Huw Jones

Computer Graphics through Key Mathematics

Springer

Huw Jones, BSc, DipEd, MSc, FSS, MBCS, CEng
Lansdown Centre for Electronic Arts, Middlesex University,
Cat Hill, Barnet EN4 8HT

British Library Cataloguing in Publication Data
Jones, Huw
  Computer graphics through key mathematics
  1.Comuter graphics - Mathematical models
  I.Title
  006.6'0151
ISBN 1852334223

Library of Congress Cataloging-in-Publication Data
Jones, H. (Huw)
    Computer graphics through key mathematics / Huw Jones.
      p. cm.
    ISBN 1-85233-422-3 (alk. paper)
    1. Computer graphics. 2. Mathematics. I. Title.
T385 .J6749 2001
006.6--dc21   2001018372

Apart from any fair dealing for the purposes of research or private study, or criticism or review, as permitted under the Copyright, Designs and Patents Act 1988, this publication may only be reproduced, stored or transmitted, in any form or by any means, with the prior permission in writing of the publishers, or in the case of reprographic reproduction in accordance with the terms of licences issued by the Copyright Licensing Agency. Enquiries concerning reproduction outside those terms should be sent to the publishers.

ISBN 1-85233-422-3 Springer-Verlag London Berlin Heidelberg
a member of BertelsmannSpringer Science+Business Media GmbH
http://www.springer.co.uk

© Springer-Verlag London Limited 2001
Printed in Great Britain

The use of registered names, trademarks etc. in this publication does not imply, even in the absence of a specific statement, that such names are exempt from the relevant laws and regulations and therefore free for general use.

The publisher makes no representation, express or implied, with regard to the accuracy of the information contained in this book and cannot accept any legal responsibility or liability for any errors or omissions that may be made.

Typesetting: Camera ready by author
Printed and bound at The Cromwell Press, Trowbridge, Wiltshire, UK
34/3830-543210 Printed on acid-free paper SPIN 10790908

Computer Graphics through Key Mathematics

Springer
London
Berlin
Heidelberg
New York
Barcelona
Hong Kong
Milan
Paris
Singapore
Tokyo

Dedicated to the memory of John Lansdown (1929-1999)

An original thinker,
an inspirational leader
and, above all, an honest and true friend

Preface

> The non-mathematician cannot conceive of the joys that he's been denied. The amalgam of Truth and Beauty revealed through the understanding of an important theorem cannot be attained through any other human activity.[1]

I have often been asked to recommend a book on the mathematics underlying the computer generation of images. Many of these are students or computer graphics users who missed out on the first chance of an extended mathematical education, having specialized in various aspects of the arts from an early age. Now, using methods that have solid mathematical foundations just below the visible surfaces of the systems they use, they want to know just what it is that prevents their systems from collapsing.

This book is intended to fill that purpose. I have taught computer graphics, particularly its mathematical underpinning, to a range of students on MSc and MA courses since the mid 1980s. These include postgraduates with first degrees in, for example, engineering, mathematics, computer science, graphic design, fashion, 3D design and fine arts. My intention has always been to give those students an understanding of how things work, to give them knowledge, as well as to develop in them the skills of doing such activities. For example, in the treatment of the calculus, as in chapter 5 of this book, the stress is on describing what differentiation and integration mean and how to interpret results of these methods, rather than to develop the very detailed skills needed for generating such solutions. Mathematics texts are usually intended to develop mathematical skills. In this book, the emphasis is on knowledge rather than skills. To this end, the description is mathematically correct in a discursive rather than a classically rigorous sense; the stress is on justifications rather than proofs. It is, however, necessary to introduce mathematical notation and at times to write down relatively complicated equations, but 'first reading' understanding of concepts should be attempted by those who find these difficult, with perhaps a later return to reinforce details. Also, knowledge is introduced on a 'need to know' priority. Topics introduced in the book are those needed for an understanding of the processes of computer graphics.

I have been rewarded by observing my students' excitement in their own successes and achievements, the sudden brightness in the eye on realizing how a desired effect can be achieved. I give just one example, which shows how Doxiadis' 'joys' in the opening quotation are equally applicable to 'she' as well as 'he'. The subject is not named to save her from embarrassment. A student who was an intelligent graduate designer but who had no previous computing experience, and only high-school level mathematics, approached me with some sketches. 'I want to do this and I want to make it move.' The sketches showed clusters of points in swirling patterns. I thought of analogies — leaves blowing in a turbulent wind, perhaps? Having thought about this for a while, I talked her through a scalar field method developed by analogy from a weather map in which atmospheric pressure creates wind in swirling forms around high and low pressure areas. In this field, particles could be dropped to be blown into the moving shapes of her illustration.

---

[1] A. Doxiadis (2000) *Uncle Petros and Goldbach's Conjecture*. Faber & Faber, page 167.

She understood my description and, with very little help, programmed the method to produce a most successful animation that closely matched her original concept. By needing to know, by putting aside any fear of mathematics, she achieved what she had set out to do and created an artistic animation with distinction.

Although most of my teaching has been to postgraduates, the subject content is not always advanced in a mathematical sense, as in many parts of this book. I have enjoyed seeing the scales fall from the eyes of those who had previously shunned mathematical study. As of my students, all I ask of readers of this book is that they approach it with a lack of fear. My intention is that those who approach this book with an open mind will gain knowledge of what is going on under the surfaces of the systems that they use day to day. By better understanding of these processes, they should at least understand their limitations and should be better able to use and enjoy them.

I started this preface with a quotation from mathematician/novelist Apostolis Doxiadis (whose work is very highly recommended). The book starts with a description of mathematical concepts, with a few insights as to the purposes of the methods. Its final three chapters describe the main processes of 'classical' computer graphics. Their 'raison d'être' can be summed up in the words of a great artist, Paul Cézanne:

> Treat nature in terms of the cylinder, the sphere, the cone, all in perspective.

Acknowledgements

This book is dedicated to the memory of John Lansdown (1929–1999), a friend and mentor over twelve years of professional development. He was a leader who inspired and encouraged people through his personal example and a true and honest friend. He was a genuine innovator and polymath; I gained so much from sharing an office with him for several years.

Many other people have helped me develop my career in the teaching of computer graphics. John Vince (now at Bournemouth University) set me going in the subject and Rae Earnshaw (Bradford University) encouraged my early development beyond the boundaries of my own institution. At Middlesex University, Gregg Moore helped me through my early years in computer graphics, and many other current and previous colleagues directly or indirectly supported my professional advancement. Those who have been directly involved in subject development or in a supervisory capacity include (in alphabetical order) Stephen Boyd Davis, Paul Brown, Aurelio Campa, John Cox, Barry Curtis, Gordon Davies, Andrew Deakin, Roger Delbourgo, Tessa Elliott, Alan Findlay, Tony Gibbs, Dick Gledhill, Jackie Guille, Avon Huxor, Hugh Mallinder, Magnus Moar, Jane Moran, Robert Myers, Martin Pitts, Dominique Rivoal, Julian Saunderson, Wally Sewell, Frank Tye and Jason White. I thank, too, those who have helped keep me sane over lunchtime conversations: Gerry Beswick, John Dack, John Lewis, Brian Peerless, Tony White and others. I give my apologies to anyone omitted by my poor memory.

I have been fortunate to have very diligent and intelligent postgraduate students over the years. I have used illustrations from the project work of several of these in the book. They are Dave Baldwin (figs 8.15 and 8.16 left), Renay Cooper (fig 8.16 right), Andrew Tunbridge (fig 8.33), Aurelio Campa (plate 8.4), Denis Crampton (plate 8.6) and Semannia Luk Cheung (plate 8.7).

My early family life gave me a strong grounding in learning and appreciation of sports and the arts. My late father Jack and my mother Megan set the frame in a house full of books; my brother John greatly influenced my early appreciation of mathematics and the visual arts. My wife Judy, with good grace, has put up with the foibles of a researcher. My son Rhodri and daughter Ceri continue to divert me, providing welcome relief from the pressures of work. Ceri, herself a professional software developer, has also given useful comment on some of the contents of this book.

Contents

Preface v

1 The Processes of Computer Graphics 1
   Introduction 1
   Object Model Building 1
   Depiction of Models 3
   Conclusion 6

2 Numbers, Counting and Measuring 7
   Introduction 7
   Natural Numbers 7
   Integers 12
   Rational Numbers 14
   Real Numbers 17
   Complex Numbers 20
   Representations of Number 25
   The Computer Representation of Number 28
   Boolean Algebra 32
   Summary 35

3 Coordinates and Dimension: Representations of Space and Colour 37
   Introduction 37
   Cartesian Coordinates 37
   Defining Space by Equations and Inequalities 41
   Angles 44
   Trigonometry and Polar Coordinates 46
   Dimension 56
   Coordinate Systems in Three Dimensions 61
   Colour and its Representation 66
   Summary 71

4 Functions and Transformations: Ways of Manipulating Space 73
   Introduction 73
   Functions as Mappings 73
   Graphs of Functions 76
   Transformations in 2D 79
   Transformations in 3D 86
   Combining Affine Transformations 89
   Inversion of Affine Transformations 94
   Inversion of Functions 95
   Shape Transformation by Function Change 100
   Conclusions 103

5 Form from Function: Analysis of Shapes 105
Introduction 105
The Straight Line 105
Drawing General Function Graphs 107
Graphs of Polynomials 107
Calculus: Differentiation 114
Calculus: Integration 124
Series Expansions 129
Calculus and Animation 134
The Exponential Function 136
The Conic Sections 136
Some Standard 3D Forms and their Equations 141
Summary 245

6 Matrices: Tools for Manipulating Space 147
Matrices in Computer Graphics 147
Definition and Notation 147
Forms of Matrices 148
Operations on Matrices: Addition 149
Operations on Matrices: Multiplication 150
The Identity Matrix 152
Matrices and Equations 154
The Inverse of a Square Matrix 159
Matrices, Transformations and Homogeneous
 Coordinates: Two Dimensions 160
Matrices, Transformations and Homogeneous
 Coordinates: Three Dimensions 166
Inverse of a Transformation Matrix 170
Perspective Projection 172
Computer Implementation of Matrix Methods 176
Summary 183

7 Vectors: Descriptions of Spatial Relationships 185
Introduction 185
Definition of a Vector 185
Notation 186
Addition of Vectors: The Parallelogram and Triangle Laws 186
Multiplication of a Vector by a Scalar 188
Examples of Vector Quantities 189
Vectors in 2D Cartesian Spaces 189
Vectors in 3D Cartesian Spaces 191
Multiplication of Vectors: The Scalar or Dot Product 193
Multiplication of Vectors: The Vector or Cross Product 194
Representation of Lines Using Vectors 196
Classification of Points against Planes Using Vectors 197
Representation of Planes in Standard Form 198
Intersection of a Line with a Plane 200
Inclusion of a Point in a Triangle 203
Reflected and Refracted Rays 205

7	**Vectors: Descriptions of Spatial Relationships** .... **continued**
	Distance between Two Skew Lines 208
	Intersection of Two Planes 211
	Summary 213

8	**Geometric Modelling and Fractals: Building Descriptions of Objects...215**
	Introduction..215
	Data Structures...215
	Geometric Modelling Systems 218
	Voxel Modelling Methods 219
	Constructive Solid Geometry (CSG)...222
	Boundary Representation (B-Rep)...227
	Isosurface Modelling..231
	Fractals...234
	Fractal Dimension 240
	Fractals Based in the Complex Plane: Julia and
	Mandelbrot Sets...243
	Fractals in Simulation of Natural Phenomena 249
	Summary 258

9	**Splines: Generation of Curves and Surfaces**..**259**
	Introduction..259
	Reasons for Splines 259
	Interpolation...260
	Bézier Splines for Curve Drawing...262
	Animation Control Using Cubic Bézier Curves...................................268
	Drawing Bézier Curves 269
	Interpolating Splines for Curve Generation 271
	Animation Control Using Interpolating Splines 274
	B-Splines..275
	Non-Uniform Rational B-Splines: NURBS 282
	Circles and Other Conic Sections 285
	Surface Construction Using Bézier Patches..286
	Surface Generation Based on Other Forms of Curve 292
	Depiction of Surface Patches..293
	Summary 294

10	**Drawing and Rendering: How to Create Pictures** **295**
	What is a 3D Drawing?...295
	Methods for Rendering 298
	Hidden Surface Removal..304
	Flat or Lambert Shading...306
	Scan Line Methods 308
	Gouraud Shading..310
	Phong Shading 313
	Exact Object Rendering 314
	Shadows...315
	Specular Highlights 317

10 Drawing and Rendering: How to Create Pictures .... **continued**
  Textures 319
  Ray Tracing 320
  Radiosity 321
  Anti-Aliasing 324
  Summary 326

Suggestions for Further Reading **327**
  Mathematical and Scientific Minds 327
  Understanding of Mathematics 328
  Computer Graphics in General 329
  Geometric Modelling 329
  Fractals and Related Issues 330
  Journals 331

Index **333**
  Names 333
  Subject Index 335

1. The Processes of Computer Graphics

Introduction

This brief resumé of the underlying processes of computer graphics is intended to overview the range of activities involved in the generation of images by computer so that detailed studies can be placed in the context of the overall picture. There is no attempt to be rigorous or exhaustive; no details of computer platforms, languages or implementation are included. The concept of computer graphics is one of 'image synthesis'; an image is built up by a sequence of activities that gradually creates the required effect. This contrasts with the related but separate subject of 'image processing', in which the starting point is an image whose properties are to be explored or amended.

There are parallels between the work of a figurative artist and the process of image synthesis. In creating a representational picture or sculpture, an artist needs to choose the arrangement of objects or the scene to be depicted and then to create an image in a selected medium and manner. The generator of computer graphic images has similarly to create a conceptual model of objects that are then depicted by a chosen method. These two stages can be designated as modelling and rendering. There is a wide variety of methods available for both these stages, so computer graphics properly includes topics such as the quest for photo-realistic images of pseudo-real scenes, non-photorealistic image generation for artistic purposes, the blending of recorded and synthetic imagery, animation and the abstract visualization of scientific data. This brief chapter overviews the processes involved in the orthodox generation of a single image using accepted mainstream methods, concentrating on the modelling–rendering concept.

Object Model Building

Directly observed models may consist of data from, for example, cartographic surveys or X-ray scans or recorded values from scientific experiments such as fluid flow dynamics studies. Similarly, direct calculations from mathematical models can yield data for visualization from simulation studies. Contour building or surface approximation methods can be used to create a geometric structure for depiction from such models. Alternatively, models of non-existent universes (whether pseudo-realistic or not) can be built directly either by programming or interactive manipulation.

The now traditional subject of geometric modelling (chapter 8) enables such direct generation. It is concerned with methods of storage, manipulation, interrogation and depiction of models of solid objects, giving an indication of the space that an object occupies as well as storing some of its properties or attributes. Modern CAD (computer aided design) systems have one or more of these modelling methods at their hearts, two types of modeller being particularly widely used.

B-Rep (boundary representation) methods hold direct information on the surfaces of objects, specifically referencing any vertices, edges and faces. Spatial occupancy is implicit through definition of the orientation (as regards inside and outside) of faces. The topology of an object (the way its structure hangs together) is defined through correct linking of the building elements of vertices, edges and faces, with its geometry (size, location, orientation) defined through reference of these elements to a mathematical description of space, typically a Cartesian coordinate system. Objects defined in this way are polyhedral (bounded by plane faces), but the domain can be extended to include curved surfaces using spline surface modelling. CSG (constructive solid geometry) systems define interior regions directly, so that categorization of a point as interior or exterior is relatively easy. However, direct depiction of the surface of a CSG object is less simple as no explicit information on the surface is held. CSG models are built through merging a range of object shapes (such as spheres, blocks, cylinders and cones) or subtracting them from each other using 'Boolean operations' of union, intersection and difference, whereas B-Rep depictions involve giving information on the location and connectivity of surface features. Typically B-Rep models are more difficult to create and take greater memory, but extraction of information for rendering is quicker than for CSG.

Other methods are less frequently used, but are becoming popular for specific applications areas. Modellers (the systems used for geometric modelling) often have hybrid features allowing models to be built according to a mix of modelling methodologies.

Spatial occupancy or voxel methods are based on a subdivision of space into small cubes, each of which can be categorized as lying inside, outside or intersecting the boundary of an object. This categorization can be specifically defined or could result from naturally occurring or mathematically calculated data. More sophisticated voxel modellers allow local surface models to be stored within boundary intersecting voxels.

Cellular decomposition methods divide objects into relatively standard shapes and are useful for creating models for finite element analysis (a major analytical tool for engineers and scientists). Swept models are creating by sweeping closed shapes (often two-dimensional) along trajectories (usually three-dimensional curves). These two methods are of relatively little importance in modern computer graphics, so are not discussed in chapter 8.

These traditional methods are supplemented by recent advances for specific and artistic purposes. 'Soft modelling' uses artificially generated mathematical scalar fields, with objects bounded by 'isosurfaces' (sets of points with the same field value, like a three-dimensional equivalent of the isobars on a weather map) to create 'blobby' objects. Developed largely by the Wyvill brothers, Geoff at Otago and Brian at Calgary, the technique has been used to produce whimsically amusing films and images, such as 'The Great Train Rubbery'.

A variety of fractal methods (L-systems, IFS and others) have become popular for generating models to simulate naturally occurring forms, such as clouds, terrain, plants and branching life forms. Pseudo-random effects are often incorporated with such models so that several instances of a model appear as different instances of the same genus, like several examples of the same type of tree (chapter 8).

This discussion covers basic methods of generating individual objects or components of objects, but the organization of these models into a coherent 'universe' is also a major topic. Complex models are often segmented or arranged in a hierarchical system.

It is clear that there is a wide variety of techniques available to the computer graphics systems creator for modelling of objects. There is no one best method. The choice depends very much on the application for which the model is to be used and the operating characteristics of the available computer platforms. Modelling methods vary as to

- the speed of depiction, manipulation and interrogation of models,
- the precision and range of models that can be generated,
- the quality of information retrieved from them (which can affect the quality of the image produced),
- the amount of computer memory required,
- their ease of use and
- their error proneness

amongst other characteristics. Often, quality in one property has to be traded off against another. For example, a model holding redundant data explicitly in a large storage space may lead to fast depiction as compared to a more compact model where explicit data must be calculated during the rendering stage. It is part of the skill of a systems developer to be able to select and implement the most appropriate method for the job in hand.

Such a developer must also have a clear understanding of the capabilities of the hardware on which the software system is to be mounted and the software base that supports graphics systems development. This involves understanding the storage and speed characteristics of the hardware as well as the peripheral devices supporting interactive activity. Human computer interaction (HCI) is also a relevant topic. Apart from using computer graphic techniques in creating GUIs (graphical user interfaces), HCI is an intrinsic part of the generation of a modelling software system, enabling the manipulation of models and selection of the mode of depiction.

Depiction of Models

This is another aspect of computer graphics that is affected by hardware. The quality of the image will depend on the spatial and colour precision of the display and the speed of processing for image generation will affect the way in which images can be presented. For the purposes of this book, we assume displays are raster devices: arrays of pixels (picture elements) in a rectangular grid, each capable of displaying a range of colours at various intensities. We also assume some simple graphical capability, such as the ability visually to join two pixels with a straight line, or to fill an enclosed region with a given colour. These are established methods, details of which are given in standard computer graphics texts. The most advanced modern systems are capable of real-time animation of complex images. When such speeds are not achievable, good animation can only be achieved by pre-calculation and storage of sequences of images for subsequent rapid replay.

Human perception is a complex issue. Our experience of the objects that surround us depends on the properties of those objects. Surfaces absorb and transmit light in different ways and that light originates from various sources; our perception of an object depends on its shape, size and location, its surface properties (such as light absorption, transparency, reflectivity) and the illumination that it receives. Our interpretation of this information, transmitted to our eyes in the form of light, depends on the physiological and psychological impact of the information. The way that humans and animals distinguish individual objects from the pattern of light that falls into the eye is so familiar that its importance is not always appreciated. The creator of computer graphic images should consider such issues in order to simulate real-world experiences or generate useful artificial worlds.

The discussion of modelling methods above concentrates on three-dimensional objects, but similar methods can be adapted to two-dimensional shapes. Much of computer graphics involves generating images of two-dimensional universes, and so-called 3D images are, in any case, produced on 2D surfaces. A 2D image is created by generating a model of a universe, selecting and cutting out (or clipping) the region of that universe to be depicted and mapping the contents of that region onto a display surface (typically a portion of a VDU screen or printer page). The image may consist of lines in one or several colours, and may have regions filled with block colours or more subtly varying tones. The methods of half toning or dithering can give the impression that more colours or shades are being used than are really available (as in old-fashioned newsprint, that gives the impression of several shades of grey by showing closely packed black dots, though using only black ink on white paper). Such methods are not considered in this work. They are currently less widely used due to the ready availability of high-integrity colour systems.

Generation of images of three-dimensional worlds involves a more complex sequence of procedures. Once the required form of universal model is created (incorporating the geometric model, surface properties and the illumination sources) in a world coordinate system, a pseudo-observer is set up. This requires definition of a location, a direction of view and viewing angles (like a photographer identifying a suitable location, pointing the camera in a particular direction and selecting a zoom setting). By transforming the space to be viewed into coordinates based on the pseudo-camera location and orientation (view coordinates), it is possible to identify portions of the image outside the viewable region (the window) for elimination by clipping. Elements of the scene that are directly visible and not hidden behind other elements are also identified (hidden surface or hidden line elimination). All visible elements (after clipping and hidden surface elimination) are transformed into a two-dimensional version of the view (projection) and are mapped onto a region of a two-dimensional viewing surface: the viewport. The mapping is called the window-to-viewport transformation, moving from view coordinates to display coordinates. Some elements of this process may be treated in different orders, but they are all present in standard three-dimensional computer graphics.

Two-dimensional images of a three-dimensional universe are familiar through paintings, photographs and television pictures. Devices are now available to simulate the effect of binocular vision by generating and viewing stereoscopic pairs of images, giving a three dimensional-effect to the pictures created. This is done by generating two images, as if from adjacent eyes or cameras, to be viewed stereoscopically with each eye being allowed to observe the relevant image only.

This can be done by projecting each image onto separate screens, each visible by only one eye, as in VR headsets. Another method links a rapidly alternating pair of images on a screen to shuttered glasses that optically open and close lenses to allow each eye access to one image.

The quality of the final image depends on the hardware platform, peripherals and the method used for rendering the image. Early display devices were only capable of displaying lines and curves (really collections of closely displayed dots) in monochrome or a limited range of colours; now it is relatively easy to fill regions with an apparently full palette of natural colours.

Rendering methods are available to match this range of display qualities. Simple line drawings can be produced with or without hidden line elimination (the latter method produces 'wire frame images'). To produce the appearance of solidity in an unambiguous way, objects need to be rendered. Simple block colour filling of polygonal regions on the screen does not lead to visually satisfying images, although 'flat shading', making use of the Lambert cosine law from physics for the relative illumination of a plane surface, can be used reasonably to depict polyhedral solids. Gouraud shading is used to give more subtle variation of perceived colour inside polyhedra and across polyhedral boundaries, giving the appearance of rounded or smooth interior edges. Even smoother appearances are created by Phong shading, also adding specular highlights to simulate the effect of reflected light from glossy or semi-glossy surfaces. As in all these methods, improvement in quality is bought only at the cost of greater complexity in the method, higher cost of display equipment and slower depiction rates. Ray-tracing is used to create very glossy images with hard-edged shadows. The increasingly popular 'radiosity' method (in which the way in which light radiates from one surface to another is simulated) generates realistic soft shadows. Hybrid methods combine the gloss of the former with the soft illumination of the latter.

Special rendering effects include the generation of surface coloured images and textures (using texture mapping or bump mapping) and the creation of spatial textures that fill an object (solid textures). These methods can produce convincing images of regularly patterned surfaces such as brickwork or randomly varying natural textures like that of marble or wood grain, usually incorporating pseudo-random features in the latter cases.

The above descriptions concern still images. Animations are created by generating and reproducing sequences of images in rapid succession. This is not a trivial process as successive images need to be properly connected to create effective sequences. Motion of the observer or camera must be smoothly generated and the required changes in the modelled environment to be observed must also be effected. Some animations are totally controlled by the systems programmer; others are directed by pseudo-physical laws that constrain the activities of objects. Such methods have been used for human body animation (constraining the motion of joints, simulating muscular forces), the flocking of birds and equivalent shoaling of fish.

It is all very well knowing the technicalities of generating images using computers, but the impact of images depends on the quality of their design as well as their production. To this end, it is worthwhile for computer graphic professionals to have at least a basic understanding of what constitutes 'good' design, a reasonable feel for 'visual literacy'.

Conclusion

This is one person's brief view of some of the knowledge involved in the generation of computer graphic images. This knowledge needs to be supplemented by the skills of creation of computer graphic systems, the skills of systems software generation and implementation on suitable platforms. Systems developers must also base their software development in the correct mathematics and physics of the objects they are simulating. These systems may be those used purely for image synthesis or special packages that use visualization as a tool for specific domain knowledge applications, such as cartography or molecular modelling.

The next six chapters describe the underlying mathematics used to create computer graphics systems of various types. Their relevant uses are pointed out as mathematical concepts are described. The kind of computer graphics is a mainstream sort, generally accepted as traditional. There are many enhancements and addenda implicit within up-to-date systems, but we concentrate on understanding of the essentials. When readers are familiar with these, they should have the ability and confidence to extend their knowledge by reading the specialist research journals in the area. A few are mentioned in the suggestions for further reading, but this concentrates mainly on books, rather than research papers, that should help to add solidity to the content of this book.

Users do not *need* to know exactly what is going on within a system, but their understanding of *how* to use it is improved by at least an understanding of the basic working characteristics of the system. The intent of this book is to give such an essential understanding of the key methods, rather than to concentrate on the skills development that is the purpose of much mathematics education.

2. Numbers, Counting and Measuring

Introduction

Mathematics does not concern itself solely with numerical problems, but number work is a large part of the mathematics that relates to computer graphics. It is worth discussing the fundamental nature of numbers to ensure that all participants in the process of creating computer graphics properly understand what they are using.

The historical development of the use of numbers is closely related to the way in which children are believed to develop their awareness of and appreciation of numbers.[1] The process builds through the identification and counting of groups to measuring, major jumps being needed whenever the system currently used becomes inadequate in the context of a particular problem. This also reflects the general history of science – major advances are often made when existing laws fail to explain an observed phenomenon.

Natural Numbers

The most basic numbers are the natural numbers

$$N = \{1, 2, 3, 4, ...\}$$

Initially, these are used merely for counting. The brilliant, but conservative, nineteenth-century mathematician Leopold Kronecker indicated their importance when he said, 'God made the natural numbers, all the rest is the work of man.'

The simple statement above has already introduced two pieces of mathematical notation. One of the major difficulties that people have in mathematics is in understanding the symbols used. They will be introduced on a 'need to know' basis throughout this text. There is a conflict between the elegance and simplicity gained by the use of symbolic notation and the level of familiarity with symbols needed to understand mathematical statements. Apart from the meanings of statements, they have syntactical laws equivalent to the laws of grammar for a language.

---

[1] Many mathematics text books discuss the development of number. A classic and popular readable general work is R. Courant, H. Robbins and I. Stewart (1996) *What is Mathematics?* (2nd edn.), Oxford University Press (original edition by Courant and Robbins dated 1941). This great book contains much that was unusual in its time, which is now very important for the development of computer graphics.

Another readable introduction to this field of mathematics is given in J. Baylis and R. Haggerty (1988) *Alice in Numberland*, Macmillan.

C.B. Boyer and U.C. Merzbach (1991) *A History of Mathematics*, Wiley, gives useful background material, as well as giving a historical overview. This takes 0 as the basis of natural numbers rather than 1, but that does not affect the development.

Mathematical statements must be properly composed, as are statements in a natural language.

The curly brackets or braces '{' and '}' indicate the contents of a set, which can be considered loosely as a collection of elements. The set of natural numbers is described above by enumeration or listing of its contents. Other sets may also be given by definitions of their contents. For example, the set S of natural numbers that are less than 5 can be defined as

$$S = \{1, 2, 3, 4\}$$

or $\quad S = \{x \in N: x < 5\}$.

The listing method simply gives all possible values in the set; the definition method tell us that the set consists of all values x that belong to N, the set of natural numbers, which have the property that x is less than 5. We are already piling up new symbols. The symbol '$\in$' indicates membership of a set.

$$12 \in N$$

states that the number 12 is an element of the set of natural numbers, so the symbol relates an element to a set. Non-inclusion of an element in a set is indicated by crossing out the inclusion symbol, so we can state, correctly with S as defined above,

$$12 \notin S,$$

meaning '12 is not an element of S'.

The symbol '<' is a comparator, meaning 'is less than'. Its use implies a sense of ordering of the natural numbers, which is indicated in the way the numbers are listed in the definition of N given above. The following symbols are often used to compare numbers:

<	is less than
>	is greater than
≤	is less than or equal to
≥	is greater than or equal to
=	is equal to
≠	is not equal to

The symbol '...' also appears above. It indicates that the process of stepping from one number to the next continues indefinitely. There is no upper bound to the process, so there is no 'largest' natural number. This concept introduces an element of challenge often encountered in mathematics. 'If you give me a natural number, I can always find a larger one.' Many mathematical texts obscure the 'my dad can fight your dad' approach by using fiercely dense symbolism. The philosophical concept of infinity is introduced by the unboundedness of the natural numbers. To say that the set of natural numbers has an infinite number of elements does *not* mean to say that infinity is a natural number. *Infinity is not a number*. This digression is beginning to lead us down a peripheral route. We have seen enough to

2. Numbers, Counting and Measuring

satisfy us for the moment, so will return to our main concern, the explanation of number.

It is very important to have natural numbers for counting and comparison purposes, but it is pretty boring to spend all your time counting and comparing. Natural ingenuity led to the development of other number tools which will be familiar to all readers. This familiarity is used as an excuse for describing them in a non-rigorous way. Some of their main properties, however, are discussed here, as familiarity can lead people into taking them for granted and even ignoring them. A simple rule gives the total number of objects in two collections using the rule of addition of natural numbers. If x and y are natural numbers, and z is their combined count or sum, this is stated symbolically as

$$z = x + y.$$

This notation is so natural that many of its properties seem obvious. Try to imagine what it is like to come across it for the first time. Such a situation may arise for other definitions later in this text, so it is useful to use this familiar case as an example. The symbol '+' is a binary operator. It represents the combination of *two* elements, both natural numbers. The result of this combination, z, is a natural number, regardless of the values of x and y. If $x \in N$ and $y \in N$, then $z \in N$. The set of natural numbers is closed under addition. It is impossible to 'escape' from N using addition alone. Addition is also commutative: the order of the two components of the binary operation has no effect on the result. Put simply

$$x + y = y + x,$$

and this is *always* true when x and y are natural numbers. This is so familiar it seems obvious, but we will later find examples of other binary operators that are not commutative.

When comparing numbers, the statement '$x > y$' could lead to the follow-up, 'by how much is x greater than y?' Finding such a difference is represented symbolically by subtraction:

$$3 = 8 - 5$$

states that 3 is the difference between 8 and 5, stated as '8 minus 5'. The binary operator for subtraction is '−', said as 'minus'. Note that subtraction is closely related to addition, for if

$$z = x - y,$$

then $\quad x = z + y.$

This can be justified by investigating examples; this text does not follow the rigorous mathematical approach of axioms and proofs. Subtraction is, in a sense, the 'undoing' of an addition: the effect of adding by 5 is 'undone' by subtracting 5 from the result.

Repeated addition gives rise to the concept of multiplication. For example, adding 5 to 5 and then adding another 5 to the result (in other words adding three 5's together or counting the total of three lots of five objects) can be represented as

$$(5 + 5) + 5 = 3*5.$$

'*' is used here as a symbol for multiplication as it is commonly used for this purpose in computer languages; a cross 'x' is more often used in handwritten work. When using characters for numbers, the symbol for multiplication may be omitted, so

$$xy = x*y.$$

xy is called the product of x and y. When used above, we assume x, y and z are natural numbers. Brackets were used in one statement:

$$(5 + 5) + 5 = 3*5.$$

Their use is rather pedantic in this case, but shows how careful mathematicians have to be in explaining simple concepts that may be over-familiar. We have only defined how to add two numbers using the binary operator '+'. The composite operation

$$(x + y) + z$$

means that x and y are added first and the result of that operation is added to z – the convention is that the contents of brackets are worked on first in arithmetic operations, the general description for symbolic combinations of this type. When x, y and z are natural numbers, the same result is achieved from

$$(x + y) + z$$

and $\quad x + (y + z),$

so this combination can be represented without ambiguity as x + y + z. This property of the addition of natural numbers is called the associative law. A little experimentation should convince readers that multiplication is also associative and commutative, as

$$(x*y)*z = x*(y*z)$$

and $\quad x*y = y*x.$

The distributive law connects multiplication with addition. For all x, y and z, we have

$$x*(y + z) = x*y + x*z.$$

If you replace x, y and z by numbers you can justify this result. This is still no proof, as proof is only achieved by exhaustively establishing the result for all

2. Numbers, Counting and Measuring

possible instances of x, y and z, but it is enough for our purposes to note that this can be proved.

Just as the result of an addition can be undone by subtraction, the result of a multiplication can be undone by division. If we create the number 15 by multiplying 3 by 5, the value 3 can be retrieved by division of 15 by 5. If 15 objects are divided into 5 equal groups, each group contains 3 objects. This is written as

$$15/3 = 5$$

or $\quad 15 \div 3 = 5,$

the division symbol '/' being most commonly used in computer languages. The symbols +, −, * and / are binary operators which are used to produce a single natural number from *two* numbers (hence the 'binary') using the relevant operation.

A further form of expression is achieved by considering repeated multiplication, as we already considered repeated addition. We can represent the result of 5*5*5 (remembering that this is unambiguous, regardless of which multiplication we perform first) by 5^3, stated as '5 to the power of 3'. x^y is the result of multiplying x by x and repeatedly multiplying the result by x so that there are y terms of x in the product. In x^y, y is the index or the power. When y takes the value two or three, special names are used. x^2 is known as 'x squared', x^3 is 'x cubed', these phrases indicating that these represent the area of a square and the volume of a cube with side length x respectively. There is a set of specific rules relating to indices (the plural of index), for example

$$x^y * x^z = x^{y+z}.$$

This means we are multiplying together y lots of x with z lots of x, so we have y + z lots of x in all. The set of all rules for manipulation of specific types of element represents the algebra of that system. The rules above begin to develop the algebra of natural numbers.

The natural numbers N are closed with respect to addition and to multiplication and both addition and multiplication are commutative operations. The question of commutativity and closure has been avoided in discussing subtraction and division. If we attempt to reverse the order of numbers in subtraction and division, we come across an intractable difficulty. The values of

$$15 - 5$$

and $\quad 15/5$

can be written down easily as 10 and 3 respectively. Reversing the order of the numbers in these operations gives

$$5 - 15$$

and $\quad 5/15.$

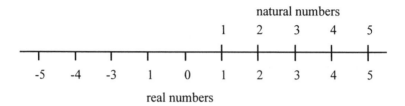

Figure 2.1 Parts of the discrete sets of natural numbers and integers on a number line

Neither of these expressions represents a value in the set of natural numbers, N. In order to find solutions, it is necessary to extend our number system. The natural numbers have been pushed as far as they will go. They must be extended to include new elements if advances are to be made. The first consideration is to devise an extended set that will allow subtraction in all cases.

Integers

The set of integers is represented symbolically as

$$Z = \{\ldots -2, -1, 0, 1, 2, 3, \ldots\}.$$

The double appearance of '...' indicates that the set is neither bounded above nor below. There is no integer that is 'the largest' and there is no integer that is 'the smallest', where size is indicated by smaller elements appearing to the left of larger elements in the set listing. The important symbol 0 ('zero' or 'nought') is included. This is called the 'identity element for addition', as adding 0 to any value leaves it unchanged, or identical to its original value. If x is any integer, then

$$x + 0 = x$$

and $0 + x = x.$

Equivalently, 1 is the identity element for multiplication, as

$$x*1 = x$$

and $1*x = x.$

The ordering of natural numbers and of integers can be illustrated by placing them on a 'number line' (fig 2.1), where successive values in each set are shown as equally spaced positions on the line. The gaps separating one number from its neighbours illustrate the discrete nature of natural numbers and integers. It is possible to name pairs of numbers such that no other number lies between them. There is no natural number between 2 and 3, no integer between -4 and -3.

2. Numbers, Counting and Measuring

It is worth noting that we have made an intuitive jump beyond the scope of the symbols we defined for natural numbers. The binary operators + for addition and * for multiplication have been used on integers without proper definition. The natural numbers N are all included inside the set of integers Z. N is a subset of Z. As a new type of number has been introduced, all the required operations for this type of number should be redefined in a 'downwardly compatible' way. If the integer addition rule is used to add two integers that happen to be natural numbers, the result should be the same as if they were added using the natural number addition rule. In a pure mathematical approach, all required operations must be redefined. This complication is avoided here, as this text concentrates on the use rather than the definition of mathematics.

One finicky rule worth stressing is that division by zero is *never* allowed. Adding together five threes makes 15. The undoing of this indicates that $15/5 = 3$. Adding together five zeros (or 15 or 1000 or ... zeros) makes zero. There is no way that repeated addition by zero can raise the sum above zero, so division by zero of any non-zero number is impossible. Division of 0 by 0 is equally problematical. We content ourselves here by stating that it is not properly defined. The difficulty is seen by considering the general division of x by y, expressed as x/y. If $x = 0$, the answer is 'always' 0. If $y = 0$, the answer is not defined; a less formal answer is that this is 'infinity'. So what happens if both x and y are 0? Should the answer be 0 or infinity? The answer depends on the circumstances: it may be zero, it may approach infinity, it may be some finite value in between. When there is an answer, it is found from the theory of limits, discussed in Chapter 5.

The symbol '–' has already been used for the binary operation of subtraction. In defining the integers, the symbol '-' is used to define some numbers. This unary minus operates on a single number, its symbol '-' being distinguished in this text as a shorter version of the binary operator '–'. Some texts raise the symbol, using '$^{-}5$' where '-5' is used here. Most texts make no distinction between the unary and binary forms; many users are unaware of the distinction and use the rules properly without problem, so this could be considered an unnecessarily pedantic distinction.

The 'new' operators have all of the old properties, such as the commutativity of addition and multiplication. New properties for general integer values x and y include

$x + 0 = x$, $\qquad\qquad x - 0 = x$,

$x*0 = 0$, $\qquad\qquad -0 = 0$,

$-(-x) = x$,

$x + (-y) = x - y$, $\qquad\qquad x - (-y) = x + y$,

$x*(-y) = (-x)*y = -(x*y)$, $\qquad\qquad x/(-y) = (-x)/y = -(x/y)$,

$(-x)*(-y) = x*y$, $\qquad\qquad (-x)/(-y) = x/y$,

$0/x = 0$ when $x \neq 0$, $\qquad\qquad x^0 = 1$ when $x \neq 0$.

No attempt is made here to justify these, but readers should make sure that they understand what they mean. Taking examples with particular values may be helpful for those unfamiliar with these rules. The final rule, for the power zero, is perhaps a bit baffling, but it produces results that 'work'.

The extension from natural numbers to integers has had one of the desired effects; the set of integers is closed for subtraction. If x and y are *any* integers, then x − y has a properly defined solution that is also an integer. The property of commutativity is still invalid for integer subtraction. In general,

$$x - y \neq y - x;$$

equality only holds if x = y.

Closure for subtraction is solved, but there is still no closure for division within the set of integers. To get over this problem, we need, once more, to move on to create an extended set of numbers.

Rational Numbers

A rational number is formed from an 'ordered pair' of integers (this means that it matters which of the pair appears first) where the first is divided by the second. This allows almost all possible divisions of integers to be expressed as a rational number. In set notation,

$$Q = \{x = p/q; p, q \in Z, q \neq 0\}.$$

Division by zero is not allowed for integers, so the case for q = 0 must be excluded. Rational numbers are more familiarly known as fractions, the character Q indicating the concept of a 'quotient', 'rational' coming from the concept of ratio. In p/q, alternatively written as

$$\frac{p}{q},$$

p is known as the numerator, q is the denominator. For example, in interpreting the fraction 3/4 or 'three quarters', we take 3 parts (the numerator indicating 'how many'), each of these parts being of a size such that 4 of them make one unit (the denominator indicating 'how big'). The more parts a unit is divided into, the smaller the individual part, so the larger the denominator, the smaller the fraction.

It is necessary to build up, once more, a useful repertoire of operations on rational numbers that include all rules defined above for integers. Without justification, we use integer values p, q, r, s, k and the rational number x to define the following new rules, with numerical examples given for each (readers may find it useful to rewrite the examples in normal fraction form):

$$\frac{p}{q}.$$

2. Numbers, Counting and Measuring

- $(k*p)/(k*q) = p/q$, when $q \neq 0$;

this is the common rule for cancelling fractions, for example,

$$8/12 = (4*2)/(4*3) = 2/3.$$

- $p/1 = p$;

this shows that integers form a subset of the rational numbers, as any integer can be expressed in rational form, for example,

$$3/1 = 3.$$

- $p/q + r/q = (p + r)/q$,

when $q \neq 0$; with the cancelling rule above, this can be used to add any pair of rational numbers or fractions, as in

$$3/8 + 1/2 = 3/8 + 4/8 = 7/8.$$

- $x - p/q = x + (-p)/q$;

this explains subtraction as an extended form of addition,

$$5/7 - 3/7 = 5/7 + (-3)/7 = 2/7.$$

- $(p/q)*(r/s) = (p*r)/(q*s)$,

when $q \neq 0$ and $s \neq 0$. This is the rule for multiplication of fractions,

$$(3/4)*(2/3) = 6/12 = 1/2.$$

- $(p/q)/(r/s) = (p/q)*(s/r)$,

when $q \neq 0$, $r \neq 0$ and $s \neq 0$, giving the rule for division of fractions as in

$$(1/4) / (2/3) = (1/4)*(3/2) = 3/8.$$

- $p^{-q} = 1/p^q$;

a rule for negative indices that makes sense when the sequence

$$\ldots p^3, p^2, p^1, p^0, p^{-1}, p^{-2}, \ldots$$

is considered, terms being successively divided by p. Remember that $p^1 = p$ and $p^0 = 1$. Thus,

$$3^{-2} = 1/3^2 = 1/9.$$

As a digression (those who want to skip ahead should do so until the symbol * later in this page) it is interesting to note that the set of rational numbers is countable. It is possible to write down a list of rational numbers (this will inevitably be of infinite length) that will include every properly defined rational number. It is fairly clear that this can be done for natural numbers (the defining set listing itself suffices). For the integers, one example of a suitable ordering for countability is

0, 1, -1, 2, -2, 3, -3, ...

In doing this, order in the list does not relate directly to size of the integer. All positive rational numbers can be reached within the following pattern

1/1	1/2	1/3	1/4	1/5	...
2/1	2/2	2/3	2/4	2/5	...
3/1	3/2	3/3	3/4	3/5	...
4/1	4/2	4/3	4/4	4/5	...
...	...	...	...	...	

Tracing a pattern of repeated diagonals through this array, the following sequence is generated:

1/1, 2/1, 1/2, 3/1, 2/2, 1/3, 4/1, 3/2, 2/3, 1/4, ...

This has repetitions (1/1 = 2/2, for example) which can be eliminated, through expressing all values in their 'cancelled down' form. The list of rational numbers can be arranged by excluding all repeats; zero and the negative rational values can also be included as in

0, 1, -1, 2, -2, 1/2, -1/2, 3, -3, 1/3, -1/3, 4, -4, 3/2, -3/2, 2/3, -2/3, 1/4,...

Inelegant as it is, this listing contains all possible rational numbers, so they are countable.

* If you didn't follow the digression, you should restart here.

In many ways, the set of rational numbers is 'complete'. It is closed with respect to the standard binary operations of addition, subtraction, multiplication and division; the combination of rational numbers using these operations is a rational number, with the exception of illegal division by zero. Rational numbers (often referred to as 'rationals') are continuous: there are no gaps between numbers. Given two rational numbers x and y, no matter how small their difference, there are other rationals lying between them: their arithmetic mean, $(x + y)/2$, is an example. The natural numbers and integers do not have this property; they are discontinuous or discrete.

This apparent completeness, however, is a sham as it can be breached in measuring geometric shapes and in some fairly easy arithmetic calculations. There are numbers and measurements that cannot be expressed as rational numbers. It is time, once more, to move on.

Real Numbers

Justification of the need for real numbers, or reals, can be from attempts to invert the power function. Suppose a rational value y is found as

$$y = x^z,$$

when x is a rational number and z is a natural (whole) number. Now attempt to recover the value of x, knowing y and z. This is known as finding the 'zth root' of y, and can be written as

$$x = \sqrt[z]{y}.$$

For example, $9 = 3^2$, so $3 = \sqrt[2]{9}$. The inverse of raising to the power of 2 (or squaring) is called the square root, and is usually written without the suffix value 2 as √9. The inverse of raising to the power of 3 (or cubing) is called the 'cube root'; subsequent roots are simple called 'fourth root', 'fifth root' and so on. The process of finding a root can always be done if it is 'undoing' the effect of the equivalent power calculation. However, it is not always possible to find a natural number that represents $\sqrt[z]{y}$ when y is any rational value and z is any natural number. This can be proved in the simplest case using the 'reductio ad absurdum' method. This supposes the required number exists, but then shows that an impossible condition arises from this assumption. (The following section, which indicates the main features of a proof, may be ignored on first reading until * on the next page.)

Suppose we wish to find a rational number x representing $\sqrt[2]{2}$, or more simply √2. Using Pythagoras' theorem (if you don't yet know of this, it will be discussed in the next chapter; we ask you to accept the result for the time being) √2 can be constructed as the longest side of a right-angled triangle with short sides 1 (fig 2.2).

Suppose that a 'cancelled down' rational form of √2 exists as p/q = √2, where p and q are integers. p and q have no common factors other than 1; there is no integer that divides exactly into both p and q.

Then $\quad (p/q)^2 = p^2/q^2 = (\sqrt{2})^2 = 2,\quad$ so $\quad p^2 = 2q^2.$

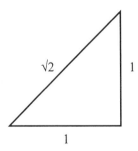

Figure 2.2 The length √2 cannot be expressed as a rational number

Thus p^2 must have a factor of 2 (it is an even number), so p must also be an even number that can be expressed, say, as 2r, where r is an integer. This gives

$$p^2 = 4r^2$$

so $\quad 4r^2 = 2q^2,$

which reduces on division by 2 to

$$2r^2 = q^2.$$

By the same reasoning as before, q also has a factor of 2. We have shown that if p/q is √2, then both p and q must have factors of 2. But we assumed that p/q was the fully cancelled down version of √2, so they cannot both have 2 as a factor. We have found the original assumption that √2 has a rational form leads to a contradiction, so no such form can exist. √2 cannot be expressed as a rational number.

This text is mainly concerned with the application of mathematics and usually follows the practice of taking things for granted, not following through proofs. However, it is worth seeing the occasional proof to understand the kind of ingenuity that extends pure mathematicians' minds. Theory and application tend to lurch forward together in the history of mathematics and science in general. Sometimes one is ahead, dragging the other forward; at other times the priorities are reversed.

* Restart here if you left out the proof.

Many other irrational values (this simply means they are not rational) or surds result from calculation and measuring problems. For example π, the length of the circumference of a circle whose diameter is one, is irrational. 'π = 22/7' or 'π = 3.142' or 'π = 3.14159265' merely give rational valued approximations to the correct value, increasingly precise as they may be. An irrational value much used in mathematics and studies of population growth is the exponential number, denoted as e and approximately equal to 2.7183. e^x has a rate of growth with respect to x which is equal to its own value. This is of interest in population studies where the growth rate of an unrestricted population is directly related to its own size. Irrational numbers such as π and e complement the set of irrational numbers from roots that have already been discussed.

When represented as decimals (once more, this concept has not been properly developed here; this text cheats by relying on prior knowledge), rational numbers will either be shown exactly to a set number of figures – for example,

$$1/4 = 0.25$$

precisely – or can be given as a repeating sequence of figures – as in

$$1/3 = 0.333...$$

or $\quad 2/7 = 0.285714285714285714...$

These last two examples of repeated sequences, which can be found by the old-fashioned skill of decimal division, can be expressed exactly as

2. Numbers, Counting and Measuring

$$1/3 = 0.\dot{3}$$

and $2/7 = 0.\dot{2}8571\dot{4},$

where the dots indicate a digit or group of digits to be repeated indefinitely, so 3 is indefinitely repeated in 1/3 and the digits 285714 are indefinitely repeated in 2/7. Numbers expressed with these repetitions are known as recurring decimals. The irrational values, however, display no repeated pattern and cannot be expressed precisely in decimal form. How, then, can we compare the values of irrationals to rational numbers? How do we know if an irrational number is greater than or less than a rational number? In relating the values of irrational numbers to rational numbers, it is possible to find a pair of known rationals that 'sandwich' the irrational to a given precision. For example, using exact rational arithmetic, it is clear that

$$1.4^2 = 1.96$$

and $1.5^2 = 2.25.$

As 2 lies between 1.4^2 and 1.5^2, then a value of $\sqrt{2}$ must lie between 1.4 and 1.5, expressed as

$$1.4 < \sqrt{2} < 1.5.$$

A similar argument shows that

$$1.41^2 = 1.9881$$

and $1.42^2 = 2.0164,$

so $1.41 < \sqrt{2} < 1.42.$

The interval of precision for approximating $\sqrt{2}$ can be narrowed down indefinitely in this way, so its value can be approximated to an arbitrary level of precision. This search can be done by trial and error or by systematically closing in on more precise interval estimates of this kind.

The last few sentences indicate that we can enclose any irrational value between an indefinitely close pair of rational values. This statement can also be turned around:

- given any arbitrarily close pair of rational values, an irrational value can be found to lie between them.

Also,

- given any arbitrarily close pair of irrational values, a rational value can be found to lie between them.

Rational and irrational values are inseparably bound together. We cannot define a range of numbers that contains only rationals or only irrationals. The rationals form a continuous set; so do the irrationals – there is no gap between successive values. What is more, these continuous sets can be considered to occupy the same space, but not the same locations in that space. They are distinct, but inseparable. Imagine taking two densely bristled hairbrushes, one white, the other black. Push the two brushes together so that the bristles are pressed between each other. Now use a pair of scissors to cut away a section purely of white bristles from the interior of this jumble. That is an easy task compared to separating overlapping sets of rational and irrational values.

Rationals and irrationals are so inextricably intermingled that they can be combined to form a single number type: the set of real numbers or reals. This set can be used to represent positions on a line, the real number line, so that all points are represented by a unique number and every real number represents a unique point, making the representation complete and one-to-one. Real numbers can be used to represent all continuous measurements, of length, angle, area and so on, of real objects.

Discussion of how the binary operators fit into this new scheme have so far been avoided. Once more, new definitions for +, –, * and / and the follow-on concepts (discussed in more detail in the section on rational numbers above) are needed that are compatible with the rules of lower-level systems already developed. These laws are carried straight over for the rational subset of reals. Laws for irrationals and for their interaction with rationals are defined by analogy from the 'thin sandwich' interval concept used above. Rational bounds on the value of the sum of (or any other calculation formed on) two irrational numbers can be found by using the standard method of addition (or equivalent calculation) on the bounds of the two numbers themselves. When developed, these laws are compatible with the familiar concepts of addition, subtraction, multiplication and subtraction for the previous number sets considered. The real numbers are closed with respect to these four binary operations. The end of this line of development appears at hand, but there is still one snag.

Complex Numbers

In the discussion on using irrational values to define roots, one important difficulty was conveniently put aside. Now it cannot be avoided. Finding a square root is equivalent to finding x if the value of x^2 is known. For example, if $9 = x^2$, then $x = 3$ is a solution, as $9 = 3^2$. This solution or answer could have been found by trial and error, guessing different answers until we find one that fits. Another possible solution is $x = -3$, as $(-3)^2 = 9$. This can be explained away by accepting that the problem has two separate possible solutions. There are several ways of crossing the English channel (by chance, this sentence was proof read by the author on the Channel shuttle, having been diverted through bad weather from a SeaCat crossing), many ways of cooking eggs. There should be no problem in accepting that there may be more than one valid answer to a mathematical question. We will visit this idea again when we discuss functions. Now consider an apparently similar problem: $-9 = x^2$. Can we find some values that satisfy this by trial and error?

2. Numbers, Counting and Measuring

x^2 represents a number multiplied by itself. From the sign rules on multiplication described above,

if $x = 0$,   then $x^2 = 0$;
if $x > 0$,   then $x^2 > 0$;
if $x < 0$,   then $x^2 > 0$.

Remember that the product of two 'pluses' is a 'plus' and the product of two 'minuses' is also a 'plus'. In multiplying a number by itself, we must be multiplying like with like, so the answer cannot be negative (it cannot have a '-' sign).

- When x is a real number, it is impossible for x^2 to take a value less than 0.

'Let's not think about such problems, then', is not an adequate response, for such problems do arise from natural circumstances. It may be useful to be able to deal with them. Remember that great advances come from overcoming difficult problems. There's no doubt that the problem in this case is difficult; no amount of ingenuity can come up with a number whose square is -9, -538 or simply -1, the easiest case. If a square root for -1 can be found, then the other cases can also be cracked.

The mathematician's answer to the dilemma is brilliant in its simplicity, but audacious in its presumption.

- If no such number exists, we will pretend that one does, and carry on working with this pretend number.

Assume that a number i exists such that

$$i^2 = -1.$$

Now all the other intractable difficulties disappear like mist clearing in the sunlight. Once more, suitable definitions of the binary operators are set up to be compatible with those for real numbers. Thus, for example,

$$\sqrt{(-9)} = \sqrt{(-1*9)} = \sqrt{(-1)}*\sqrt{(9)} = 3i,$$

as $\sqrt{(-1)} = i$ and $\sqrt{(9)} = 3$. This assumes we take 3 rather than -3 for $\sqrt{9}$; note that -3i is also a correct alternative solution to this problem. Roots of any negative values can be found in terms of i. As the existence of i has been assumed, it is only fair to give it a name. i, or any multiple of i, is called an imaginary number. Once more, there is a beautiful simplicity – not only do mathematicians come up with something that doesn't exist, they admit it openly. Engineers find the same concept useful particularly in dealing with electrical circuits, but they habitually use the character i to represent electric current; they represent the mathematician's i by the character j.

Imaginary numbers are intrinsically different from real numbers. They do not mix easily, as do rationals and irrationals. No 'sandwiches' can be made; it is not possible to compare real numbers and imaginary numbers for size using the familiar comparators of <, =, > and so on. Even when attempts are made to force them together, they remain as essentially separate as oil and water. The failure of such attempts to mix unlike quantities results in the definition of a hybrid and clumsy beast called a complex number. All complex numbers z can always be expressed uniquely in the form

$$z = x + iy,$$

where x and y are real numbers. In this structure, x is the real part and y is the imaginary part. Note that both the real and imaginary parts are real numbers. z could be expressed as an ordered pair of reals, (x, y). The superstructure of $x + iy$ is normally retained, however, as it gives an easy reminder of the sort of number that is being dealt with.

Laws of algebra for complex numbers can now be defined to be compatible with their real number equivalents. The standard binary operations of addition, subtraction, multiplication and division are worked through as follows.

- $$\begin{aligned} z_1 + z_2 &= (x_1 + iy_1) + (x_2 + iy_2) \\ &= (x_1 + x_2) + i(y_1 + y_2); \end{aligned}$$

- $$\begin{aligned} z_1 - z_2 &= (x_1 + iy_1) - (x_2 + iy_2) \\ &= (x_1 - x_2) + i(y_1 - y_2); \end{aligned}$$

- $$\begin{aligned} z_1 z_2 &= (x_1 + iy_1)(x_2 + iy_2) \\ &= x_1 x_2 + i x_1 y_2 + i x_2 y_1 + i^2 y_1 y_2 \\ &= (x_1 x_2 - y_1 y_2) + i(x_1 y_2 + x_2 y_1). \end{aligned}$$

- $$\begin{aligned} z_1/z_2 &= (x_1 + iy_1)/(x_2 + iy_2) \\ &= \{(x_1 + iy_1)(x_2 - iy_2)\}/\{(x_2 + iy_2)(x_2 - iy_2)\} \\ &= \{(x_1 x_2 + y_1 y_2) + i(-x_1 y_2 + x_2 y_1)\}/(x_2^2 + y_2^2) \\ &= (x_1 x_2 + y_1 y_2)/(x_2^2 + y_2^2) + i\{(x_2 y_1 - x_1 y_2)/(x_2^2 + y_2^2)\}. \end{aligned}$$

*

In each case the result is given in '$x + iy$' form, where the real part and the imaginary part can easily be identified. Some ingenuity is needed to achieve this in the cases for multiplication and division. Both incorporate $i^2 = -1$ with the standard real number rules for use of brackets and commutativity of multiplication. The trick in establishing the form for division (line *) is to multiply the numerator and denominator by $(x_2 - iy_2)$, known as the complex conjugate of the denominator $(x_2 - iy_2)$. The complex conjugate $\bar{z} = x - iy$ of a complex number $z = x + iy$ is found by changing the sign of the imaginary part. It has the property that

$$z\bar{z} = (x + iy)(x - iy) = x^2 - i^2 y^2 = x^2 + y^2,$$

which is a real number. Thus, the denominator of the division is reduced to a real number, so it is easy to find the real and imaginary parts of the result of the division by ordinary real number division.

When all definitions have been made, the set of complex numbers can be shown to be closed with respect to addition, subtraction, multiplication, division and root operations. A more general problem that underlies the search for completeness in the

2. Numbers, Counting and Measuring

preceding number types can also be challenged. A polynomial of order n in x is expressed in the form

$$a_n x^n + \ldots + a_3 x^3 + a_2 x^2 + a_1 x + a_0$$

where a_0, a_1, a_2, a_3 and a_n are real constants (fixed real values) and x is a real variable. It can be shown that the equation

$$a_n x^n + \ldots + a_3 x^3 + a_2 x^2 + a_1 x + a_0 = 0$$

has *at most* n values of the variable x that satisfy it, or n roots. (This meaning of 'root' as a value of a variable that satisfies an equation is distinct from the 'square root' or 'cube root' definition.) Such equations may have no real valued roots; for example,

$$x^2 + 1 = 0,$$

in which $n = 2$, $a_2 = 1$, $a_1 = 0$, $a_0 = 1$, has no real roots, as it reduces to the form

$$x^2 = -1,$$

which has two solutions; $x = i$ and $x = -i$. Polynomial equations may have fewer than n real roots; for example, we can set

$$n = 3, a_3 = -1, a_2 = 1, a_1 = -1, \text{ and } a_0 = 1$$

to give the equation

$$-x^3 + x^2 - x + 1 = 0,$$

which has only one real valued root $x = 1$, although $n = 3$ (its other two roots also happen to be i and -i, as in the above example). We do not explain how these values can be found, but the result can be justified by showing that $x = 1$ (or i or -i) makes the equation 'balance'. The equation cannot always be solved to 'its fullest extent' using real numbers alone.

If the problem is recast in terms of complex numbers, a complex polynomial equation of order n has $a_n, \ldots a_3, a_2, a_1, a_0$ as complex constants and z as a complex variable to give

$$a_n z^n + \ldots + a_3 z^3 + a_2 z^2 + a_1 z + a_0 = 0.$$

The fundamental theorem of algebra states that such an equation has exactly n roots $z_1, z_2, z_3, \ldots z_n$, all of which can be expressed as complex numbers, so that it can be expressed in the form

$$a_n(z - z_1)(z - z_2)\ldots(z - z_n) = 0.$$

Casting the two examples above in the form of complex problems, we can re-express the equation

- $z^2 + 1 = 0$

as $(z + i)(z - i) = 0,$

giving roots -i and i;

- $-z^3 + z^2 - z + 1 = 0$

is expressed as

$(-1)(z + i)(z - i)(z - 1) = 0,$

giving roots -i, i and 1. Again, we do not show how these expressions are found (using an algebraic process known as factorization), but results can be checked by multiplying the separate brackets.

This important theorem confirms the completeness of the complex number set. For this class of problems, the complex number set produces a complete set of solutions. This indicates that we have come to the end of the line in our search for new forms of number. The complex number set is enough to satisfy a broad range of problems.

The development of number systems is upwardly compatible. New systems are enlargements; they contain previous systems without destroying them. The laws of sub-systems remain valid in the larger systems. The complex number set has a sense of completeness. There are no numbers outside this set – the exterior of the complex number set is null, a set with no elements (fig 2.3). This also shows the way in which the development of number builds by expansion while still containing the previous system within the expanded version.

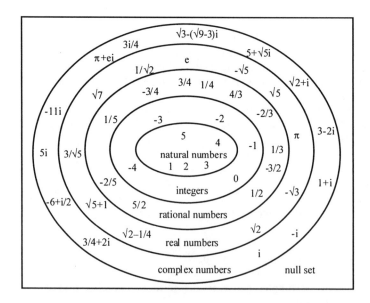

Figure 2.3 Number systems and some examples of numbers

Representations of Number

In the section above, the familiar denary or decimal representation of numbers has been used. Roman numerals are also regularly encountered on watch faces, page numbers or dates. The year 1977, to take a relatively rich example, is represented in Roman numerals as MCMLXXVII, interpreted in denary form as the sum of 1000 (M), 100 less than 1000 (CM), 50 (L), 20 (XX), 5 (V) and 2 (II). If a lower-value symbol appears immediately before one of the next higher status, the implication is that the lower is subtracted from the higher. Symbols always represent the same numerical value regardless of their positions, although their positions relative to each other indicate whether they are to be added or subtracted. Even fairly simple arithmetic tasks are difficult to perform in the Roman system, so it has no more than a decorative role today, as in film or book copyright notices. Number systems based on absolute position, where symbols for digits are interpreted in terms of their location within a number representation rather than relative position, have proved more easy to manipulate arithmetically. The denary systems is an example – the different digits of the number 11.1, or 'eleven point one', for instance, indicate that the number comprises 1 ten, 1 unit and 1 tenth.

Universality of the denary system leads to its acceptance as the 'natural' system. It is a good system, but it is not the only useful one. It is probably an accident of nature that a number base of ten is used, as early civilizations may have learned to count on their ten fingers and thumbs. The denary system has ten symbols for digits 0, 1, 2, 3, 4, 5, 6, 7, 8 and 9; the word 'digit' also means 'finger'. Advanced arithmetic systems have been built on other bases. The Mayan culture of central America used a number base of twenty – they probably wore open-toed sandals which extended their basis for counting. Babylonians used a sub-base of tens and sixties to create an eventual basis of 360.

The fact that use of the denary system has persisted indicates that it is not a bad system for humans. The balance between how many symbols are used and the relative length of numbers seems to be suitable for human memory. A lower number base has fewer symbols, but requires longer strings of symbols to represent values, vice versa for higher number bases. Imagine trying to remember a few telephone numbers in binary (base 2) systems as a string of zeros and ones, or having to identify 100 symbols representing digits in a base 100 system. These two cases are extreme; number bases at about ten would probably also have persisted had they been naturally suggested. One difficulty is that ten has relatively few natural divisors: only 1, 2 and 5. With a decimal system it is difficult to divide sums of money, say, into thirds, although the author is *not* suggesting a return to the previous British systems which had a multiplicity of number bases. There were 12 pence in each shilling, 20 shillings in each pound for money calculations. In terms of weight, there were 16 ounces to each pound, 14 pounds to each stone, and so on. Length had 12 inches to one foot, 3 feet to one yard, 22 yards to a chain, and so on. There is much to be said for standardization.

However, if humans had one extra finger per hand, we could well count in a base 12 system, with divisors 1, 2, 3, 4 and 6, giving fewer recurring numbers on division (these are not 'recurring decimals', as 'decimal' implies using base 10).

Table 2.1 Counting in denary and binary

Denary	Binary	Denary	Binary	Denary	Binary
0	0	8	1000	16	10000
1	1	9	1001	17	10001
2	10	10	1010	18	10010
3	11	11	1011	19	10011
4	100	12	1100	20	10100
5	101	13	1101	21	10101
6	110	14	1110	22	10110
7	111	15	1111	23	10111

Other number representations occur in computer systems. Digital computers use, internally, binary arithmetic with a number base of two.[2] In the usual base 10 method, digits are represented by one of ten characters: 0, 1, 2, ... 9. In a binary system there are only two binary digits (abbreviated to bits): 0 and 1. These can be considered as equivalent to the absolute absence or presence of a current or voltage in an electrical circuit, giving guaranteed levels of precision in a discrete system compared to the imprecise measurement of continuous current used in analogue computers, a process that is prone to measuring error. A bit represents the basic minimum item of information (on or off) and this can be identified with some certainty compared to attempting to assess the 'amount' of a current. To count in the binary system, start with 0, alternate 0 and 1 in the first position, then pairs of 0s and 1s in the second, sets of four 0s and 1s in the third, sets of eight 0s and 1s in the fourth, and so on. Table 2.1 shows this cycle compared to the denary equivalents, with leading zeros suppressed (0011 is the same as 11).

A group of people standing shoulder to shoulder in line can act as a binary counter using a simple set of rules. Each person is allowed to raise or lower the left arm (indicating a digit one when raised, zero when not), and must nudge the person to the right when the left arm is lowered. When a nudge is received, a person must change the state of the left arm (raise a lowered arm, or lower a raised arm with the consequential nudge being invoked). Starting with all arms down, when a stimulus in the form of a series of nudges is applied to the leftmost person, the row's state counts the number of stimulus nudges received, up to the limit of the number of people in the row. Practically, this usually ends in disarray, as humans find it difficult to follow such boringly simple instructions. However, computers can be

---

[2] The alternative to a digital computer is an analogue computer, now infrequently used except as a special-purpose machine. These work directly on a continuous 'analogue' of the object being studied. For example, an electronic circuit can be set up so that the equations governing its behaviour are equivalent to those for a problem in structural design, say. By monitoring currents or voltages from the circuit, the behaviour of an analogous structural system can be forecast, without having to construct the structure itself. It is far cheaper to patch together an electronic circuit than to build a bridge, or even a model of a bridge. Also, the equations can be scaled to accelerate the effects of time. An analogue computer has mechanisms enabling such circuits to be set up and monitored to conform to particular equation sets, but the devices have limited accuracy. Powerful numerical analysis packages on digital computers have all but replaced analogue devices.

2. Numbers, Counting and Measuring

constructed to undertake such mindless activity without error; at a basic level, they can be considered as fast-working morons, following simple instructions without question. Ingenious programs do build the complexity of activity beyond such simple repetition, but this is still the basis of modern digital computers.

This method can be used to count on the fingers – with ten fingers, it is possible to count up to 1023, with an extended finger for 1 and a folded finger for 0, as

$$1023_{10} = 1111111111_2$$

suffices indicating the number base. This is more easily calculated by noting that

$$1111111111_2 = 10000000000_2 - 1_2$$

and that

$$10000000000_2 = 2_{10}^{10} = 1024_{10}.$$

The latter result is easy to establish from the positional nature of binary numbers. The digit one in position 11 with ten zeros to its right represents 2^{10} in denary. This positional property enables relatively easy conversion of binary numbers to their denary equivalent values. For example, 10111_2 can be reconstructed in denary form by 'peeling off' digits from right to left, giving

$$(1*1) + (1*2) + (1*4) + (0*8) + (1*16)$$
$$= 1 + 2 + 4 + 16 = 23.$$

The suffix 10 indicating that these are denary values is left out as this should be clear from the context. Each binary digit position represents a value twice as large as the one to its right (in the denary systems, each denary digit position represents a value ten times as large as the one to its right). Evaluating a binary number in denary form involves repeated multiplication by 2, so it is not surprising that the reverse process involves repeated integer division by 2, at each stage holding the remainder term. The remainder on division of 23 by 2 is 1, as

$$23 = 2*11 + 1.$$

The remainder must be less than the dividing value (or divisor), so division by 2 can only leave remainders 0 or 1. For example, to interpret 23 as a binary number,

23/2 = 11,	remainder 1;
11/2 = 5,	remainder 1;
5/2 = 2,	remainder 1;
2/2 = 1,	remainder 0;
1/2 = 0,	remainder 1.

The binary interpretation 10111 of the denary number 23 can now be read off the list of remainders from bottom to top. As we noted above, although fewer symbols are used in binary form, more digits are used to represent a number compared to the denary form.

Early computers were 'programmed' by hard wiring sets of binary commands; operating nodes were linked by soldered wires or through jack plugs. Primitive programs usually consisted of instructions in the form of long strings of 0s and 1s. These were easy to misinterpret, so methods more easily recognized and remembered by humans were developed. The octal and hexadecimal systems (with number bases 8 and 16 respectively) are easily transformed to and from binary systems as their number bases are powers of two. The octal digits 0 to 7 represent all combinations of 3 bits (binary digits), as shown in the first column of table 2.1, if we insert leading zeros to 'pad out' each number to 3 bits length. The hexadecimal digits whose decimal values are 0 to 15 (usually written as 0, 1, ..., 9, a, b, c, d, e, f) represent all combinations of 4 bits, as shown in the first two columns of table 2.1 (now padding with zeros to 4 bits length). Numbers written in octal or hexadecimal systems are more easily remembered or interpreted by humans and can readily be reinterpreted in the binary form that is more digestible to digital computers. They are not developed further here, but are mentioned as they may be encountered in supporting literature for computer graphics. This is also a part explanation of why powers of two occur so frequently in computer specifications. Memory is often sold in 'chunks' of 32K or 64K, where K represents $1024 = 2^{10}$ (and remember that $32 = 2^5$ and $64 = 2^6$). Each memory location has to be given an individual identity, its 'address' in the computer's own binary coding. It is easier if this address can be identified as a set number of bits, which can be achieved by clustering memory in powers of two.

The simplicity of arithmetic manipulation is one of the main reasons for using binary representation of numbers in digital computers. The rules for addition and multiplication of single bits are easily expressed:

$0 + 0 = 0,$ $0*0 = 0,$
$0 + 1 = 1,$ $0*1 = 0,$
$1 + 0 = 1,$ $1*0 = 0,$
$1 + 1 = 10,$ $1*1 = 1.$

A product (indicated using *) takes the value 1 only if both initial bits are 1. The addition rule is slightly more complicated. The 'sum bit' is 1 when both initial bits are different; the 'sum bit' is zero when both initial bits are the same, but there is an extra 'carry bit' of 1 when both initial bits are 1, otherwise the carry is taken as 0, and is not shown in the examples above, just as you would write 'twenty' as 20 rather than 020. Logical circuits such as adders and binary counters are devised to perform such operations on the internal representations of number. These can then be constructed from electronic components and incorporated in digital computers. Designers of such logical circuits make use of Boolean algebra, discussed later in this chapter.

The Computer Representation of Number

To state that computers represent numbers in binary form is a gross simplification. The basic storage element in a digital computer is a 'bit', representing binary or logical value 0 or 1. The internal representation of numbers is related to the pure binary form used by mathematicians, but is a distorted version to enable most efficient use of the limited availability of computer memory. Memory is organized hierarchically as bits, bytes and words. Typically, a byte is 8 bits, a word is 4 bytes or 32 bits. Numbers are stored within these structures as patterns of 0s and 1s; individual manufacturers may use different ways of arranging these patterns and there are two IEEE standards for number representation (IEEE 754 and IEEE 854). However, there are some basic similarities that are discussed here using typical forms of representation.

We consider first a few definitions of the ways in which numbers can be represented, taking numbers in the denary system as examples. Numbers resulting from calculations may be described as 'correct to four significant figures', or 'correct to four decimal places'. Consider how we would describe the number π in this way; its value is approximately 3.14159265. The first significant digit is the leftmost non-zero digit in the representation. In this case, it is 3. If we start from this digit and only use four digits to represent the number as accurately as possible, we have 3.142 as the representation (the rule is that the leftmost figure must not be zero). Clearly, 3.14159265 lies between 3.141 and 3.142. Consider the digits that come after 3.141: as they start with the digit 5 or more, the value of π must be closer to 3.142 than 3.141, so 3.142 is chosen as the most accurate representation correct to four significant figures. Representing π correct to four decimal places, by a similar argument, counting four digits from the decimal point, we have 3.1416 as the appropriate value. Both these examples have 'rounded up', but this is not always the case.

If the 'remainder' is exactly halfway between the two contending approximate values, one of several conventions must be adopted consistently. A typical way is to 'round up' whenever the remainder is midway between the two possible values, but this imposes an upward bias on approximations. An alternative, spreading some errors upwards, others downwards, is to 'round to even'. For example, if we have to choose between rounding towards a number ending in either 7 or 8, we choose 8.

Representing numbers correct to four significant figures controls errors relative to the size of the representation; there are always four significant figures, so this controls the maximum percentage error. Suppose we have a number between 1000 and 9999 inclusive when represented to four significant digits. The error in this representation must be between -0.5 and 0.5. The greatest relative error is found by combining the largest possible numerical error with the smallest possible value represented (the value 999.5 is rounded up to 1000). In this case, the relative error is $(0.5/999.5) \approx 0.0005 = 0.05\%$ (the precise figure is -0.00050025 or 0.050025%). When quoting a number to four significant figures, you are guaranteeing an error of no more than about 0.05% or one twentieth of a percent. Representing numbers to four decimal places controls the absolute error; it must always lie between 0.00005 and -0.00005, regardless of the size of the number. Thus, for example, the error in representing 0.00005 as 0.0001 is limited to 0.00005 absolutely, but is $0.00005/0.00005 = 1 = 100\%$ in relative terms. The analysis of errors is properly the domain of numerical analysis. Within permitted ranges (see below), computers store integers to absolute precision and real (or floating point) numbers to relative precision. With complicated sequences of calculations in which intermediate results are used to calculate subsequent values, care is needed to ensure errors do not proliferate beyond acceptable levels. The subject of numerical analysis deals with this topic.

We consider just one simple problem to illustrate the concept. Suppose we know that numbers a and b are stored in approximate form as $\alpha = (a + \delta)$ and $\beta = (b + \gamma)$, where δ and γ are small absolute errors. If a and b are multiplied, the approximate value found will be

$$\alpha\beta = (a + \delta)(b + \gamma) = ab + a\gamma + b\delta + \delta\gamma,$$

the absolute error being $(a\gamma + b\delta + \delta\gamma)$. If δ and γ are small compared with a and b, the value of $\delta\gamma$ is extremely small compared with other terms in this expression, so it can be approximated as $(a\gamma + b\delta)$. The relative error is found by dividing the error by the value that *should* have been found. The relative error in calculating the product of two imprecise values is approximately

$$(a\gamma + b\delta)/(ab) = (b\delta/ab) + (a\gamma/ab) = \delta/a + \gamma/b.$$

Note that δ/a is the relative error in the original value of a and γ/b is the relative error in the original value of b. Thus, in multiplying two values with small relative errors, the maximum relative error in the product is approximately equal to the sum of the relative errors in the original values. In practice, we do not know ab, but do know its approximate value $\alpha\beta$. A good estimate of the error can be found using α and β in place of a and b. There are many other results of this form that are useful in checking potential problems in errors of evaluation. Here, we merely point out that they exist and that care needs to be taken, particularly with relatively long strings of calculations on imprecise values.

Computer languages can define different forms of number. We consider four essential number types: short and long integers, floating point numbers (reals) and double precision floating point numbers. The integers can also be categorized as 'signed' or 'unsigned'.

A 'short' integer is typically held in 2 bytes, so has 16 bits available. These can range from 0 000 000 000 000 000 to 1 111 111 111 111 111 in binary number form or 0 to 65 535 in denary[3] (the upper limit is calculated as $2^{16} - 1$). If negative integers are needed, one of these bits (typically the first) is used to indicate the sign of the value encoded, leaving 15 bits for numerical information. There is one ambiguous value here, as +0 and -0 mean the same thing. Most manufacturers use this to allow an extra value in the negative range, using a 'hidden bit' to interpret -0 000 000 000 000 000 as -10 000 000 000 000 000. This gives a range from -10 000 000 000 000 000 to +111 111 111 111 111 (-2^{15} to $+2^{15} - 1$), -32 768 to +32 767. This range can also be achieved by interpreting a signed short integer to take the value equivalent to the unsigned version minus 32 768; this is known as the 'biased form' of representation.

---

[3] The SI standard shows four-digit numbers without separation, as 2562. For more than three digits, a space is used as separator rather than the common comma. The SI method is used here, so one million is written as 1 000 000 rather than 1,000,000. To avoid numbers being split across lines at a separator, internal spaces can be typed as 'control space'. Many countries use a comma instead of a decimal point, so 12,352 could be ambiguously interpreted as 12 352 or 12.352.

2. Numbers, Counting and Measuring

A 'long' integer is stored in 4 bytes or 32 bits, and also has unsigned and signed form. These allow storage of numbers from 0 to $2^{32} - 1$ or 4 294 967 295 in unsigned form or -2^{31} to $2^{31} - 1$ (-2 147 483 648 to 2 147 483 647) in signed form. Integer values outside these ranges cannot be stored in digital computers other than by using special routines to merge the contents of more than one standard integer memory location. Real numbers allow a greater range, so users may want to use these in extreme situations, although their manipulation is slower due to greater complexity and they are not as precise.

The storage of real or 'floating point' numbers is more complicated. They are stored in standardized 'base-exponent' form. It is useful to consider denary equivalents first. The number 2573 represents exactly the same value as (2.573×10^3), (25.73×10^2) and (0.2573×10^4). Similarly, 0.00034 is the same as (3.4×10^{-4}), (34×10^{-5}) and (0.34×10^{-3}). Multiplication by 10^n moves the decimal point n places to the right; multiplication by 10^{-n} moves it n places to the left. In the representation (2.573×10^3), 2.573 is known as the 'significand' or 'mantissa' and 3 is the 'exponent'. A 'normalized' representation has one non-zero digit to the left of the point in the significand for all numbers other than zero, so our two examples are given in normalized form as (2.573×10^3) and (3.4×10^{-4}). These are sometimes written as 2.573E3, or 3.4E-4, the character E (or even e) indicating the exponent.

Similar definitions can be made in binary numbers, this time using a binary point in which each shift to the right represents multiplication by 2, and each shift to the left represents division by 2. Thus, binary numbers -10100 and 0.0000011 can be represented in standardized form as -1.01×10^{100} and 1.1×10^{-110} respectively in binary notation; here the '10' is a binary number, equivalent to the denary number 2. These have values $-1.25 \times 2^4 = -20$ and $1.5 \times 2^{-6} = 0.0234275$ in denary form. The digit 1 *must* appear to the left of the decimal point in the binary standardized form for all numbers other than zero, which is again the anomalous value. In the computer representation of numbers in 'base-exponent form', using the binary number base 2, standards specify how many bits are available to the significand and the exponent. The IEEE single precision definition of a floating point number uses 4 bytes or 32 bits, allocating one bit for the sign of the significand, 23 bits for its numerical content and the remaining 8 bits for the exponent. IBM's System/370 version uses one fewer bit for the exponent and one more for the significand, thus reducing the range of numbers that can be represented. Also, the ordering of these components of a number can vary from computer manufacturer to computer manufacturer. A typical arrangement sees the significand sign bit as the leftmost, the next 8 bits represent the exponent in biased form, the remaining 23 bits contain the numerical content of the significand, with the most significant bit to the left. Other systems have different internal word organization.

Consider the representation of floating point numbers in the IEEE specification. We have not mentioned a sign bit for the exponent. This is usually stored using a 'displacement' method, so values available in an 8-bit unsigned number are 'shifted down' by subtraction to give an exponent range from -126 to 127; this is the biased form mentioned above. This is not the full range available in an 8-bit value, but other exponent values are reserved for some special quantities such as +0 or -0. Also, a 'hidden bit' is normally used in the significand. For all numbers (other than zero), the 23-bit numerical content of the significand is considered to have '1.' as a prefix, thus adding one significant bit to the representation. This has a minor effect on the range of values allowed, but has more effect on the precision of the depiction, allowing more significant figures. The numerically largest significand is approximately 1.111... in binary form, a little less than 2 in denary. The outer limits of numbers available in single precision floating point is thus from approximately (-2×2^{127}) to (2×2^{127}) or (-3.4×10^{38}) to (3.4×10^{38}). Figures that fall outside these values may give rise to a 'floating point overflow' error. The related error, 'floating point underflow', arises when values fall between the inner limits of the representation close to zero, from (-1×2^{-126}) to (1×2^{-126}) – remember that the numerically lowest allowed value of the significand is 1 using the 'hidden bit'. In denary form, this is approximately (-1.2×10^{-38}) to (1.2×10^{-38}).

We have so far ducked the explanation of how to deal with zero using the 'hidden bit' that forces an implicit minimum numerical value of 1 onto the significand and therefore does not allow the number represented to take value zero. The IEEE standard reserves some special bit patterns for +0 and -0, although it does not discriminate in value between these. Exponents with denary values 127 and 128 are used for this purpose, as well as for other special values like the 'non numbers' +∞ and -∞. With all these complexities, it is clear that computer arithmetic must differ from pure binary arithmetic, due to the unnatural ways in which numbers are encoded. However, it does follow correct procedures from these encoded representations, providing correct results within the limits of precision allowed.

Double precision floating point numbers are held in two words of 4 bytes each, comprising 64 bits. In the IEEE standard, this comprises a discriminand sign bit, an 11 bit exponent and a 53 bit representation of the numerical value of the discriminand. Limits for the exponent are set from -1022 to 1023, allowing values from (-2×2^{1023}) to (2×2^{1023}) to be represented: approximately (-1.8×10^{308}) to (1.8×10^{308}). Although technically limited in range, these are enormous numbers, and most realistic problems can be solved within these allowed limits. A further extension in some systems is to allow 'long double' or 'double extended' within a 79 or 80 bit range depending on implementation, with exponents of at least 15 bits.

Because of the organization of floating point numbers, the possible values that can be represented are not evenly spread along the number line. As the exponent increases, values become multiplicably larger, and separations between successive values within the range of the significand become wider. Thus, there is higher density of values close to zero than there is far from zero. This is not a significant problem in terms of relative error, so using sensible algorithms based on sound principles of numerical analysis, relative error limits can be bounded within acceptable ranges in all other than exceptional cases.

Boolean Algebra

Operations on binary numbers can be defined in terms of Boolean algebra. This is an abstract technique that has many potential uses in computer graphics and has a range of parallel (and meaningful) methods that are its exact equivalents. Only the two values 0 and 1 are allowed in Boolean algebra. These can be interpreted as 0 for 'false' and 1 for 'truth', and the laws of Boolean algebra used to analyze the truth or falsity of combinations of statements, given the truth or falsity of the component statements. This interpretation of Boolean algebra is a form of symbolic logic.[4] Another interpretation in set theory interprets 0 to indicate that a point or value lies outside a particular set, with 1 indicating that the point lies in the set. All these meanings are potentially useful in computer graphics, so it is worth considering all three alongside each other. Boolean variables are combined using binary operators (this does not mean they combine binary numbers, but that they combine *two* values to give one) of Boolean addition and multiplication (table 2.2). To define addition, the sum is one if either the first combining digit *or* the second *or* both have value one. In multiplication, the product is one if both the first *and* the second digit have value one. Logical operators are **or** (indicating that 'either one statement *or* the other *or* both are true') and **and** (indicating that 'both the first *and* the second statement is true'). Sets are linked by union (symbolic representation ∪, meaning that a point is in either one *or* the other of the sets being linked, which includes being in both) and intersection (represented by ∩, meaning that the point is in both the first set *and* the second).

The equivalence of the three interpretations is established by noting that the three parts of table 2.2 have identical body structures using different headings and symbols. The Boolean algebra tables are abstractly defined laws. In symbolic logic, the statement 'P **or** Q' is true if either P or Q (or both) are true statements; 'P **and** Q' is true only if both P and Q are true – these are the occasions when the 'truth value' is set to one. In set theory, the 'union' of sets A and B (indicated as A∪B) is the set of all points that are either in A or in B or in both, including overlaps. The 'intersection' of A and B is the set of all points which are in both A and B (indicated as A∩B). The set interpretation will be used later to combine various sets of points to create complex shapes from simpler 'primitive' forms. Boolean variables are used in computing to evaluate the truth (value 1) or falsity (value 0) of comparisons; for example the comparison

$$5 > 3$$

has Boolean value 1, as it is true, but

$$8 > 11$$

has Boolean value 0, as it is false.

Two other Boolean operators may be used in computer graphics. The unary operator (it operates on only one variable) of negation, (the complement or **not**),

Table 2.2 Equivalent operations in Boolean algebra, symbolic logic and set theory

Boolean algebra				Symbolic logic				Set theory			
Digits		+	*	Truth value		or	and	Point pos'n		∪	∩
0	0	0	0	false	false	false	false	out	out	out	out
0	1	1	0	false	true	true	false	out	in	in	out
1	0	1	0	true	false	true	false	in	out	in	out
1	1	1	1	true	true	true	true	in	in	in	in

[4] In symbolic logic, only 0 for 'false' and 1 for 'true' can occur. Fuzzy logic allows 'degrees of certainty' values between 0 and 1. The method has been applied to a number of applications, but is not of major interest for this text.

which reverses the value of a given variable, is a fundamental or axiomatic definition. The symbolic Boolean form of complement, represented here by a bar above the value to be complemented [5] gives

$$\bar{x} = 1 \text{ when } x = 0;$$

$$\bar{x} = 0 \text{ when } x = 1.$$

Its logical interpretation is equivalent to saying that **not**(true) is false, **not**(false) is true. In set theory, $\bar{A}$ is the set of all points lying wholly outside set A.

The difference (representing Boolean subtraction) is not an axiomatic definition, as it can be derived using Boolean product and complement. Symbolically,

$$x - y = x * \bar{y} = x\bar{y}$$

(the symbol for logical product may be left out, its presence being implicitly understood as in 'normal' number algebra). This means that $(x - y)$ takes value 1 only when x is 1 and y is 0. The logical difference is written as 'P **and not** Q'; in set theory $(A - B)$ is the set of all points in A which are not in B.

From the axioms and definitions given here, structures of theorems can be devised. The logical interpretation of two famous laws due to de Morgan are

not(P **and** Q) = (**not** P) **or** (**not** Q),

not(P **or** Q) = (**not** P) **and** (**not** Q).

These expressions are sometimes referred to as 'P **nand** Q' and 'P **nor** Q', using **nand** for (**not** ... **and** ...) and **nor** for (**not** ... **or** ...). It can be shown that all logical manipulation of bits required in computer systems can be expressed solely in terms of **nand** or **nor** operators. These operators can be reproduced readily using electronic components, so are much used in the murky interiors of digital computers. One last special Boolean operator will be mentioned: the 'exclusive or', or **xor**. This takes the value 1 when either of the initial bits is 1, but not when both are 1. The exclusive or gives the 'sum bit' in Boolean addition and may also be used in computer graphic image manipulation.

Boolean operations can be illustrated using Venn diagrams. These have relatively obvious interpretations in the set theory interpretation and can be used to illustrate Boolean expressions for up to three variables. Some simple examples are given in fig 2.4, where sets are depicted as oval shapes and the 'universe' of all points under consideration is depicted as a surrounding rectangle. Use of Venn diagrams can help with understanding of Boolean algebra and with the development of logical expressions.

This is a very brief excursion into the area of Boolean algebra. Readers who may not have encountered the topic before should not be alarmed if they find this cursory

---

[5] The 'tilde' symbol '~' appearing before the value to be complemented may also be used to denote 'not'. Some mathematical texts use the symbols ∨ for **or**, ∧ for **and** and ¬ for **not**.

2. Numbers, Counting and Measuring 35

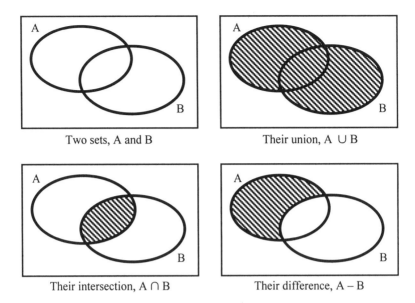

Figure 2.4 Venn diagrams used to represent Boolean operations in the set theory sense

treatment bewildering. The topic will be revisited when it is needed and its use should be more understandable when linked to concrete applications.

Summary

This chapter has described the basic features of number systems, and has delved into the initial concepts of Boolean algebra. Many computer users are unaware of the nature of these, as many car drivers are unaware of what is going on beneath the bonnet. However, a rudimentary understanding can lead users to 'drive' better.

Essentially, after reading this chapter, you should understand that there are different types of numbers. Those used in computer systems are integers: discrete numbers used for counting; and real numbers: continuous numbers used for measuring. The computer representation of a number is to a limited number of bits, thus limiting the precision of the number, so 'real' numbers or 'floating point' computer representations of reals are incomplete. They only represent a limited set of rational values which have upper and lower bounds, and which cannot represent irrational numbers. This limit on the length of representation means they can be expressed exactly as fractions. These values are accurate within the precision of the machine, but two types of error can occur in computer calculation.

One is in attempting to represent a number that is larger than or smaller than the limited memory can store. A common instance of this is in the representation of colour to limited precision. If calculations create a value larger than the largest colour limit, the image can be 'washed out' with most intense whites or, in alternative systems, colour can 'wrap around' giving dark central patches to regions which should be bright.

The other form of inaccuracy is due to accumulation of 'rounding error'. Measurements may be stored in slightly inaccurate form due to the limited precision, and the error is considered insignificant. However, if many repeated calculations are performed on such numbers, and this is typically the case in computing or synthesizing a computer graphic image, these rounding errors can comprise a significantly large fraction of the required value. Sophisticated techniques can be used to minimize this effect, and properly devised programs should reduce the chance of such problems. We will not look in detail at these, but readers should be aware that the apparent infallibility of the digital computer can be breached by certain circumstances, for example when dividing by a number close to zero.

We took the trouble to describe the complex number system, not yet explained in this summary. Its influence in computer graphics is mainly in the creation of certain forms of fractal form, particularly those related to the Mandelbrot set. However, a complete understanding of complex numbers is not really necessary to create these forms, as the concept can be approached from another angle.

Boolean algebra is an intrinsic part of most computer languages. This also has uses in computer graphics in combining various structural forms to produce more complex structures for depiction. For example, the essential structure of the tower of Big Ben can be represented as the union of a rectangular block with a suitably placed pyramid. What about all the wrinkly bits of detail, you ask? Have patience, let's just start with the easy stuff!

3. Coordinates and Dimension: Representations of Space and Colour

Introduction

Computer graphics is involved with space and its occupancy. From a mental concept of a set of objects in some kind of abstract space, we define and locate those objects in a specific place. Then, we must perform calculations on the objects to be able to show their images in a display space, typically a computer monitor screen.

This chapter introduces ways of defining, measuring and 'controlling' two- and three-dimensional spaces. As is the habit in this book, concepts are more important than specifics, so although examples are used, and readers new to these may find the details a little difficult, they are encouraged to consider the underlying ideas as most important. Displaying images also involves interpreting their colours. A short section on colour is included as it involves a coordinate-like representation.

Cartesian Coordinates

Real numbers can be represented as points on a line in a one-to-one and complete relationship. This is a two-way relationship, meaning that every point represents a number and every number is represented by a point on the line (fig 3.1). A range of uninterrupted values on this line is called an interval. A closed interval, one which contains its boundary points, is written as [a, b], expressed as

$$\{x: a \leq x \leq b\}$$

in set notation, 'the set of values x such that x is greater than or equal to a and x is less than or equal to b'. A more fussy formulation would have insisted on defining x, a and b as real numbers such that a is not greater than b, but this is already formal enough for our purposes. An open interval does not contain its boundary points, written as (a, b) or]a, b[in some texts, and expressed using 'strictly less than' inequalities as

$$\{x: a < x < b\}.$$

The inequalities here do not allow the inclusion of boundary values. Sometimes,

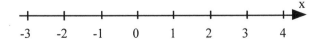

Figure 3.1 A number line

semi-open intervals are found, with one boundary point included but the other excluded, as in [3, 5).

For those with sporting interests, a soccer pitch or tennis court is closed, as the ball is still in play if it touches the boundary line, it is only out of play if the ball completely crosses over the line. Rugby and American grid iron pitches are open, as the ball is out of play if it merely touches the boundary – the line is outside the playing area. American footballers talk of 'breaking the plane'.

Ideas of intervals can be extended to higher-dimensional spaces with two, three and even more dimensions, particularly spaces expressed in terms of Cartesian coordinates.

Points in a two-dimensional (2D) space (we rely on prior experience at the moment to consider exactly what 2D means before discussing it later) have a one-to-one and complete relationship with ordered pairs of real numbers. Any point in a 2D space (typically a plane) can be represented by a pair (x, y), and every pair of real numbers (x, y) has an equivalent point in a plane. The most common method of identifying points is through Cartesian coordinates,[1] which refer points to a framework of axes defined by the user. 2D Cartesian coordinates have two mutually orthogonal axes (they are at right angles to each other): an x-axis and a y-axis. Each axis is a one-dimensional number line drawn so that the zero points coincide at the origin, usually denoted O, conventionally with an x-axis pointing horizontally to the right and a y-axis pointing upwards on a page or screen (fig 3.2). This orientation is known as a 'right handed' set of axes. The point (x, y) is found by moving x units to the right from the origin (in the x-direction) and y units upwards (fig 3.2).

There is nothing absolute about Cartesian coordinates – a set of axes forms a convenient frame of reference. Any particular set of axes is just one choice; the origin, x-direction, units of measurement could all be changed, but consistently used to define exactly the same points and objects in space as another axis configuration.

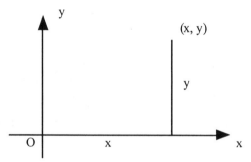

Figure 3.2 The point (x, y) in 2D Cartesian coordinates

---

[1] 'Cartesian' is written with a capital letter in honour of the philosopher and mathematician Réné Descartes, who published the method as the appendix *La Géometrie* to his 1637 *Discours de la Méthode* (this is the shortened version of a very long title). Two other appendices were *La Dioptrique*, the first published explanation of the diffraction of light, and *Les Méthéores*, the first proper explanation of rainbows. Subsequent editions of the book were published without these appendices as the publishers did not consider them particularly important.

3. Coordinates and Dimension: Representations of Space and Colour

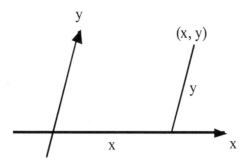

Figure 3.3 The point (x, y) in a non-Cartesian coordinate system

If different people choose to describe the same planar structure in terms of a Cartesian system, each will do so relative to coordinate axes that may be differently defined. It is unlikely that they would both choose the same origin and, unless there are directional preferences inherent in the structure, the same directions for axes. It is even possible that they would choose different measuring units (such as centimetres and inches). When such anomalies exist, it may be important to be able to transform from one description to another. That is a major feature of computer graphics systems, when more than one coordinate system may be in consideration at any time.

For example, a designer has a mental concept of 2D space from which he or she creates precise definitions of objects in a Cartesian 'world' plane. Creating an image involves transforming this 2D space into that of a graphics screen, with coordinates defined by a screen-addressing system. The conversions from concept space to 'world' space to 'screen' space (the latter is known as a window to viewport mapping) will be discussed later in discussing functions, matrices and perspective projection. Such transformations are at the heart of modelling and viewing systems in computer graphics.

Points in the plane could be defined with respect to any two number lines with a non-zero angle between them (fig 3.3), but calculations such as distance finding are easier in orthogonal systems, so the Cartesian system is the norm. It is even more difficult to find distances between points when the axes are not straight lines, but coordinate systems in which points are related to well-defined curves (fig 3.4) are possible, even sometimes desirable. This may seem odd until you consider how we

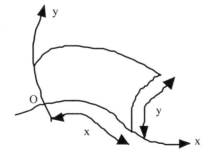

Figure 3.4 2D points referred to curvilinear coordinates

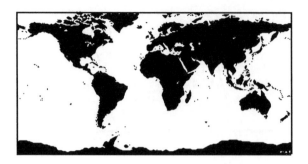

Figure 3.5 A Mercator projection; curvilinear coordinates converted to linear

look at maps of the earth. The Mercator projection distorts curvilinear coordinates of latitude and longitude to linear axes (fig 3.5). Difficulties in distance interpretation are shown when the shortest distance between two towns on the globe, following the arc of a great circle, appears as an odd curve in an in-flight magazine map.

Use of a right angle in Cartesian systems makes it relatively easily to calculate distances between points. Pythagoras' theorem relates the lengths of sides of a right-angled triangle (it is assumed readers have experienced this in another mathematical existence; this development sacrifices rigour in favour of explanatory power). If the distance between two points (x_1, y_1) and (x_2, y_2) is d, (fig 3.6), d is the hypotenuse of a right-angled triangle with the other two sides $(x_2 - x_1)$ and $(y_2 - y_1)$, giving

$$d = \sqrt{\{(x_2 - x_1)^2 + (y_2 - y_1)^2\}}$$

With different orientations of the triangle, either or both of $(x_2 - x_1)$ and $(y_2 - y_1)$ could be negative, but the formula still 'works', as squaring these produces the equivalent positive squared value. This is not a particularly pleasant formula, but is much more convenient than for non-orthogonal systems, such as that of fig 3.3. Note that d is always taken as the non-negative square root, the distance between two distinct points is always considered to be positive, regardless of the direction of travel. Clearly, d is zero only when both points (x_1, y_1) and (x_2, y_2) coincide. This only happens when both

$x_1 = x_2$ and $y_1 = y_2$.

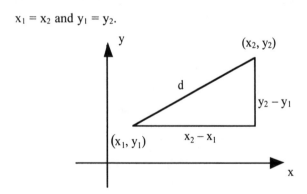

Figure 3.6 The distance d between two points in 2D Cartesian coordinates

3. Coordinates and Dimension: Representations of Space and Colour   41

In mathematics, other 'distance metrics' may be adopted for specific purposes; we merely mention the 'Manhattan grid' distance (imagine how you would get from A to B in such a regularly laid out town) as

$$|x_2 - x_1| + |y_2 - y_1|.$$

The symbols '| ... |' indicate that the magnitude (also known as modulus or absolute value) of whatever is between them is taken – the 'non-negative equivalent' of the value. For example, $|5| = 5$ and $|-5| = 5$. If a minus sign exists, it is eliminated in the 'absolute value'. The distance is the same, whichever direction you walk.

Defining Space by Equations and Inequalities

A relationship between the values of x and y expressed as an equation can be illustrated as a curve called a locus (plural loci). Figure 3.7 shows how all points (x, y) which satisfy the equation

$$y = 2x + 1$$

lie on a straight line, whereas all points (x, y) that satisfy

$$\frac{x^2}{4} + y^2 = 1$$

lie on an oval shape known as an ellipse. A number of equations representing specific shapes will be introduced in the next two chapters. The purpose of this section is to explain the concept of representing forms by equations, rather than to examine specific forms in detail, so these shapes should be taken on trust for the time being.

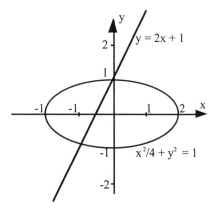

Figure 3.7 Graphs of equations in the form of a straight line and an ellipse

A typical way of establishing such forms is to plot many values of y for given values of x, and visually to assume smooth connections between the established points. For example, if we find values of y from the formula

$$y = 2x + 1$$

for x = -2, -1, 0, 1, 2, we obtain (x, y) number pairs (-2, -3), (-1, -1), (0, 1), (1, 3) and (2, 5). Plotting these points (this means marking the positions of these points typically with a cross) in a Cartesian plane, it appears that they all lie on the straight line shown in fig 3.7, so the inference is that *all* points satisfying the equation lie in this line. Inference is notoriously unreliable; just because a system does something for a few instances does not mean it is always going to behave in the same way. However, more sophisticated mathematical analysis to be discussed in chapter 5 shows it to be correct in this case, as for the illustration of the ellipse shown in the same figure. These graphs do what is expected of them and are said to be 'well behaved'.

The essential ellipse shape could be found by plotting a number of points that fit the equation, and assuming that if they are joined smoothly the new points formed in this joining process will also satisfy (or fit) the equation. This is not as easy as for the line, partly because the ellipse equation cannot be easily rearranged into a formula giving y directly. For example, inserting the value x = 0 into the ellipse equation

$$\frac{x^2}{4} + y^2 = 1$$

gives

$$y^2 = 1.$$

This has two possible solutions at y = -1 and y = 1 as both $(-1)^2$ and $(1)^2$ are equal to 1. Thus, for x = 0, there are two points on the curve: (0, -1) and (0, 1). This should be clear from the shape of the ellipse (fig 3.7), representing the two points on the ellipse's oval shape above and below the x-axis. All other values of x in the open interval (-2, 2) give similar results, there being two square rooted values of y for the particular value of x. When x is -2 or 2, the equation reduces to

$$1 + y^2 = 1$$

or

$$y^2 = 0.$$

This has only one solution, y = 0, as there is no difference between 0 and -0. If we give x any value outside the closed interval [-2, 2], there is no real solution for y, as the equation will give y^2 a negative value. When x = 4, the equation becomes

$$4 + y^2 = 1,$$

which reduces to

$$y^2 = -3.$$

3. Coordinates and Dimension: Representations of Space and Colour

This has no real valued solution, as explained in chapter 2. A complex valued solution for y could be found, but this cannot be plotted against the real valued y-axis in Cartesian coordinates. Figure 3.7 shows that the graph of the ellipse does not appear outside values of x in the interval [-2, 2]. The line in fig 3.7, on the other hand, can be extended for any value of x, it is theoretically limitless, although the size of our or any practical illustration means we can only draw a part of it.

As inequalities in the number line represent intervals or ranges of numbers, inequalities in 2D coordinate systems represent regions of the 2D space. Figure 3.8 shows how the line and curve of fig 3.7 are replaced by regions with these forms as boundaries, sometimes called half-spaces, when equal signs are replaced by inequalities. As illustrated, the region

$$y > 2x + 1$$

lies to the dotted side of the line

$$y = 2x + 1.$$

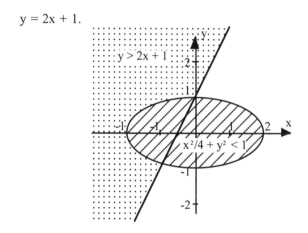

Figure 3.8 Regions represented by inequalities based on the equations of fig 3.7

Taking the example point (-1, 1), it is clear that the inequality $y > 2x + 1$ is true. (When $x = -1$, $2x + 1$ takes the value -1, so $y = 1$ is greater than this.) Consider a line or any curve joining (-1, 1) to a point for which the inequality is false, for example (1, 1). As we move away from the starting point (-1, 1) the inequality remains true until the curve reaches the line $y = 2x + 1$; the status of the inequality can only change at points that make both sides equal. Thus all points to the same side of the line as (-1, 1) make the inequality true, as it is possible to join them without changing the sense of the inequality.

Similarly, all points for which

$$\frac{x^2}{4} + y^2 < 1$$

lie inside the ellipse $x^2/4 + y^2 = 1$ (illustrated by hatching in fig 3.8). The test point (0, 0) makes the inequality true (as 0 is less than 1), so all points to the same side of the boundary curve (in this case inside it) also satisfy the inequality. This is also considered as a 'half-space', although it cannot be considered to have cut off 'half' of the (x, y) plane. There is an an intuitive idea that this is less than a half, but the philosophical question of fractions of infinity will not be expanded on here.

The inequality used, '<', indicates that only points inside the ellipse and not on its boundary are to be included (the set is 'open'). The boundary can be included by using '≤' as the inequality (a 'closed' set). The difference is like observing a cut-out ellipse (closed) and the shape seen through the cut-out hole that this leaves (open).

Angles

The concept of angle, representing change of direction, is familiar to most readers, who will be able mentally to visualize angles of 60°, 90° and other measures. The concept of one right angle being 90° and a complete revolution being 360° is well known. This evaluation of angle may be due to an historical accident and inaccuracy, that one day's worth of the sun's annual travel round the earth (which was the accepted concept until the days of Galileo[2] and for some time after for those who didn't believe the 'heretic') was equivalent to one degree.[3] This measure has persisted because it is essentially a good one. 360 is a reasonably sized number (not too big, not too small) meaning that 'ordinary' sized angles have understandable values (for example, 30°, 60°). 360 is divisible by many values (2, 3, 4, 5, 6, 8, 9, 10, 12…), so many fractions of a full revolution can be expressed as integer degree values. However, although it is familiar, it is not the preferred method for mathematical development, where radian measure is used.

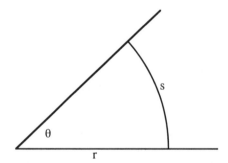

Figure 3.9 The angle θ is calculated in radians as s/r

---

[2] The story of Galileo is explained in Arthur Koestler's *The Sleepwalkers*, Penguin, 1964.
[3] The delightful illustrated children's book, *Man Must Measure* by Lancelot Hogben, Rathbone, London, 1955, has a very readable explanation of this. Hogben also wrote two classics of popular science and mathematics: *Mathematics for the Million*, and *Science for the Citizen*.

3. Coordinates and Dimension: Representations of Space and Colour

To measure an angle θ in radians, draw a circular arc of radius r, where r is greater than zero, with centre at the vertex of the angle (fig 3.9). The Greek character θ is often used in mathematics to denote an angle.) Measure the length s of the arc cut off by the two 'arms' of the angle. The radian measure of θ is the fraction of s divided by r:

$$\theta = (s/r)^c.$$

The symbol ^{c} for radians looks like a superfix 'c', but is supposed to represent the arc of a circle, like that used to measure the angle. The symbol is often omitted, the context usually indicating the type of measure used. Note that this measure remains the same for a given angle regardless of the length r chosen, as if we change r by a set proportion the length of s would be changed by the same proportion, which would be 'divided out' in finding the ratio s/r.

$$c = 2\pi r$$

is a familiar formula from elementary geometry giving the circumference of a circle in terms of its radius r and the familiar constant π (the Greek character 'pi', one of the irrational numbers that contributes to the set of real numbers). This means that the familiar angle 360°, representing a complete rotation, has radian measure

$$(s/r)^c = (2\pi r/r)^c = 2\pi^c.$$

By considering fractions of this, we can form a table of comparisons of familiar angles in the two measures (table 3.1).

Note that values of familiar angles in radians are expressed in terms of the constant π. This is often the contextual clue that identifies the form of measure; if an angle contains π, then it is almost invariably measured in radians, so the symbol for radians is often left out when the context is clear. From the table, we can derive an easy formula for conversion of degrees to radians and vice versa.

$$r = (\pi/180)d, \quad d = (180/\pi)r.$$

Table 3.1 Comparison of degrees and radians

Degrees	Radians
0	0
1	π/180
30	π/6
45	π/4
60	π/3
90	π/2
180	π
270	3π/2
360	2π

An angle of one radian is a little smaller than 60°. For those familiar with working in degrees, this can be helpful in visualizing the approximate size of an angle given in radians. In fact, one radian is

$180/\pi° = 57.2958°$ (to four decimal places).

The real reason for the mathematical adoption of radians as an angle measure is not solely because mathematicians want to show off their knowledge of Greek characters (they do, anyway), but in more advanced mathematical analysis radians make calculations easier by avoiding the use of multiplying factors. In certain applications, radians give a multiplying factor of 1, whereas degree measure needs a $\pi/180$ factor. Conversion formulae are useful in computer use, as computer formulae depend on degrees expressed in radians. If you want to input an angle in degrees, the first formula above will effect the required conversion.

Trigonometry and Polar Coordinates

Cartesian coordinate systems are very useful for many purposes, but are not the only systems used. For example, if a navigator wishes to guide a ship from position A to position B, it is more useful to know what direction the ship should be steered in and how far the ship will have to travel (to answer questions like, do we have enough fuel?). The turtle graphics system which is a subpart of the computer language LOGO, uses this method – a line is defined by the distance to be travelled from a starting point and its direction. Figure 3.10 shows a point expressed in Cartesian coordinates (x, y) and in polar coordinates (r, θ). r is the distance from the point O (equivalent to the origin of Cartesian coordinates) and θ is the angle of the line OP to a fixed direction from O (equivalent to the x-axis in Cartesian coordinates). The conventional positive direction of θ is anti-clockwise from the given axis.

The transformation to polar from Cartesian coordinates and vice versa makes simultaneous working in both systems possible, enabling the benefits of both to be used. This can only be achieved through use of the trigonometric functions sine, cosine and tangent, abbreviated to sin, cos and tan. Figure 3.10 can be taken as a defining diagram for these functions, with

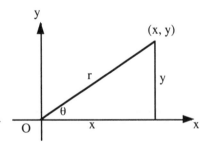

Figure 3.10 A 2D point in Cartesian coordinates (x, y) and polar coordinates (r, θ)

3. Coordinates and Dimension: Representations of Space and Colour

$$\sin(\theta) = y/r,$$
$$\cos(\theta) = x/r$$
and $\tan(\theta) = y/x$ (provided $x \neq 0$).

Note that these are definitions, sometimes called 'trigonometric ratios' to indicate their fractional form. To find sin, cos and tan for a particular angle θ, draw a line outwards from the origin at angle θ from the x-axis of a Cartesian system. For any point P on the line at a positive distance r from the origin, measure the Cartesian description (x, y) of P. Then, the trigonometric ratios of sine, cosine and tangent are defined as above. Note that these are fixed for a given value of angle and are hence functions of the angle (functions are discussed in more detailed in the next chapter; this is one of those occasions where we assume a little knowledge in order to progress). If r is changed in fig 3.10 without changing θ, the values of x and y change in the same proportions as the change of r, so the ratios y/r, x/r and y/x will not change. It is worth noting that the definition for tan(θ) excludes x = 0, when θ would take the value 90° or -90°, representing a vertical line upwards or downwards. The value of tan(θ) is not defined for this value of θ; we discuss this concept later in this chapter and in the section on functions. Fortunately, we do not have to go through this laborious drawing and measuring process each time we need to use trigonometric functions, as someone has already done this. Originally these values were to be found in tables hand calculated by teams of people; now they are readily available from calculator– and computer–based functions.

Many readers will have become familiar with – or perhaps baffled by – the functions sin, cos and tan as related to right-angled triangles. If one of the angles of a right-angled triangle is labelled as θ, its sides can be labelled by their locations with respect to θ (fig 3.11).

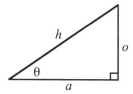

Figure 3.11 The configuration of a right-angled triangle to define trigonometric ratios

The hypotenuse h is the longest side, opposite the right angle, the adjacent side a is next to angle θ, and side o is opposite angle θ (fig 3.11). The familiar schoolbook definition of the trigonometrical functions is

$$\sin(\theta) = o/h,$$
$$\cos(\theta) = a/h$$
and $\tan(\theta) = o/a$ (provided $a \neq 0$).

It should be clear from the layouts of figs 3.10 and 3.11 that for any angle θ in a right-angled triangle the two formulae are equivalent (as o replaces y, a replaces x and h replaces r). There is nothing *wrong* with the schoolbook definition other than its limitation. If θ is an angle in a right-angled triangle, then of necessity its value lies in a range from 0° to (π/2)°, or 0° to 90°. However, if θ is a general angle, it could take values outside this range (we can consider negative turns or corners of greater than 90°). The definitions of sin, cos and tan taken from the relationship with polar coordinates allow any size of angle to be considered, as the signs of x and y change depending on the **quadrant** of the plane within which the angle lies.

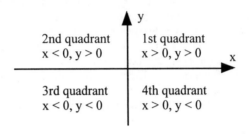

Figure 3.12 The four quadrants of a 2D Cartesian system

Figure 3.12 shows that in the first quadrant θ lies between 0° and 90°, the second quadrant has θ between 90° and 180°, the third quadrant has θ between 180° and 270° and the fourth quadrant has θ between 270° and 360°. Angles do not need to be restrained to the range from 0° to 360°. Angles between -90° and 0° are in the fourth quadrant and those between 360° and 450° are in the first, and so on

As r is always positive, this gives the following sign properties of the trigonometric functions; if we know the signs of sin and cos, we can determine the quadrant in which the angle lies:

- First quadrant, $0° < θ < 90°$:
  $\sin(θ) > 0; \cos(θ) > 0; \tan(θ) > 0$.

- Second quadrant, $90° < θ < 180°$:
  $\sin(θ) > 0; \cos(θ) < 0; \tan(θ) < 0$.

- Third quadrant, $180° < θ < 270°$:
  $\sin(θ) < 0; \cos(θ) < 0; \tan(θ) > 0$.

- Fourth quadrant, $270° < θ < 360°$:
  $\sin(θ) < 0; \cos(θ) > 0; \tan(θ) < 0$.

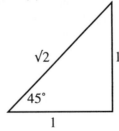

Figure 3.13 Side lengths of a 45° right-angled triangle

3. Coordinates and Dimension: Representations of Space and Colour

Some particular values of sin, cos and tan can be found from familiar right-angled triangles, for example 45° and 60° triangles (as seen in the old fashioned drawing tools called set squares). For fig 3.11,

$$h^2 = a^2 + o^2,$$

from Pythagoras' theorem (which should at least be familiar as a distant memory from the incantation, 'the square of the hypotenuse is equal to the sum of the squares of the other two sides'). For a 45° right-angled triangle with short sides length 1 unit (fig 3.13), this gives the hypotenuse as

$$h = \sqrt{(1^2 + 1^2)} = \sqrt{2}.$$

From the triangle definition, we then have

$\sin(45°) = 1/\sqrt{2},$
$\cos(45°) = 1/\sqrt{2},$
$\tan(45°) = 1.$

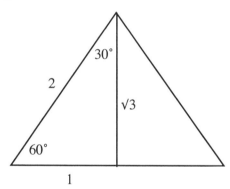

Figure 3.14 Construction of a 60° right-angled triangle

A 60° right-angled triangle can be drawn as half an equilateral triangle (all sides have the same length), which has angle 60° (fig 3.14). If the equilateral triangle has side length 2 units, the half side has length 1, and Pythagoras' theorem again is used to find the third side as $\sqrt{3}$, which is about 1.732. The correctness of this can be checked from the theorem as

$$a^2 + o^2 = 1^2 + (\sqrt{3})^2 = 1 + 3 = 4,$$

which is the same as $h^2 = 2^2$. Again using the triangle formulae for sin, cos and tan,

$\sin(60°) = \sqrt{3}/2,$
$\cos(60°) = 1/2,$
$\tan(60°) = \sqrt{3}.$

As a bonus, this triangle also gives the sin, cos and tan of 30°. Note from fig 3.14 that if 30° is the 'featured angle' the hypotenuse h is still 2, but the side adjacent to the angle is now $a = \sqrt{3}$ and the opposite side is $o = 1$. Inserting these into the triangle formulae for sin, cos and tan gives

$$\sin(30°) = 1/2,$$
$$\cos(30°) = \sqrt{3}/2,$$
$$\tan(30°) = 1/\sqrt{3}.$$

This begins to hint at ways in general that we can relate sines and cosines of various angles. Two of these, demonstrated by example but not proved for all values of θ by the above, are

$$\sin(\pi - \theta) = \cos(\theta),$$
$$\cos(\pi - \theta) = \sin(\theta).$$

It should be clear from the use of π in these relationships that the angles are defined in radian measure for this purpose. There are many such 'trigonometric identities' linking sines, cosines and tangents in the advanced study of trigonometry; they will not be considered in detail here. However, it is worth noting the use of 'identity' rather than 'equation'. An equation is only valid for certain values of its variables. An identity is true for all values of variables concerned; it is a kind of universal law. Establishing the proof of an identity involves showing that all possible values 'work'. This is a central topic for mathematicians; the concept of mathematical proof has obsessed mathematical theory at least since the time of Euclid. The two identities given above are valid for *any* value of θ whereas the equations used to define graphs earlier in this chapter themselves defined a limited set of points (x, y) that made the equations 'true'. Properly, we should use the symbol '≡' (with three bars) instead of '=' to distinguish an identity from an equation, but use of the equals sign '=' has become quite common for both meanings.

Table 3.1 can be extended to a wider range of values and to include values of sines, cosines and tangents as in table 3.2, which shows the wave-like behaviours of sine and cosine functions. Note the way the sin and cos values rise and fall in cycles. We will discuss the concept of graphs and their manipulation more generally in chapters 4 and 5, but figs 3.15–3.17 show the cyclic variations in graphs of sin, cos and tan, all shown on the same page for comparison. One complete period of sin and cos is represented in table 3.2; the cycle repeats itself indefinitely as the angles rotate around further, for example the angles -270° and 450° are effectively the same as 90°.

3. Coordinates and Dimension: Representations of Space and Colour

Table 3.2 Trigonometric ratios for some standard angles between 0° and 360°

Degrees	Radians	sin	cos	tan
-180	$-\pi$	0	-1	0
-150	$-5\pi/6$	-1/2	$-\sqrt{3}/2$	$1/\sqrt{3}$
-135	$-3\pi/4$	$-1/\sqrt{2}$	$-1/\sqrt{2}$	1
-120	$-2\pi/3$	$-\sqrt{3}/2$	-1/2	$\sqrt{3}$
-90	$-\pi/2$	-1	0	∞
-60	$-\pi/3$	$-\sqrt{3}/2$	1/2	$-\sqrt{3}$
-45	$-\pi/4$	$-1/\sqrt{2}$	$1/\sqrt{2}$	-1
-30	$-\pi/6$	-1/2	$\sqrt{3}/2$	$-1/\sqrt{3}$
0	0	0	1	0
30	$\pi/6$	1/2	$\sqrt{3}/2$	$1/\sqrt{3}$
45	$\pi/4$	$1/\sqrt{2}$	$1/\sqrt{2}$	1
60	$\pi/3$	$\sqrt{3}/2$	1/2	$\sqrt{3}$
90	$\pi/2$	1	0	∞
120	$2\pi/3$	$\sqrt{3}/2$	-1/2	$-\sqrt{3}$
135	$3\pi/4$	$1/\sqrt{2}$	$-1/\sqrt{2}$	-1
150	$5\pi/6$	1/2	$-\sqrt{3}/2$	$-1/\sqrt{3}$
180	2π	0	-1	0

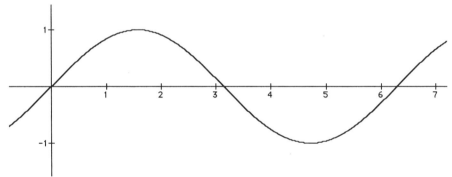

Figure 3.15 The graph of the sine function

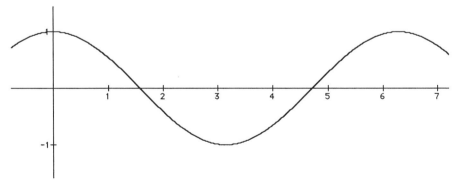

Figure 3.16 The graph of the cosine function, a displacement by $\pi/2$ to the left of the sine (fig 3.15)

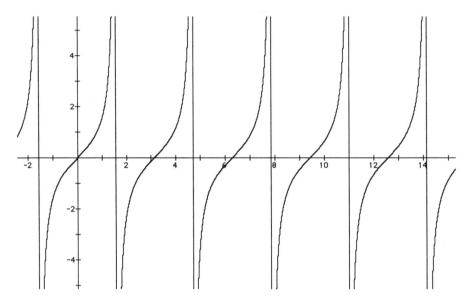

Figure 3.17 The graph of the tangent function; its approach to infinity within each cycle of length π can only be hinted at by extending the graph to the edge of the box. Vertical lines indicate where the graph extends to infinity and are not part of the function graph

The shapes of sine and cosine graphs (generically known as 'sine curves') make them useful in modelling the shape of deep-water ocean waves. These forms are distorted to produce breakers in shallow water, but the effect of wind on the deep water creates essentially long waves with sine curve-cross sections. With the angle interpreted as time, standard oscillations such as pendulum swings and spring vibrations are described by sine curves. Sound waves of all types can also be synthesized from combinations of sine wave forms.

Two complete cycles of the tan function appear in table 3.2. The tangent has no value for 90° and 270°; the symbol '∞' shows that its value becomes very large as the angle gets close to 90° or 270°. The formula for tan involves dividing by x, which is zero for these angles, and is why 'x ≠ 0' and 'a ≠ 0' appear in relevant definitions above, as we would be attempting the invalid division by zero (chapter 2). The correct way to state this is that the value of tan(θ) approaches infinity (∞) as θ approaches 90° or 270°.

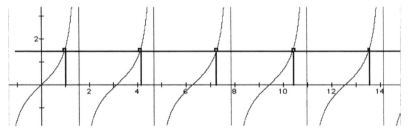

Figure 3.18 A choice of possible angles with the same tangent value

3. Coordinates and Dimension: Representations of Space and Colour

We can find values of sin, cos and tan for any angle in our extended domain, but finding the angle for a particular value of, say, a known tan value is not so straightforward. Figure 3.18 shows that many angles can be found to have the same value of tan; five cases are shown in the limited range of fig 3.18. If the range is limited to one revolution, 0° to 360°, there is still ambiguity. Consider a specific case: if we know that the tan of an angle is $1/\sqrt{3}$, how can we tell if the angle is 30° or 210° (table 3.2)? The full range of angles can be reduced to the interval [0°, 360°) or [0, 2π), as angles of more than one revolution or negative angles can be replaced by the equivalent value in this range. When an ice skater spins through, say, ten and a half turns, the *positional* effect is equivalent to just half a turn, although the effect on the skater's balance would be greater! Effectively, we start counting again when the angle passes 360° or 2π. This still leaves a choice, for tan(x) = $1/\sqrt{3}$, of two possible answers in this interval: 30° or 210°.

The value of tan will usually have been found by dividing a y value by an x value (fig 3.10), that is, from

$$\tan(\theta) = y/x.$$

By returning to the separate values of y and x, we can determine the quadrant in which θ lies (fig 3.12). Suppose our tan was calculated as

$$\tan(\theta) = (-1)/(-\sqrt{3}).$$

Noting that the 'y' value is -1 (negative), the angle must lie in the third or fourth quadrant. The sign of the tan is positive, so the angle must lie in the first or third quadrant. These two conditions mean the angle must be in the third quadrant, from 180° to 270°, so we can choose the correct value of θ as 210° rather than 30°. In most languages, the computer function arctan reverses the effect of a tan function, but returns values between -90° and 90°. Some offer an alternative of form arctan(y, x); this correctly allocates values in the range -180° to 180° having taken into account the signs of x and y. Computer trigonometric functions work in radian measure, so we should switch here to radian description of the range as -π/2 to π/2 for arctan and -π to π for arctan2. Thus, if we wish to find the true value of an angle, it is not enough to know its tangent, we must also know the sign of at least one of the (x, y) coordinates. Given real (or floating point in computer terminology) values of x and y and assuming the constant π is stored as 'pi', the following instructions give the effect of the arctan2 function, so it can be recreated if not available:

```
theta = arctan(y/x);
if (y < 0)
    theta = theta + pi;
```

This gives a result in a full cycle range of answers from -π/2 (-90°) to 3π/2 (270°). This could be corrected to 0° to 360° if required, but the important thing is that a full revolutionary cycle of angles has been allowed.

Inverse functions, which reverse the effect of another function, are discussed in chapter 4. Now, it is enough to know that 'arctan' (often written 'tan^{-1}') reverses the effect of performing the function 'tan'.

If a point (x, y) can be drawn in Cartesian coordinates, then its position can be measured in polar coordinates and vice versa (fig 3.10). Transformation from one system to the other is practically achievable by drawing and measurement. From the principle, 'if I can measure it, I can calculate it', it must be theoretically achievable as well. Given that the functions sine and cosine are now well established (through tables, calculators or computers), conversion from polar (r, θ) to Cartesian (x, y) is straightforward by rearranging the definitions for cos and sin:

$$\cos(\theta) = x/r \qquad \sin(\theta) = y/r.$$

By making x and y the subjects of these formulae, conversion from polar to Cartesian form is given as

$$x = r\cos(\theta); \qquad y = r\sin(\theta).$$

Conversion the other way, finding (r, θ) from (x, y) depends on Pythagoras' theorem to find the hypotenuse r from x and y and the inverse of tan to express θ in terms of x and y, as $\tan(\theta) = y/x$. Thus,

$$r = \sqrt{(x^2 + y^2)}; \qquad \theta = \arctan(y/x).$$

In the formula for θ, it is assumed that the decision of which quadrant the result lies in, as discussed above, is correctly made.

Trigonometry is initially concerned with angles in right-angled triangles, typically used to solve problems such as the steeple height in fig 3.19. Suppose the distance along the ground from an observer to the centre of the steeple's base is 20 m, the steeple has unknown height h and the angle of the observer's line of sight to the horizontal is 35°. From the triangle definition of tangent, we see that

$$\tan(35°) = h/20.$$

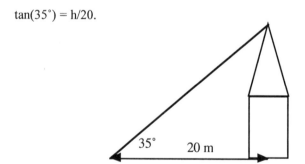

Figure 3.19 A simple steeple problem

Simple rearrangement and a few calculator key presses gives us the height of the steeple, from the definition of tangent, as

$$h = 20\tan(35°) = 14.00 \text{ m (correct to 2 decimal places).}$$

The answer here is given to a limited precision. As the value of tan(35°) is irrational, we cannot write it down exactly. If the answer is given as '14', we know

3. Coordinates and Dimension: Representations of Space and Colour

only that the true solution is somewhere between 13.5 and 14.5. By writing the solution ('solution' is a mathematician's term for 'answer') as '14.00', we know that the answer lies in the more precise range of 13.995 to 14.005.

This is a restricted form of problem, but the subject of trigonometry was developed to solve a much wider range of questions, particularly in the development of cartography and surveying. Accurate maps were created by the process of triangulation, linking together structures of triangles from known positions to establish the location and heights of other positions. The triangles used were not usually right angled, so rules for finding side lengths and angles of all forms of triangle were needed. The two essential laws, stated here without any attempt at proof, for triangles with any side lengths a, b and c, and angles opposite to these as A, B and C respectively (fig 3.20), are the cosine rule and the sine rule

$$a^2 = b^2 + c^2 - 2bc\cos(A);$$

$$\frac{a}{\sin(A)} = \frac{b}{\sin(B)} = \frac{c}{\sin(C)}.$$

The cosine rule can be used to find side a when two other sides and the included angle (b, c and A) are known. This is a generalization of Pythagoras' theorem, as if A = 90°, cos(A) = 0 (from table 3.2), so the cosine rule reduces to the familiar

$$a^2 = b^2 + c^2.$$

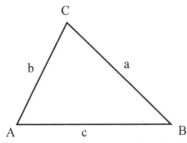

Figure 3.20 Labelling of the sides and angles of a general triangle

The sine rule can be used if the angles and the length of one side are known. These techniques were developed to a high level of sophistication to help surveyors create precise maps. Lengths were measured using chains of known size. The old British measurement of one chain, 22 yards, was a standard, and incidentally was used to measure the wicket-to-wicket length of the normal cricket pitch. Surveyors' assistants were commonly known as 'chain men', as part of their job was to carry the heavy chain. Angles were measured using precision instruments such as theodolites, and the process of mapping a region was known as 'triangulation', as the heights and locations of ground points were generated by 'covering' the surface with a mesh of triangles (as is now frequently done to define surfaces in computer graphics). Locations were found from the sine and cosine formulae. Knowing the locations of one triangle, one of its sides could be used as the base for positioning another point from the angles subtended at the ends of this line. This repetitive process did generate some inaccuracy, but by measuring some convenient lengths

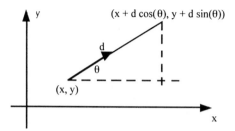

Figure 3.21 The forward draw command 'F' in turtle graphics

using the chain corrections could be put in place every now and then. The concept of triangulation is one used for modelling objects for computer depiction (see chapter 6), so the ancient art is regenerated.

Trigonometry is used in computer graphics for identifying relative position. In the previously mentioned turtle graphics system, a 'turtle' in 2D is defined by its location (x, y) and direction θ as an angle from the x-axis. Drawing commands are executed relative to the current position and orientation. To draw a line forward a length d from the current position, a new turtle position is calculated as

$$(x', y') = (x + d \cos(\theta), y + d \sin(\theta))$$

(fig 3.21) and a line is drawn from (x, y) to (x', y'). The current position of the turtle is then updated to (x', y') without change of θ to complete the process. The full range of turtle commands includes turns (changing θ), forward and backward moves with 'pen down' to draw or 'pen up' to move without drawing, all relative to the current turtle definition. Implementation of these in display screen coordinates is achieved by simple trigonometric commands like the forward line drawing command shown here. The turtle state [(x, y), θ] is held in computer memory; drawing involves adjusting this and creating the relevant effects on screen.

Dimension

We have discussed ways of describing points and spaces in 2D (Cartesian and other coordinates) and 1D (the real line) without really considering what we mean by 'D' or dimension. As we approach a discussion of 3D, and possibly higher orders, it is useful to reflect on just what dimension means.

There are many possible approaches to this concept. In their classic book, Courant and Robbins[4] consider an approach based on a 1912 argument of Poincaré, which is related to counting as used in the development of natural numbers, starting with a concept of zero dimension and building sequentially. First, we need to accept

---

[4] R. Courant, H. Robbins and I. Stewart (1996), *What is Mathematics? (2nd edn)*, Oxford University Press, pp 248–251 (1941 first edition by Courant and Robbins). This highly recommended classic book is a most readable exposition of the true nature of mathematics.

3. Coordinates and Dimension: Representations of Space and Colour 57

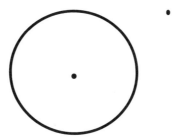

Figure 3.22 Separation of two points in 2D by a 1D curve

that a single point has dimension zero – this is one of the mathematical definitions that has to be taken as a basic 'truth' or axiom before we can build from it; a fundamental concept. Of course, to indicate such a point in a drawing, we must give it some size, so the blobs of fig 3.22 are approximations of the ideal theoretical points. What is the dimension of a line? We can 'separate' a distinct pair of points on a line by removing a single point from the line between them. A loose consideration of 'separation' is that the pair of points cannot be joined within the space of the line without leaping this gap; this would be much more rigorously defined in a pure mathematical text. This makes the line, or indeed any curve that it can be bent into, one-dimensional, one greater than the dimension of the space needed for separation. Two distinct points in a plane can be separated by a one-dimensional curve, such as a circle with radius half the distance between the points centred at one of the points (fig 3.22). This makes the plane, or any surface that it can be bent into, two-dimensional. It is impossible to join the two points without crossing the circle, which as a 'bent line' is one-dimensional. As the space is separated by a one-dimensional object, its dimension is two. Separation of points in the space that we live in can be achieved by a two-dimensional surface – imagine placing a sphere or balloon of sufficiently small radius not to engulf the other around one of the points. Thus, this space is three-dimensional. And so on ...? Could we extend this concept to higher dimensions? This will be considered after we look at another approach, using a method of geometric construction rather than separation or deconstruction.

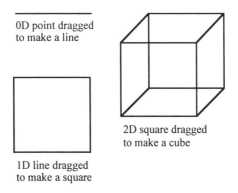

Figure 3.23 'Dragging' increases dimensionality

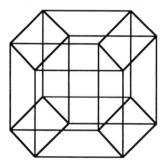

Figure 3.24 A 3D cube dragged to make a 4D hypercube

We start once more with a single point or vertex of zero dimension. This again begs the question of what we mean by a point, which is taken for granted here. We must start somewhere. If we drag this point to create a line (or a curve if the hand is allowed to wobble), the line is one-dimensional. It has one edge and two vertices (end points). Dragging this line out of its own structure then produces a two-dimensional shape, which could be a square if our hand is steady, the drag is in a specific direction and of a particular distance (fig 3.23). The resulting 2D square has one face, four edges and four vertices. Further dragging this two-dimensional shape out of its own form creates a three-dimensional structure, for example a cube, a 3D object that has six faces, 12 edges and eight vertices. Note that the 'purest' form of drag in these cases is at right angles to (or orthogonal to) the previous directions. Although three non-orthogonal directions could be used, as we can use non-orthogonal axes to define a 2D space, the orthogonal form is somehow 'neater', and distances are much easier to calculate. The minimum requirement is for the drag to leave the space of the previously generated object.

So why stop at three? Now we need a fourth dimension that moves outside the space of the three dimensions already generated; in the purest form, we need a direction at right angles to all the three previous orthogonal lines. Within the physical limits of our 3D universe, we cannot build such a line, but there should be no limit to our mathematical imagination. A 2D representation of the result of dragging the final cube of fig 3.23 along an axis into the fourth dimension gives an image like fig 3.24. All eight vertices of the cube have been displaced without reorientation (this is known as a translation), and all eight joins of new and old vertices completed as straight lines. The new object, a tesseract that is the four-dimensional version of a hypercube, has 16 vertices. It also has 32 edges: each of the 12 edges of the two contributing cubes plus the eight linking edges. As to faces, there are two lots of six from the contributing cubes plus another 12 created by dragging the original cube's edges, making 24 in all. This four-dimensional object is also 'bounded' by eight 3D cubes. The original cube plus the one it is dragged into are augmented by the cubes created by dragging the previous six boundary squares, as emphasized in fig 3.25. An object of n dimensions is bounded by (n – 1)-dimensional objects, a line by two points, a square by four edges, a cube by six squares, a tesseract by eight cubes. We see that there is an increase of two at each stage for creation of a 'hypercube'. Suppose an (n – 1)-dimensional hypercube has k (n – 2)-dimensional boundaries. In creating an n-dimensional hypercube, we drag these k objects to create k (n – 1)-dimensional boundaries, and add the two extra

3. Coordinates and Dimension: Representations of Space and Colour

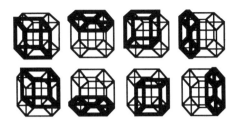

Figure 3.25 The eight 3D boundaries of a 4D hypercube or tesseract

'boundaries' from the start and end locations of the dragged (n − 1)-dimensional object. A hypercube of n dimensions has 2n boundary objects of (n − 1) dimensions. We can argue this with confidence, but the mental process of visualizing just what this means is much more difficult.

How can we draw something that can't exist within our universe? We can depict a 3D object (such as the cube of fig 3.23) in the 2D universe of a piece of paper. This is done by projection, a technique developed by artists and mathematicians.[5] Mathematical projection is a technique for reducing the dimension of an object, typically from 3D to 2D, but the concept once more can be extended to higher dimensions. In fig 3.24, we have taken a 4D object and reduced it to 2D in order to depict it. There's no need to stop at 4D, either.

The options were beautifully described in Edwin Abbott's 1884 book *Flatland*, published under the pseudonym, 'A Square'.[6] Abbott's moving point creates a line, a moving line creates an square and a moving square creates a cube. He describes how, using analogy, a 'divine Cube' creates a 'still more divine Organisation' with 16 points in a fourth dimension. He then fantasizes about leaving this 'blessed region' to enter five-, six-, seven-, eight- and so on, dimensional worlds.

We are limited by our inability to find a new direction. A one-dimensional line is dragged at right angles to itself to create a square, the square at right angles to itself to create a cube, the cube at right angles ... . This is where we step outside the familiar three-dimensional universe that allows us to draw only three mutually orthogonal lines. This is not an unfamiliar situation: we can draw only two mutually orthogonal lines on a piece of 2D paper, but create '3D' drawings.

A third, and very common, approach to dimension is through the concept of coordinates. If we have a one-dimensional space, which can be represented as a line or curve, any point within it can be determined by a single value, representing, for example, its distance along the curve from a given origin or fixed point. In 2D spaces, we need two numbers or coordinates, (x, y) in a Cartesian description, to fix a specific point. In 3D we use three coordinates (x, y, z) in a Cartesian system to fix a point (fig 3.26). Dimension is the least number of values needed to locate a point.

What, then, is the dimension of a piece of string? If we measure location as the distance along the string from one end, it is one-dimensional. Suppose it is rolled

---
[5] R. Courant, H. Robbins and I. Stewart (1996) *What is Mathematics? (2nd edn)*, Oxford University Press, p 167; Brunelleschi (1377–1446) was a pioneer; da Vinci (1452–1519) and Dürer (1471–1528) were early users of perspective.
[6] E.A. Abbott (1978), *Flatland: A Romance of Many Dimensions (2nd edn)*, Basil Blackwell (first published in 1884), pages 88–89. This is a delightfully whimsical classic on the nature of dimension as seen by inhabitants of 2D and other worlds.

up tightly in a ball, and we can't really track along the string without changing the ball's structure. It is reasonable to consider the ball as three-dimensional; a point within it could be identified by three coordinates. If the string were knitted to form a scarf, we might have a case for considering it to be two-dimensional. This should teach us to be wary of making instant interpretations of dimensionality, which will be reinforced when we consider fractal objects in a later chapter.

We have used Cartesian coordinates here to illustrate the argument, but other systems also satisfy this condition. In 2D polar coordinates, we need to identify the two values of a distance and an angle. Figure 3.4 shows how the original concept of a Cartesian space can be warped, so that axes need not be straight to enable the location of points. Thus, although lines and planes are natural objects of one or two dimensions respectively, they can be distorted into general curves and surfaces without losing their dimension. Consider, for example, the surface of a sphere, or even a distortion from the sphere such as the surface of the earth. We are used to locating points on the earth's surface by two numbers: latitude and longitude. Thus, the surface is two-dimensional. We also recognize 'mappings' of the surface onto a plane, which we easily accept as a 2D space, as in the Mercator projection (fig 3.5).

To identify a point on the globe, we need only two coordinates, so the surface is two-dimensional. However, in order to create an adequate model of the globe itself, we must have a three-dimensional universe. The globe's surface is a two-dimensional sub-space within this three-dimensional universe. In the same way (remember Abbott) we can consider that our three-dimensional universe may be a sub-space of higher-dimensional worlds. There are other possibilities, dimensions between dimensions, but this must serve only to whet your appetite for the later discussion of fractals.

The concept of dimension as the least number of values needed to identify position is closely related to the concept of 'degrees of freedom' in physical systems. The number of degrees of freedom of a system can be thought of as the number of measurements of a system that you can freely define. For example, if a structure is made of iron, we cannot separately define its volume and its mass. Once the volume is given, the mass is automatically known from the density of iron and vice versa. Some defined concepts can be linked in this way; others are independent. In creating an iron block, fixing the length of the block does not prevent me from choosing the width and height freely, so length, width and height are independent in this context. The number of independent variables that may be freely chosen for a particular system is the number of 'degrees of freedom' of the system.

To define precisely the location and attitude of an aeroplane, we need six coordinates, each of which can be defined freely, so the aeroplane has six degrees of freedom. These could be three spatial coordinates to define location, for example latitude, longitude and height above mean sea level, together with three to identify the orientation of the aeroplane. Three angles are used to define the horizontal direction in which it is heading, the angle by which its nose is tilted up or down from this and the angle by which the wings are rotated from the horizontal about a line from tail to nose of the craft. These are related to the 'Euler'[7] angles that denote

---

[7] Leonhard Euler (1707–1783) was a prolific mathematician born in Basel, Switzerland. His work permeates this book. He popularized the symbol π and defined the use of 'i' for complex numbers and the exponential constant 'e'. His name is also remembered in a major theorem of topology (chapter 8).

object orientation, representing the heading, pitch and roll of the aircraft. This assumes that we consider the aeroplane as a basic rigid body. The number of degrees of freedom becomes enormous if we want to define every item that defines the plane's state, such as the angles of wing flaps, by what angle seat 27B is tilted, whether the rear port lavatory door is shut, and so on.

If we have a system that is described by six coordinates, like the position and attitude of an aircraft, and there is a defined relationship between some of these coordinates that links them by an equation, this dependence reduces the degrees of freedom of the system by one. Degrees of freedom represent the number of values that can be independently chosen. If we choose five of the six values, and there is an equation linking the sixth with one or more of the others, we no longer have the freedom to choose this sixth value, as it is determined by the equation. The equation or condition is sometimes known as a constraint. If we have a system defined by n values subject to r different constraints, it has n − r degrees of freedom. This is an important concept in the consideration of the meanings of equations, discussed later in this chapter.

Physical systems of great complexity, and problems in other disciplines such as financial markets, can be described as high-dimensional lists of values, which may be considered as coordinates. A specific situation can then be identified as a point in a multidimensional space. Ways of reducing the dimensionality of these sets by projections to give meaningful images of their main features are used in multi-dimensional analysis, a branch of statistics. Projections are ways of reducing the dimension of a set, the most familiar being the way in which perspective projection creates 2D images of 3D objects.

The abstractions of four and higher-dimensional spaces have meaningful purpose in other than directly spatial descriptions. In developing ideas of computer-generated spaces to depict three dimensions, we limit ourselves to conventional interpretations, but it is useful to understand that more abstract interpretations may prove to be useful in unexpected ways. This is useful in the branch of computer graphics known as 'visualization' deals with visual interpretation of data sets, for example scientific or financial data, to illustrate and elucidate their properties.

Coordinate Systems in 3 Dimensions

The way that three-dimensional spaces can be generated by dragging in three mutually orthogonal directions is the basis of the Cartesian method of describing 3D space (fig 3.26). The z-axis is drawn orthogonal to the x and y coordinates used in 2D. Normally, this is done to create a 'right-hand set', in which a 'right-hand screw' (that which drives in a normal screw) in the direction of the z-axis turns in the shortest way from the x-axis towards the y-axis. Gardner[8] has an amusing and

---

[8] Martin Gardner (1978) *The Ambidextrous Universe: Right, Left and the Fall of Parity (2^{ne} edn)*, Sribner, New York. Gardner discusses how you could distinguish left from right handedness to intelligent beings from another galaxy, using only coded radio transmissions. He concludes that it can be done, but it is about the most difficult concept to describe in this way.

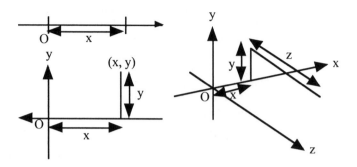

Figure 3.26 Definition of points in 1D, 2D and 3D Cartesian systems

interesting discussion on the nature of handedness. At the risk of dislocating readers' wrists, an alternative way of considering this is to point the thumb of the right hand in the z-axis direction. If the fingers of the right hand curl in the natural 'x to y' direction, the axes are right handed. You should confirm this, carefully, on the axes of fig 3.26. Because information is inevitably lost in projection from 3D to 2D, the illustration of 3D coordinates in fig 3.26 is potentially ambiguous. Its interpretation is that if x points towards the right and y points towards the top of the page, then z points out of the page towards the reader. Reversing the direction of any axis, for example the z-axis, creates a left-handed set of axes. These are used in some computer graphics systems, where the normal z-direction is reversed to create a left-handed set of axes in view coordinates, discussed later. The question of left and right handedness is largely a question of personal choice or convention, as long as formulae are correctly applied and distances correctly evaluated in the circumstances.

It is interesting to note that most mathematical textbooks illustrate the x and y axes to be horizontal, with the z-axis vertical, whereas computer graphics texts often show the z-axis coming horizontally towards the reader, with the y-axis upwards. This may be because the mathematician's 2D universe is a horizontal piece of paper on a table or desk, whereas computer graphics practitioners are used to working with vertical 2D display screens.

When our axes are properly defined, a point's position is fixed by the ordered triplet (x, y, z), indicating distances to be traversed from the origin in the x, y and z axis directions respectively to reach the point (fig 3.26). As in 2D systems, we need to devise working concepts, such as a definition of distance, within this space. The distance between two general points (x_1, y_1, z_1) and (x_2, y_2, z_2) is an extension of the law in 2D quoted earlier in this chapter, derived from Pythagoras' theorem,

$$d = \sqrt{\{(x_2 - x_1)^2 + (y_2 - y_1)^2 + (z_2 - z_1)^2\}}.$$

We shall need this result when discussing vectors, when many relationships in 3D will be explored.

We discussed the importance of equations as constraints on the degrees of freedom of a system. Equations similarly affect the dimensionality of a system. We saw above that equations involving 2D coordinates (x, y) define lines or curves. What does an equation in a 3D system define? Figure 3.27 shows a surface generated from the equation

3. Coordinates and Dimension: Representations of Space and Colour

Figure 3.27 The surface defined by $y = (x^2 + z^2)/4$ for $-2 < x < 2$, $-2 < y < 2$

$$y = (x^2 + z^2)/4.$$

Is there an inconsistency here in that equations in 2D give us curves, whereas an equation involving 3D coordinates gives us a surface? This can be explained by the interpretation using degrees of freedom. In a 3D system, we can choose our location freely by varying all the three coordinates. However, if an equation or constraint is applied, only two of these can be defined independently, thus reducing our degrees of freedom or dimension to two and defining a surface. If we choose x and z each to take values, say, 1, then we have no option other than to take $y = 1/2$. The constraint gives us the freedom to choose only two values. Similarly an equation in a 2D system will reduce the dimensionality by one, leaving us with a one-dimensional curve.

This argument can be extended further. A 2D equation in (x, y) can be taken as a particular case of a more general equation in (x, y, z) with a constraint that $z = 0$. This interpretation takes a 2D curve as that part of a surface in 3D that passes through the 2D plane that has $z = 0$, which is the x-y plane. The apparent inconsistency has been explained.

As equations represent surfaces in 3D, then inequalities (as we discussed in the 2D cases above) represent half-space regions in 3D space, or volumes with these surfaces as boundary curves known as half-spaces. We will not consider this further here, but the concept will be looked at again when we discuss geometric models.

We mentioned polar coordinates in 2D, which were instrumental in the development of the trigonometric functions sine, cosine and tangent. Two major types of polar system are commonly used in 3D: cylindrical polar coordinates and spherical polar coordinates.

Figure 3.28 shows a point P, (x, y, z) in 3D Cartesian coordinates with axes in the usual computer graphics orientation. The point P', (x, y, 0), lies in the (x, y) plane. Suppose (r, θ) is the polar identification of P' (x, y) within this plane (fig 3.10), where r and θ are found from formulae associated with fig 3.10. The cylindrical polar coordinates of P are then defined as (r, θ, z).

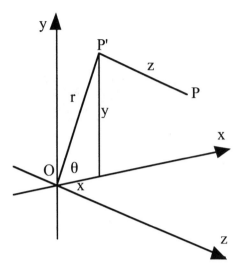

Figure 3.28 A point P referred to Cartesian coordinates (x, y, z) and cylindrical polar coordinates (r, θ, z)

Formulae for switching from Cartesian to polar forms are useful. It should be clear, in comparison with the 2D case, that conversion formulae from cylinder polar to Cartesian coordinates and vice versa are as follows:

$x = r\cos(\theta);$ $\quad y = r\sin(\theta);$ $\quad z = z.$

$r = \sqrt{(x^2 + y^2)};$ $\quad \theta = \arctan(y/x);$ $\quad z = z.$

The only difference between the 3D case is the addition of the trivial 'z = z', as this performs the same role in both Cartesian and cylindrical polar systems. The name of the system relates to the surface defined by the equation

$\quad r = $ constant.

All points for which r is fixed will lie the same distance from the z-axis; for example, all points on the line PP' (fig 3.28) have the same value of r. As θ changes, this line sweeps out the surface of a horizontally oriented cylinder.

Spherical polar coordinates (fig 3.29) have the property that all points

$\quad r = $ constant

lie on the surface of a sphere. Consider the same point P with Cartesian coordinates (x, y, z); x is not shown in fig 3.29 to avoid hiding the label for φ. The angle θ is defined exactly as for cylindrical polars, the angle between the x-axis and the line OP'. This time, r is the direct distance from the origin of the axes to the point P. This is what defines the 'spherical' property: all points with the same value of r lie at the same distance from O, and hence on the surface of a sphere. The third coordinate is the angle φ (pronounced 'phi') between the z-axis and the line OP,

3. Coordinates and Dimension: Representations of Space and Colour   65

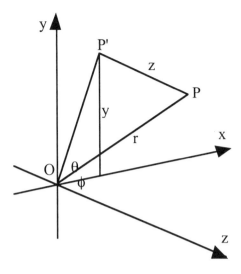

Figure 3.29 A point P referred to Cartesian coordinates (x, y, z) and spherical polar coordinates (r, θ, φ)

where O is the origin. Figure 3.29 can be visualized as a book with its spine along the z-axis, originally lying in the x-z plane. A page of the book is opened until it reaches the point P, turning through an angle θ. On this page, the line from O to P can be drawn, length r, and the angle this makes with the book spine can be measured as φ.

Once again, conversion formulae are useful. In the particular configuration given in fig 3.29 (spherical polars may also be defined in other contexts with the 'book spine' against other axes), we have:

$$x = r\cos(\theta)\sin(\phi); \qquad y = r\sin(\theta)\sin(\phi); \qquad z = r\cos(\phi).$$

More confident readers may find it an interesting exercise to extract the triangles from fig 3.29 that generate these formulae; others may want to take this on trust. The reverse formulae, to find spherical polar coordinates from Cartesian coordinates, are:

$$r = \sqrt{(x^2 + y^2 + z^2)}; \qquad \theta = \arctan(y/x); \qquad \phi = \arccos(z/r).$$

The formula for r is the length of the line from O to P. The value of θ is as for cylindrical coordinates. The 'arccos' in the formula for φ is the reverse of the formula for cos, so φ is the angle whose cosine is z/r. This cheats a little, as we should give formulae for r, θ and φ in terms of x, y and z alone. We note that r has already been given in terms of x, y and z, so we could have quoted the less elegant version

$$\phi = \arccos\{z/\sqrt{(x^2 + y^2 + z^2)}\}.$$

Cylindrical and spherical polar coordinates are sometimes useful in describing objects that fall naturally into the shapes implicit within the systems, such as cylinders and spheres. In such cases, it is necessary to transform these into Cartesian coordinates in which the standard computer graphics methods are based. These will not be used too often, so the reader who is a little confused by this should move on, and accept that they exist.

Colour and its Representation

Colour is included in this chapter as it can be represented in terms of three coordinates, although the coordinates do not share the spatial characteristics of those discussed earlier in thie chapter. The ways in which we understand colour are part of the study of perception, which looks at the psychological ways we interpret clues received by physiological means. If our eyes 'work' properly, light rays from a specific point on an object can be made to impinge on a specific point on the retina, the light-sensitive surface at the back of the eye-ball (fig 3.30). Muscles surrounding the transparent lens squeeze it to change its characteristics so that we can focus light in this way; other muscles surrounding the eyeball move the whole structure so we can 'point' our eyes at a required object.

The retina is covered more or less densely with light-sensitive cells capable of feeding signals to the optic nerve, which transmits these signals to the brain, where the psychological component of vision takes over to interpret these signals. In broad terms, the reception of light and conversion of them to 'signals' comprises the physiological element. There are two types of retinal cell, known as rods and cones because of their approximate shapes. Rods give us most of our vision in dim light; they 'fire' with greater sensitivity. It requires more light energy or intensity to make a cone transmit a signal. The signal from rods is interpreted as a range of intensities devoid of colour, so our vision in faint light is broadly monochrome, like a 'black and white' film.

The nature of light is studied in quantum physics. The smallest 'packet' of light is a tiny particle called a photon, but light also shows wave-like behaviour, with wavelengths between about 400 nm and 700 nm being visible. (A nanometre, nm, is 10^{-9} metres; there are one thousand million nanometres in a metre.) This range represents the 'spectrum' of light. Light waves at 700 nm are observed as red; as the wavelength decreases, we see the spectral range of orange, yellow, green, blue and indigo, reaching violet at 400 nm. This packs about 2.5 million violet wavelengths

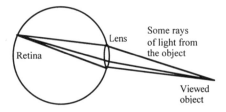

Figure 3.30 Rays from an object are focused on the retina by action of the lens

3. Coordinates and Dimension: Representations of Space and Colour

and about 1.4 million red wavelengths into one metre. The phenomenon of light exists outside this range, but humans cannot perceive ultraviolet or infra-red light.

Ultra, meaning more than', and 'infra', meaning below, refer here to the light's frequency of oscillation, found as c/w, where w is the wavelength and c is the constant speed of light. With c at approximately 3×10^8 metres per second, this gives the frequency of oscillation of violet light as 7.5×10^{14} hertz and that of red light as about 4.3×10^{14} hertz (one hertz is one oscillation per second).

Pure colour information is carried by light of a single wavelength. We seldom experience such purity, such as in the yellow light of a sodium lamp created by pure oscillation of a single type of atom; most light is a mix or 'spectrum' of wavelengths. A sonic analogy is to compare the purity of tone of a flute, which delivers a near single-wavelength sound, with that of a cello, which delivers a more complex mix of wavelengths.

Retinal cones transmit colour information; for this three different types are known to transmit more strongly on reception of red, green and blue light. The blue cones give their maximum response at about 440 nm, green at about 545 nm, red at about 580 nm. These maxima are surrounded by largely overlapping ranges, for example at the green maximum, red will fire at about half its maximum, and blue has just about tailed off to zero, its range being between about 400 nm and 540 nm. Blue is also the least sensitive, absorbing only about 2% of the colour reaching it, the red and green responses being close to 20%. The eye is most sensitive to light at about 550 nm, perceived as a yellow-green colour. The combination of signals from all three types of cone is interpreted as colour.

The reproduction of colour uses this 'tristimulus' response by feeding the eye with three colour components: red, green and blue. We consider first the additive colour system, created by adding light to the environment, as in a television or computer display screen. These start as essentially colourless (black), with colour created in the form of glowing red, green and blue spots (look at a television screen or computer monitor with a magnifying glass if you're not convinced). The human perception system interprets a close juxtaposition of glowing dots as a single colour. Colour is represented as a fraction of the maximum display level of each component, red, green and blue (r, g, b), so this has the characteristics of a coordinate representation. However, the concept of distance is not constant; we have already been told that the eye is less responsive to blues than to yellows and greens, so the sense of a 'distance' between colours is very difficult to define, and is beyond the scope of this text. The coordinate analogy stops here. This form of additive colour can be represented in an (r, g, b) cube, where each point in a unit cube is coloured by its colour representation. The cube at the top left of plate 3.1 has the red axis towards the lower right, green axis upwards and blue axis towards the lower left.

An (r, g, b) representation of (0, 0, 0) means no colour is added, so this represents black (hidden at the back of plate 3.1). The maximum of (1, 1, 1) is interpreted as bright white (seen towards the front in the full cube of plate 3.1). Equal colour components (f, f, f), where $0 < f < 1$, are interpreted as greys, becoming brighter as f nears the value 1. Greys occur along the black–white diagonal of the (r, g, b) cube. The partial cubes in plate 3.1 show slices at 0.75, 0.5 and 0.25 through the green axis, to show how internal colours are distributed. The vertices of the (r, g, b) cube have the following colour interpretations.

Black	(0, 0, 0)	White	(1, 1, 1)
Red	(1, 0, 0)	Cyan	(0, 1, 1)
Green	(0, 1, 0)	Magenta	(1, 0, 1)
Blue	(0, 0, 1)	Yellow	(1, 1, 0)

Cyan, magenta and yellow are formed from mixes of primary additive colours as green–blue, blue–red and red–green respectively. Plate 3.2 shows the way in which colours in the additive colour system mix; this could be reproduced by shining three torches onto an otherwise non-illuminated white surface (the author has seen this figure so many times, but knows of no-one who has performed this actual physical experiment).

The colour sensitivity of computer displays is related to the number of bits used to represent each pixel on a display screen; we should now realize that a pixel is a triad of coloured red, green and blue dots. Most high-quality colour systems allocate a minimum of 24 bits of memory to each pixel: eight bits each for red, green and blue. This allows a total of 2^8 levels for each of the primary colours, giving a total of $2^8 \times 2^8 \times 2^8 = 2^{24}$ different colours to be displayed (this is almost 16.8 million colours). This is adequate to fool the human perception system that it sees a continuous display of colour. Although it is convenient to consider the colour component as a value in the range 0 to 1, most systems deal with an integer representation. So, each colour component may be represented as an unsigned integer in 0 to $2^8 - 1$, or 0 to 255 . Less precise colour systems, such as an 8 bit colour display that can show only $2^8 = 256$ different colours, may show visible bands of colour rather than continuous tone variation. Such systems are a distant memory for many users. Even with the full colour range, some colours that may be observed from real objects cannot be displayed correctly in an (r, g, b) universe. This is because the full complexity of a spectrum cannot be created from a combination of three primitive 'spot' values. However, the available range is enough to satisfy most television viewers, for example.

The interpretation of an (r, g, b) colour is sometimes difficult. For example, just what colour does (0.2, 0,8, 0.6) give you? Adding a few lines to a bar chart of the colours can help with the interpretation (fig 3.31). The lower band shows when all

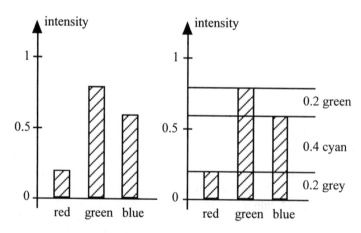

Figure 3.31 Interpretation of (r, g, b) colour components

3. Coordinates and Dimension: Representations of Space and Colour

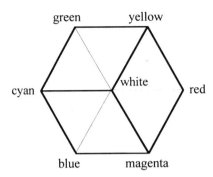

Figure 3.32 The (r, g, b) cube viewed along the black–white diagonal

three primaries are in action, giving a level of grey (from white at 0 to black at 1). Increasing this element mixes in more white, to give more of a pastel or washed out effect. The central band has two of the primaries in action; in this example the mix of green and blue gives cyan. The upper band shows how much single primary (in this case green) is mixed in. The mix of 0.2 grey, 0.4 cyan and 0.2 green gives a not too 'washed out' aquamarine, slightly more green than blue (plate 3.3).

In spite of this method of interpretation, designers who need to specify colours have found another colour representation, the (h, s, v) method, valuable. Plate 3.1 is observed almost along the direction of the white–black diagonal. Colours seen around the border of this image, known as pure hues, appear to be spread around this in rotation, from red through yellow, green, cyan, blue, magenta back to red (fig 3.32). Hue can be represented as an angle around white, from 0° for red, 60° for yellow, and so on, back to 360° for red. Now effectively forget the cube ever existed, and take fig 3.32 as a flat representation of all hues mixed with white. Distance from the white centre to the outer hues can be considered as 'saturation': the amount of pure hues mixed in with white. A pure hue is a 'fully saturated' colour. Darker colours can be represented by mixing hues with greys rather than whites; this is visualized by gradually darker layers of this colour hexagon lying underneath the original. These gradually become smaller, in recognition of the way that human colour perception is less precise for darker colours. The 'darkness' of the colour is represented as a 'value' from 0 for black to 1 for white or fully saturated colours. Black is a single colour, and is therefore represented as a single point, so the (h, s, v) hexcone (fig 3.33) results. This can be further transformed to a cone, as shown in plate 3.4, where black lines indicate where pure hues of red, green and blue can be seen at 0°, 120° and 240° respectively.

Many designers find it easier to specify a hue h as an angle referred to a colour reference circle, and then 'mix in' levels of white to 'wash out' the colour (more white gives lower saturation s). The level of darkness is a value v from 0 for black to 1 for full brightness. A colour defined as (h, s, v) can be transformed automatically into an alternative description for depiction on screen – typically (r, g, b) – or on paper, as described in the next paragraph.

The representation of colour on the printed page operates on a different principle: that of subtractive colour. To see a blank page, it must be illuminated, and a white sheet of paper diffusely reflects most of the light cast upon it. Thus, if we shine a white light on white paper, all wavelengths of light are reflected by the paper, and

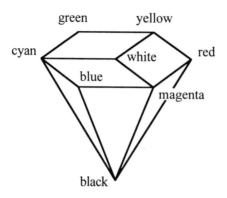

Figure 3.33 The (h, s, v) hexcone

we see it as white. By putting paint on the paper, we trap some of the wavelengths of light (the light energy so absorbed is converted to a small amount of heat). A yellow paint or ink allows only wavelengths close to 560 nm to escape, trapping all others, so the perceived colour is yellow. This is the subtractive light model: the paint subtracts light from the environment. The primary components of subtractive colour are cyan, magenta and yellow, the 'complements' of the additive primaries – just change all zeros to one in the (r, g, b) description to get from a colour to its complement. Colour print is formed from a combination of small adjacent specks of paint in these primaries. In (c, m, y) coordinates, (0, 0, 0) represents white (no colour subtracted) and (1, 1, 1) represents black (all the colour subtracted); this illustrates 'complementarity' with the (r, g, b) system. In practice, a full mix of cyan, magenta and yellow gives a dark browny colour, near black, so high-quality printers usually include a fourth black ink to give 'true blacks'. All (c, m, y) mixes of primaries are shown in plate 3.5. Coordinate levels in (c, m, y) have the following interpretation:

White	(0, 0, 0)	Black	(1, 1, 1)
Cyan	(1, 0, 0)	Red	(0, 1, 1)
Magenta	(0, 1, 0)	Green	(1, 0, 1)
Yellow	(0, 0, 1)	Blue	(1, 1, 0)

Above, we stated blandly that (h, s, v) values can be converted automatically to (r, g, b) and (c, m, y) forms for depiction using different media. This is valid up to a point. In theory, a colour converted from one schema to another should appear identical when compared; a printed page held next to a monitor should show indistinguishable colours if the conversions have been properly calculated. The practical result is different. Although there are industry standards, manufacturers use different forms of glowing phosphor to produce their colour effects, and triads of colour units are differently organized on different screens. Users also adjust colour displays to their own tastes, so the balance of manufacturers' settings may be disturbed. This can be confirmed by standing in front of a display bank of televisions in a department store. For print media, the quality of paper, differences between primary coloured inks and the way in which ink specks are distributed on the page can affect the appearance of colour. The author's 'photo-quality' colour printer gives superb results, but red is accentuated compared to screen-based images. Manufacturers strive to standardize, but in this case we must accept that we live in an imperfect world.

Summary

So what should you have learned by the time you get here? You should understand how several types of coordinate systems exist, with Cartesian coordinates being those of major interest for us. Angles and the general trigonometric functions have use in computer graphics, in transforming objects (chapters 4 and 6), manipulating space (chapter 6), creating models of objects (chapters 8 and 9) and lighting methods (chapter 10), moving things about in space (chapters 7 and 8) and creating fantastic and natural objects (chapter 8). Even if you don't feel confident you could calculate things using trigonometry, you should carry with you the concept of its usefulness, as it will crop up in most of the areas to be considered later. You should also have some idea of the meaning of dimension, as we'll visit that again in a more perplexing light when we discuss fractals (chapter 8).

We have also discussed the representation of colour, although it does not properly belong as a mathematical coordinate system with the usual considerations of distance, and so on. It is, however, convenient to address it here as a phenomenon capable of description in a coordinate-like way, and it is clearly of central importance to the creation of images (chapter 10).

Being able to understand rather than to 'do' is the purpose of this chapter, but it should leave you with a reasonable overview of the importance, influence and potential applications of these methods.

4. Functions and Transformations: Ways of Manipulating Space

Introduction

We have seen examples of functions, such as sine, cosine and tangent, in the previous chapter, without discussing the concept of function. We also illustrated graphs of functions. This chapter explains their meanings as essential mathematical devices, and shows how they are particularly important in the generation of computer graphic images. In image synthesis, from modelling to image generation, several coordinate systems are used. The process of transforming between these is done by special functions that act as 'mappings' from one space to another. Several of these forms of mapping are introduced later in this chapter.

Functions as Mappings

The equation

$$y = x^2$$

defines a function rule. It represents a mapping from a set of values of x, called the domain of the function, to a set of values of y, which lies in the codomain of the function. For a particular value of x in the domain, such as

$$x = -3,$$

the rule gives *the* corresponding value of y, in this case

$$y = (-3)^2 = 9.$$

We are free to choose any value x from the domain in order to find its corresponding y value. In this example, x is the independent variable and y is the dependent variable. It is important to realize that any function is 'single valued'. For any value of the independent variable x from the domain, the rule gives one and only one value of the dependent variable y. However, it may be the case that several values of x may give the same result when entered into the rule. For

$$y = x^2,$$

both $x = -3$ and $x = 3$ map to the same value of y. This shows an intrinsic difference between the independent and dependent variables. The independent variable is an input that can be freely chosen from the function domain; the dependent variable is an output that is determined by the function.

The concept of input and output is helpful in comparing the mathematical definition of a function to the definition of a function in a computer language such as C or Pascal or Java. For those familiar with a procedural computer language, standard functions which do not have type 'void' are equivalent to mathematical functions – they have domains through the type definitions of their arguments, codomains through the type definition of the value returned by the function and function rules operated by the body of the function. For any particular set of values of the arguments, the same value will be returned by the function every time it is called with these particular argument values. In C and other languages, functions declared as void differ in that no value is returned – this form of function is identified as a procedure in, for example, Pascal and is designed to undertake a specific set of operations without necessarily having a 'value'. This is not permitted in mathematical functions; all functions 'return' or define a value dependent on permitted 'input' values from the function's domain.

The concept of a function as a mapping from one region to another is illustrated in fig 4. 1. The function's independent variable is selected from a domain of values; the function in some sense 'rules' over this domain. Using a predefined function rule, signified here as f(.), any value x in the domain is associated with a single well-defined value f(x) from the codomain of the function. The rule is a formula involving x. In the specific case above we defined a function rule as

$$y = x^2,$$

so in the general case we can define a function rule to have form

$$y = f(x),$$

where x is an element of the domain and y is its 'image' in the codomain. From the single-valued nature of a function above, it should be clear that no two arrows of fig

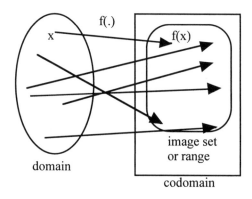

Figure 4.1 The normal concept of a function as a mapping from a domain to a codomain

4. Functions and Transformations: Ways of Manipulating Space

4.1 can leave from the same x, but more than one arrow may arrive at the same resulting value.

The codomain may be defined in a more or less vague way – it just specifies a broad target region into which the function values will fall. The image set or range of the function is a more precise concept. If every value x of the domain is considered, and the function rule is operated on these to produce all possible values of f(x), this comprises the image set or the range, the set of all possible 'outputs' of the function. This may lie inside the codomain as illustrated in fig 4.1 or may be equivalent to the codomain.

Strictly, the domain and codomain should be specified in defining a function, particularly if the domain is limited in some way. If not given, the codomain is generally taken to be the most widely available set of values that it is sensible to use. For the rule

$$y = x^2,$$

x could be taken to be any real number, so the domain is the set of all real numbers. In this case, as y is the result of squaring a real number, it must take a non-negative real value, so the image set or range is the set of non-negative real numbers. As the codomain does not have to be precisely specified, it is only necessary to give a set that *includes* all possible outputs from the function; it would be perfectly legitimate to take the codomain as either the more broad set of all real numbers or the less vague set of all non-negative real numbers.

In many cases, a function is simply specified by giving its rule – interpretation of the domain and codomain is left to the common sense of the user. This convention will be adopted here. Although important for the mathematical definition of a function, the domain will only be mentioned when there is particular need to draw attention to it. In using the rule

$$y = 1/x,$$

for example, we must exclude $x = 0$ from the domain, otherwise the 'zero divide' fault may be invoked.

We have used the notation

$$y = f(x)$$

for a general mapping from x to y, where x is the independent variable which can be chosen freely from the domain, y is the dependent variable calculated using the rule and f represents the function rule. f(x) is said as 'f of x'. The characters used here are common, but arbitrary. For example, if we want to express a velocity **v** as a function of time t, we could use

$$\mathbf{v} = \mathbf{g}(t),$$

or even

$$\mathbf{v} = \mathbf{v}(t).$$

g(t) or **v**(t) represents the activity of applying the controlling function rule to the independent variable t, giving a single value as the result. The use of bold characters for **v** and **g** indicates that they are vector functions, to be explored in chapter 7, indicating that velocity has direction as well as value.

These are cases when the codomains are not restricted to simple real values; there are also cases of non-real valued domains. The distance r of a point (x, y) from the origin of a Cartesian system (chapter 3) is given by

$$r = \sqrt{(x^2 + y^2)}.$$

This could be represented as

$$r = d(x, y),$$

where the domain is the set of all pairs of real values (x, y), the range or image set r is the set of all non-negative real values (as the positive square root is always taken) and the rule 'take the square root of the sum of the square of x with the square of y' is interpreted in shorthand form as d(.).

Graphs of Functions

Figure 4.1 gives a reasonable interpretation of the general concept of a function, but it is an unsatisfactory representation of any specific function as there is no interpretation of the domain, codomain and function formula in a realistic sense. Figure 4.2 is an attempt to depict the specific function

$$y = x^2,$$

linking corresponding values of the domain x and codomain y. This again is unsatisfactory; the connections from x to y overlap (we can see here how some values of y have more than one corresponding value of x), only a few values of x are chosen (the integer values in this case), and the image does not give any real concept of what is happening.

The frequent use of x for the independent variable and y for the dependent variable may suggest some connection with Cartesian coordinate systems. This is

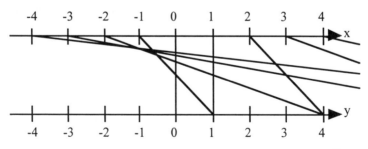

Figure 4.2 Mapping from real value x to real value y using $y = x^2$

4. Functions and Transformations: Ways of Manipulating Space

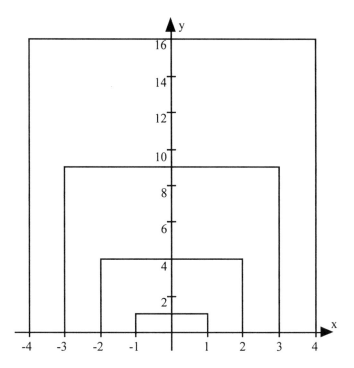

Figure 4.3 Rearrangement of fig 4.2 with the y-axis vertical and links as vertical/horizontal 'hinged' lines (note re-scaled y-axis)

no accident. We can rearrange fig 4.2 so that the y-axis is vertical and the links between corresponding values are drawn as 'hinged' vertical/horizontal lines (fig 4.3). This figure 'cheats' in that the scale for the y-axis is half that of the x-axis; this should be noted by clear labelling of axes (as in fig 4.3) whenever it is more convenient to illustrate images this way. This figure improves on previous attempts to show just what is going on in this particular function. The values of y or f(x) can be related to chosen x values from the domain, the integers from -4 to 4, through the corners of the 'hinged' links. For example, for x = 3, we follow the link up from the x-axis, the location of the corner setting it at the correct level to identify the function value at that level on the y-axis. However no values are illustrated for those between the discrete chosen set of x values. If we attempt to enter more x values, the links will overlap and become an uninformative smear.

Figure 4.4 is derived from fig 4.3 by eliminating the linking lines, retaining only the 'hinge' points, each marked with a cross. For reference, the full 'hinged link' for x = 3 is shown. These crosses are seen to lie in a smooth form which may be linked by a curve, known as the graph of the function. There is a major assumption here, that all values between those directly plotted as crosses will behave in such a 'reasonable' way. We surmise that for any value of x in the domain for which the graph is shown, we can find the corresponding value of y by moving up from the x-axis until we reach the curve, then moving horizontally to the y-axis to read off the corresponding y value on the scale. We cannot justify this faith in the 'good sense' of the function here, but the subject of calculus does give some

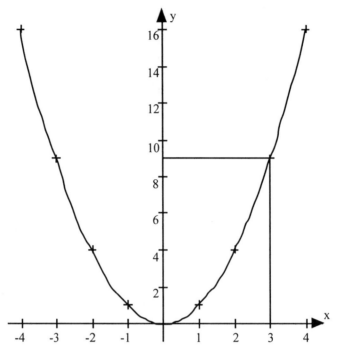

Figure 4.4 The graph of the function $y = x^2$ for the domain $-4 \leq x \leq 4$, joining link corners smoothly

reassurance that most functions do behave in such a way. If the individual points plotted are not so far apart as to hide some detailed feature of the function behaviour, a good approximation to the graph can be found by joining those points plotted with a smooth curve.

Conventionally, the independent variable (often x) is plotted on the horizontal axis, with the dependent variable (often y) on the vertical axis. If the rule is given as

$$y = f(x),$$

the curve is the 'graph of y = f(x)' or the 'graph of f(x)'. Due to the strict functional rule, graphs may not 'double back' on themselves, as there can only be one value of y for a particular value of x. If we want to draw a general curved shape, this limitation can be overcome using 'parametric functions' to define graphs. Two functions x(t) and y(t) are defined on an independent variable t. Linked pairs of values (x(t), y(t)) then define points on the graph. This will be discussed more thoroughly in the next chapter. We consider only basic functional forms in the discussion that follows. By looking at the graphs of functions already shown in chapter 3 (sine, cosine and tangent, for example), readers should reassure themselves that, although some odd things seem to be happening in the case of tangent, all graphs shown depict valid functions. The study of graphs and their forms for particular sorts of function is a major part of mathematics. Readers will become familiar with some standard forms as the book develops.

Transformations in 2D

Transformations are important throughout the creation of computer-generated images. From the original mental concept, objects are moved from easily defined positions in *canonical space* to new positions in *world space*, allowing the accumulation of objects into a *scene*, as discussed later in chapter 8. When the scene is displayed on a graphics device, such as a display screen, that portion of the scene to be depicted in world space (the contents of a *window*) has to be transformed into a specified part of the display space (called the *viewport*) using a *window to viewport transformation*; this will be described in chapter 10. If the scene is created in 3D space, the lowering of dimension necessary to produce a 2D image will be achieved through a *projection transformation*, typically a *perspective projection* or *parallel projection*. At first we consider purely 2D transformations, but will develop their 3D equivalents later.

We concentrate here on affine transformations, which have the defining property of maintaining parallel lines. If an object or drawing is transformed, any lines in the original that are parallel will remain parallel to each other in the affine transformed version. In practice an affine transformation can be created from any combination of translation, scaling and rotation.

In its most useful form for computer graphics, a 2D affine transformation transforms an original point to a new point, so it is a mapping from the domain of the 2D plane into the codomain of the 2D plane. It takes in a 2D point, manipulates it, and outputs a 2D point. If points are given in Cartesian coordinates, the function rule can be written as

$$(x', y') = f(x, y),$$

indicating that the point (x, y) has been transformed into the point (x', y'). All affine transformations can be defined as equations of form

$$x' = ax + by + c,$$
$$y' = dx + ey + f,$$

where a, b, … f are constant values. These are particularly simple forms of equation, known as 'linear equations'. Each of the three main forms of affine transformation,

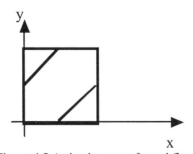

Figure 4.5 A simple untransformed figure

translation, scaling and rotation, are considered below in turn, showing the effect of each on a simple image (fig 4.5) comprising a unit square with the mid points of two sides joined.

Translation

This simply moves the object without changing its size, shape or orientation. Each point of the original figure has to be moved in this way, so if an original point is (x, y), this can be achieved by

$$x' = x + t_x,$$
$$y' = y + t_y,$$

where t_x and t_y are constants (just numbers) giving distance to be moved in the x and the y direction. Note this is equivalent to the general affine form given above, with a = e = 1, b = d = 0, c = t_x and f = t_y.

Consider the effect on the particular point (x, y) = (1, 1) when t_x = 1 and t_x = 0.5. The equations give

$$x' = 1 + 1 = 2,$$
$$y' = 1 + 0.5 = 1.5,$$

given the 'output' point (x', y') = (2, 1.5). The effect of applying this to all points of the shape in fig 4.5 is shown in fig 4.6. The shape is seen to have moved one unit to the right and half a unit upwards. Values of t_x and t_y less than zero move the object to the left and downwards respectively.

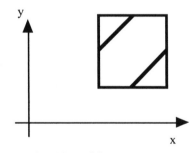

Figure 4.6 The object of fig 4.5 translated by (1, 0.5)

Scaling

Scaling changes the shapes of objects by enlarging or contracting them in the x and/or y directions. Constant scale factors s_x and s_y can be used to control these according to the following equations.

$$x' = s_x x,$$
$$y' = s_y y.$$

4. Functions and Transformations: Ways of Manipulating Space   81

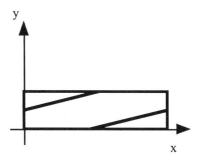

Figure 4.7 A scaling of (2, 0.5) on the object of fig 4.5

Taking (x, y) = (1, 1) again as an example point, with $s_x = 2$ and $s_y = 0.5$ this gives (x', y') = (2, 0.5). Simply, the value of the x coordinate is doubled and that of the y coordinate is halved. The effect on the object of fig 4.5 is shown in fig 4.7.

Comparing fig 4.7 with the original shape in fig 4.5, it should be clear that only one point, (0, 0), remains in its original position; all other points in the image are relocated whenever both s_x and s_y are not equal to 1. Scaling should really be identified as 'scaling about the origin'; it is as though the image is drawn on an elastic sheet, a pin is used to fix the point (0, 0) and all stretching or contraction is done about this point. In the trivial case when both s_x and s_y are 1, there is no change to the image and all points (including the origin) are fixed. When $s_x = 1$, all points on the x-axis are unchanged; when $s_y = 1$, all points on the y-axis are unchanged. Of course, the origin is part of both x- and y-axes, so it is the only point that remains unchanged in all possible cases.

Scalings with negative constants create mirror images or reflections. With $s_x = 1$ and $s_y = -1$, x coordinates are unchanged and y coordinates are multiplied by -1, 'flipping' the image about the x-axis. This is not the same as translating the image to this new position: a genuine mirror image is created with the expected phenomenon of left and right side inversion, as seen in fig 4.8. Scalings with negative values other than -1 perform size changes as well as reflection.

Viewing your own face in a double mirror reflection can be quite disconcerting:

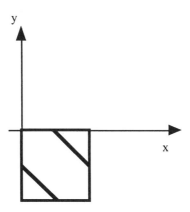

Figure 4.8 Reflection of the object of fig 4.5 using scale factors (1, -1)

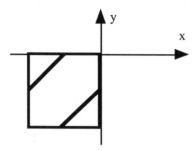

Figure 4.9 Double reflection of the object of fig 4.5 using scale factors (-1, -1)

you see your face as others see it and not in the reflected form that you normally see. Two joined vertical mirrors at right angles to each other enable you to see yourself in this way, centred at the join of the mirrors. Facial asymmetries, such as positions of hair partings, are shown to you as others see them, not as you usually see. A double reflection is, in that sense, not a reflection: the left–right inversion is 'cancelled' so a positive image is seen. The same is observed in a scaling with factors s_x and s_y both set to -1 (fig 4.9). This effect is equivalent to a rotation of 180° about the origin, which makes a good lead in to the next section.

Rotation

The third standard affine transformation is rotation about the origin. As in the case of scaling, the origin (0, 0) is a fixed point, as if a drawing on a piece of paper has been pinned down at the point, and the paper rotated about the pin. Figure 4.10 shows the effect of rotating our simple shape about the origin by 30° in the positive sense of rotation. For every point (x, y) defining the original shape, we need to find its equivalent position after rotation, (x', y'). Consider the effect of rotation through angle α on a general point (fig 4.11).

Working in polar coordinates, the solution is very easy. If Cartesian point (x, y) is given as (r, θ), the new point is (r, $\theta + \alpha$). However, it is not very easy to interpret this transformation in Cartesian coordinates as

$$x' = r \cos(\theta + \alpha),$$
$$y' = r \sin(\theta + \alpha).$$

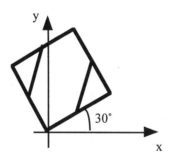

Figure 4.10 The object of fig 4.5 rotated by 30°

4. Functions and Transformations: Ways of Manipulating Space

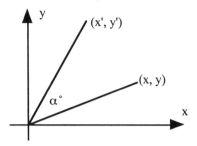

Figure 4.11 Rotation of a general point through α°

Working directly from the diagram, the secret is to rotate not just the radial line from the origin to (x, y) as in fig 4.11, but to rotate the triangular 'superstructure' connecting (x, y) to the axes (fig 4.12). Now we have some right-angled triangles with edges of known length x and y, and we can use our knowledge of trigonometry on these edges. In fig 4.17 the triangle with edges length x, y and hypotenuse OP (left image) has been rotated so that its hypotenuse is in position OP' (right image). This means that OT = x, TP' = y and ∠OTP' is a right angle. If OP has been rotated through α from its original direction, then the line OT will also have been rotated by that amount from its original direction of the x-axis. By a similar argument, the line P'T in the right image is an angle α from the vertical direction, as it was originally vertical in the left image. This means that angles ∠SOT and ∠UP'T are both α. From the right-angled triangle OST, we can now extract the lengths

and
$$OS = x \cos(\alpha)$$
$$ST = x \sin(\alpha)$$

(remember that OT = x, ∠SOT = α and ∠OST is a right angle). Similarly, from triangle UP'T,

and
$$UT = y \sin(\alpha)$$
$$UP' = y \cos(\alpha)$$

(using TP' = y, ∠UP'T = α and ∠P'UT = 90°). We are trying to determine the

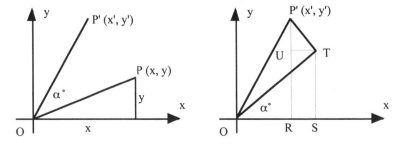

Figure 4.12 Rotation of the supporting triangle to position OTP'

position of P' (x', y'). Note that x', the horizontal displacement of P' from the origin, can be found from

$$x' = OR = (OS - RS) = (OS - UT),$$

as RS is the same length as UT. We know both OS and UT, so

$$x' = x \cos(\alpha) - y \sin(\alpha).$$

Similarly for y', the vertical displacement of P' from the origin,

$$y' = RP' = RU + UP' = ST + UP',$$

giving us the formula for y' from known sides of triangles as

$$y' = x \sin(\alpha) + y \cos(\alpha).$$

We have achieved the required result, finding the position of a general point (x, y) when it has been rotated by α about the origin. This is probably the most difficult geometric proof in this book. Those who have found it difficult to follow should just accept the important result, repeated below for convenience.

$$x' = x \cos(\alpha) - y \sin(\alpha),$$
$$y' = x \sin(\alpha) + y \cos(\alpha).$$

Thus, for example, applying a rotation of 30° to (1, 1), the upper right vertex of our original shape (fig 4.10), its location (fig 4.15) moves to

$$(\sqrt{3}/2 - 1/2,\ 1/2 + \sqrt{3}/2)$$

using the values $\cos(30°) = \sqrt{3}/2$ and $\sin(30°) = 1/2$ from chapter 3. These are not very pleasant numbers to deal with, but computers cope with these without distinction from the kinds of values we humans prefer, so there is no pain in computer implementation of such tricky formulae.

A trigonometric spin-off

In the last section, we commented that the polar coordinate version of a rotation was easy to derive. Rotating the point (r, θ) through angle α gives the point (r, θ + α). Suppose (x, y) is equivalent to the original point. This makes

and
$$x = r \cos(\theta)$$
$$y = r \sin(\theta),$$

from the formulae of chapter 3. Also, the transformed point (x', y') is equivalent to the polar point (r, θ + α), so

and
$$x' = r \cos(\theta + \alpha)$$
$$y' = r \sin(\theta + \alpha).$$

4. Functions and Transformations: Ways of Manipulating Space

We have already established formulae for x' and y' in terms of x, y and α as

$$x' = x \cos(\alpha) - y \sin(\alpha),$$
$$y' = x \sin(\alpha) + y \cos(\alpha).$$

Substituting the polar forms of x, y, x' and y' into these equations gives

$$r \cos(\theta + \alpha) = r \cos(\theta) \cos(\alpha) - r \sin(\theta) \sin(\alpha),$$
$$r \sin(\theta + \alpha) = r \cos(\theta) \sin(\alpha) + r \sin(\theta) \cos(\alpha).$$

Now we can divide by r, as r appears in all terms, so

$$\cos(\theta + \alpha) = \cos(\theta) \cos(\alpha) - \sin(\theta) \sin(\alpha),$$
$$\sin(\theta + \alpha) = \cos(\theta) \sin(\alpha) + \sin(\theta) \cos(\alpha).$$

These important trigonometric formulae are identities; in other words they are true for any possible values of θ and α. Their proof is a relatively straightforward 'spin-off' from the rotation formulae. Other forms are easily derived from these. For example, if α is set to equal θ, we get the 'double angle' formulae,

$$\cos(2\theta) = \cos^2(\theta) - \sin^2(\theta),$$
$$\sin(2\theta) = 2\cos(\theta) \sin(\theta),$$

where $\cos^2(\theta) = \cos(\theta)\cos(\theta)$ and $\sin^2(\theta) = \sin(\theta)\sin(\theta)$. Also, using the rules that

$$\cos(-\alpha) = \cos(\alpha) \quad \text{and} \quad \sin(-\alpha) = -\sin(\alpha)$$

from chapter 3 (this can be established from the defining equations of cos and sin), we can replace α by $-\alpha$ to get

$$\cos(\theta - \alpha) = \cos(\theta) \cos(\alpha) + \sin(\theta) \sin(\alpha),$$
$$\sin(\theta - \alpha) = \cos(\theta) \sin(\alpha) - \sin(\theta) \cos(\alpha).$$

These identities are very important in the development of trigonometry. Although not of great use in computer graphics, they are mentioned here as they can be so easily derived from one of the methods that is central to computer graphics.

Shear

Shear is not one of the 'core' affine transformations, as it can be performed by a combination of rotations and scaling. However, we show it explicitly here as it can be performed by particularly simple linear equations. If rotations and scaling are used to obtain the same effect, the calculation of the appropriate rotation and scaling constants can be quite difficult. The defining equations for shear in the x direction are

$$x' = x + k_x y,$$
$$y' = y.$$

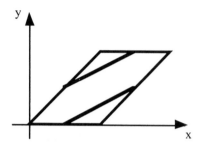

Figure 4.13 A shear of factor 1 in the x direction on the object of fig 4.5

Note that y values are unchanged, but x values are changed more when y is greater, when the point is further from the x-axis. The effect of this on the point (1, 1) is to transform it to the point $(1 + k_x, 1)$. With the shear constant k_x set to 1, this is the point (2, 1), the location of the top right corner of the shape of fig 4.13.

A similar 'shear in the y direction' (not illustrated) is performed using

$$x' = x,$$
$$y' = y + k_y x.$$

In this case, x is unchanged; y values are changed more for points more distant from the x-axis. Readers may want to sketch their own figure for this case, to ensure that they understand the process.

It was stated earlier that a defining property of affine transformations is that parallel lines are preserved. Several examples of transformations of a basic image containing three pairs of parallel lines (fig 4.5) have been shown (figs 4.6–4.10, 4.13). Some of these have changed the location, the orientation and the shape of the original form, but in each case it should be clear that the transformed versions of the three pairs of parallel lines are still parallel. This is not a proof that the defining property is maintained, but to the eye what has been achieved seems consistent with this property.

Transformations in 3D

The extension to 3D is straightforward for translation and scaling, so we merely state the equations as extensions of the 2D case into the 3D Cartesian world where points have coordinates (x, y, z).

Translation

$$x' = x + t_x,$$
$$y' = y + t_y,$$
$$z' = z + t_z.$$

4. Functions and Transformations: Ways of Manipulating Space

When applied to all points of a 3D object this moves it without change of shape or orientation by t_x units in the x direction, t_y in the y direction and t_z in the z direction.

Scaling

$x' = s_x x,$
$y' = s_y y,$
$z' = s_z z.$

This stretches or compresses a 3D object by factors s_x in the x direction, s_y in the y direction and s_z in the z direction with the point (0, 0, 0) held fixed.

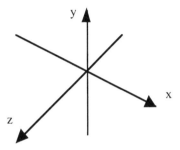

Figure 4.14 A standard set of 3D Cartesian axes

Rotation

Now some complications start to occur. In 3D, three forms of rotation are defined, about each of the coordinate axes in the x, y and z directions (fig 4.14). Consider rotation about the z-axis. If we observe fig 4.14 directly from the positive end of the z-axis, we see the x- and y-axes as in fig 4.15, shown in a familiar orientation with x pointing right and y pointing up. This is exactly the set up for 2D rotation, so the formulae are almost the same as in the 2D case,

$x' = x \cos(\alpha) - y \sin(\alpha),$
$y' = x \sin(\alpha) + y \cos(\alpha),$
$z' = z.$

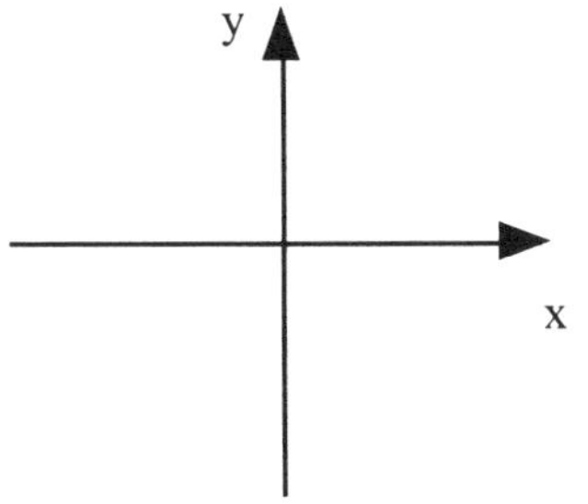

Figure 4.15 The x- and y-axes as seen from the positive z-axis of fig 4.14

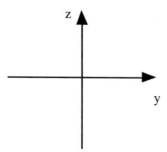

Figure 4.16 The y- and z-axes as seen from the positive x-axis of fig 4.14

The one 'extra' is the trivial case z' = z, as the z value is unchanged by rotation about its own axis. Rotation of an object about the z-axis (all points on this axis will remain unchanged) turns by the required angle 'from x towards y' as seen in fig 4.15. This is a right-hand screw in the z direction.

Now consider rotation about the x-axis. If we view the y- and z-axes from the positive x-axis, they can be seen in an orientation as in fig 4.16. This is no longer the same set-up as for 2D rotation, but we see that y has taken the former x position and z has taken the former y position. So, rotation around the x-axis is equivalent to rotation from y towards z, with x remaining unchanged, just as rotation around the z-axis is rotation from x towards y with z remaining unchanged. We can take the formulae we already have, changing x to y, y to z and z to x, as we are solving the same problem with relabelled axes. This gives

$$y' = y \cos(\alpha) - z \sin(\alpha),$$
$$z' = y \sin(\alpha) + z \cos(\alpha),$$
$$x' = x.$$

Restating these in a more usual alphabetic order,

$$x' = x,$$
$$y' = y \cos(\alpha) - z \sin(\alpha),$$
$$z' = y \sin(\alpha) + z \cos(\alpha).$$

Now it should be clear how we obtain the equations for rotation about the y-axis. Viewing our axes of fig 4.14 from the positive y-axis, the z- and x-axes are observed as in fig 4.17. This rearrangement puts z and x in place of the original x and y in

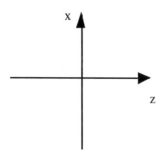

Figure 4.17 The z- and x-axes as seen from the positive y-axis of fig 4.14

our 2D rotation formulae, which are equivalent to rotation about the z-axis. By appropriate replacements into this rule, we get

$$z' = z \cos(\alpha) - x \sin(\alpha),$$
$$x' = z \sin(\alpha) + x \cos(\alpha),$$
$$y' = y.$$

Again, rearranging in alphabetical order x, y and z,

$$x' = x \cos(\alpha) + z \sin(\alpha),$$
$$y' = y,$$
$$z' = -x \sin(\alpha) + z \cos(\alpha).$$

These will be discussed again in chapter 6, when a matrix method for implementing these effects is considered. At first sight, there seems to be an asymmetry about the forms of equations for rotations about the different axes, but that is because we have arranged them in alphabetical order, beginning with x. There is an elegant cyclic symmetry within the equations, with the sequence 'x to y to z to x ...' at its core. This is demonstrated by rearranging the equations (yet again) as

Rotation about x:
$$x' = x,$$
$$y' = y \cos(\alpha) - z \sin(\alpha),$$
$$z' = y \sin(\alpha) + z \cos(\alpha).$$

Rotation about y:
$$y' = y,$$
$$z' = z \cos(\alpha) - x \sin(\alpha),$$
$$x' = z \sin(\alpha) + x \cos(\alpha).$$

Rotation about z:
$$z' = z,$$
$$x' = x \cos(\alpha) - y \sin(\alpha),$$
$$y' = x \sin(\alpha) + y \cos(\alpha),$$

when the cyclic structural similarities should be clear.

Combining Affine Transformations

The affine transformations are very important in manipulating computer graphics 'objects', but are seldom used in their individual restricted forms. By repeatedly applying combinations of transformations to an object, a much wider variety of controlled effects can be created. To illustrate this, we consider a simple 2D example. Suppose we want to reflect an L-shaped object defined by the joins of vertices (0, 0), (0, 3), (1, 3), (1, 1), (2, 1), (2, 0), (0, 0) (fig 4.18(a)) in a line at 45° to the x-axis passing through the point (2, 1). This can be achieved by the sequence of transformations illustrated in fig 4.18(b)–(f). The L shape in its original location is shown in each of these frames as a shaded shape. The sequence is not unique; the

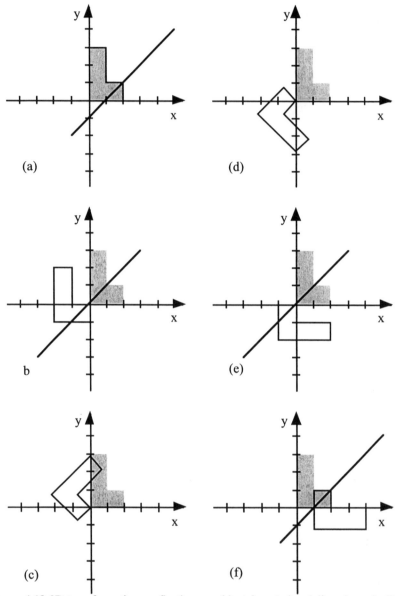

Figure 4.18 2D transformations reflecting an object in a 'mirror' line through (2, 1) at 45° to the x-axis

effect could have been achieved in other ways – some readers may want to devise their own alternative combination of transformations. The stages of fig 4.18(b)–(f) are described in turn below, each effect being shown on the L shape and its 'mirror line' in the figure.

4. Functions and Transformations: Ways of Manipulating Space

(b) The point (2, 1), a known point on the mirror, is moved to the origin using a translation with constants (-2, -1). This makes the mirror line pass through the origin.

(c) The object is rotated by -45°, to align the mirror line with the x-axis.

(d) The object is reflected in the x-axis, using a scaling with factors (1, -1) to 'flip' the y values.

(e) The object is rotated by 45°, to reorient the mirror at 45° to the x-axis, its original alignment.

(f) Finally, the object is translated (2, 1) to replace the mirror line in exactly its original position.

The effects of these transformations, in sequence, on the vertex at the top right of the vertical stem of the L shape, (x, y) = (1, 3), are given below.

(b)   Translation with $t_x = -2$, $t_y = -1$

$$x_b = x + t_x = 1 - 2 = -1,$$
$$y_b = y + t_y = 3 - 1 = 2,$$

gives $(x_b, y_b) = (-1, 2)$, where suffix b is used to indicate this is the location of the transformed point in fig 4.18(b).

(c)   Rotation with $\alpha = -45°$

$$x_c = x_b \cos(\alpha) - y_b \sin(\alpha) = (-1)(1/\sqrt{2}) - (2)/(-1/\sqrt{2}) = 1/\sqrt{2},$$
$$y_c = x_b \sin(\alpha) + y_b \cos(\alpha) = (-1)(-1/\sqrt{2}) + (2)(1/\sqrt{2}) = 3/\sqrt{2},$$

gives $(x_c, y_c) = (1/\sqrt{2}, 3/\sqrt{2})$. Trigonometric values established from chapter 3 are used, $\cos(-45°) = 1/\sqrt{2}$, $\sin(-45°) = -1/\sqrt{2}$.

(d)   Scaling with $s_x = 1$, $s_y = -1$

$$x_d = s_x x_c = (1)(1/\sqrt{2}) = 1/\sqrt{2},$$
$$y_d = s_y y_c = (-1)(3/\sqrt{2}) = -3/\sqrt{2},$$

gives $(x_d, y_d) = (1/\sqrt{2}, -3/\sqrt{2})$.

(e)   Rotation by angle $\alpha = 45°$

$$x_e = x_d \cos(\alpha) - y_d \sin(\alpha) = (1/\sqrt{2})(1/\sqrt{2}) - (-3/\sqrt{2})/(1/\sqrt{2}) = 2,$$
$$y_e = x_d \sin(\alpha) + y_d \cos(\alpha) = (1/\sqrt{2})(1/\sqrt{2}) + (-3/\sqrt{2})(1/\sqrt{2}) = -1,$$

gives $(x_e, y_e) = (2, -1)$. From chapter 3, we use $\cos(45°) = 1/\sqrt{2}$, $\sin(45°) = 1/\sqrt{2}$.

Also, remember the meaning of $\sqrt{2}$ implies that $(\sqrt{2})(\sqrt{2}) = 2$, so, for example, it is easy to evaluate $(1/\sqrt{2})(1/\sqrt{2}) = 1/2$.

(f)   Translation with $t_x = 2$, $t_y = 1$

$$x_f = x_e + t_x = 2 + 2 = 4,$$
$$y_f = y_e + t_y = -1 + 1 = 0,$$

gives $(x_f, y_f) = (4, 0)$.

The effects of these can be tracked in fig 4.18 through the points $(1, 3)$ in frame (a), $(-1, 2)$ in (b), $(1/\sqrt{2}, 3/\sqrt{2})$ in (c), $(1/\sqrt{2}, -3/\sqrt{2})$ in (d), $(2, -1)$ in (e) and $(4, 0)$ in (f). For convenience in tracking, it is easier to identify points in (c) and (d) by their approximate values $(0.71, 2.12)$ and $(0.71, -2.12)$ (correct to two decimal places).

To find the final location of the L shape, we apply all transformations to each of the shape's six defining vertices. This is a tedious process – even more so in a computer graphics application when a shape may contain several thousand vertices. It would be more convenient to put all the five transformations together into one. This can be done working back through the formulae of stages (f) to (b) given above. So, for example,

(f) $x_f = x_e + 2,$
         $y_f = y_e + 1,$

uses x_e and y_e from

(e) $x_e = x_d/\sqrt{2} - y_d/\sqrt{2},$
         $y_e = x_d/\sqrt{2} + y_d/\sqrt{2},$

so we can substitute these expressions for x_e and y_e to find x_f and y_f as

$$x_f = x_d/\sqrt{2} - y_d/\sqrt{2} + 2,$$
$$y_f = x_d/\sqrt{2} + y_d/\sqrt{2} + 1.$$

Then, using forms x_d and y_d from

(d) $x_d = x_c,$
         $y_d = -y_c,$

we get x_f and y_f as

$$x_f = x_c/\sqrt{2} + y_c/\sqrt{2} + 2,$$
$$y_f = x_c/\sqrt{2} - y_c/\sqrt{2} + 1.$$

Two more stages to go. Now we substitute for x_c and y_c from (c)

(c) $x_c = x_b/\sqrt{2} + y_b/\sqrt{2},$
         $y_c = -x_b/\sqrt{2} + y_b/\sqrt{2},$

4. Functions and Transformations: Ways of Manipulating Space

to give x_f and y_f as

$$x_f = (x_b/\sqrt{2} + y_b/\sqrt{2})/\sqrt{2} + (-x_b/\sqrt{2} + y_b/\sqrt{2})/\sqrt{2} + 2,$$
$$y_f = (x_b/\sqrt{2} + y_b/\sqrt{2})/\sqrt{2} - (-x_b/\sqrt{2} + y_b/\sqrt{2})/\sqrt{2} + 1.$$

Some algebraic manipulation (this is one task that is perhaps easier for a human than a computer, although there are ingenious computer algebra systems available today) simplifies this dramatically to

$$x_f = y_b + 2,$$
$$y_f = x_b + 1.$$

Readers should note that this is a particularly friendly example. Most sequences of transformations lead to ever more complex sets of equations: a simple case has been chosen here to illustrate a point. Finally, we use

(b) $\quad x_b = x - 2,$
$\quad\quad y_b = y - 1,$

to find (x_f, y_f) in terms of the starting point (x, y) as

$$x_f = y + 1,$$
$$y_f = x - 1.$$

Extracting this formula has not been easy, and the particular example chosen has made the situation much simpler than in most arbitrary cases, but at the set-up cost of developing the formula, transformation of all vertices of the shape can now be done more easily. It is fairly straightforward to see that the original object defined by points

$$(0, 0), (0, 3), (1, 3), (1, 1), (2, 1), (2, 0)$$

is transformed into the final object defined by points

$$(1, -1), (4, -1), (4, 0), (2, 0), (2, 1), (1, 1).$$

The point used to work out our example above has been italicized, $(1, 3)$ is transformed to $(4, 0)$. It is also worth noting that the point $(2, 1)$, a point on the mirror line, has not been changed by the overall transformation.

This has been a long and tedious process; computers work well in evaluating formulae, rather than manipulating them, so such a method would be difficult to implement in a computer system. However, in situations where several thousands of points may have to be manipulated in 2D or 3D, there are obvious benefits of being able to put together, or to 'concatenate', sequences of affine transformations. If we can reduce the point manipulations by a factor of 5 through reduction from five to one transformation equation, it is worth more than a little set-up cost. Chapter 6 gives a method for doing this in a way more easily performed in computer systems, based on matrix manipulation. It will start from the base point of the defining affine transformation equations given in this chapter.

Inversion of Affine Transformations

We have looked at a number of affine transformations that act as a function on points in 2D or 3D space to create new points. They have domain as all points in 2D (or 3D) and codomain as points in 2D (or 3D as appropriate). We saw examples in the last section of cases when it was necessary to undo the effects of certain transformations, and this was easy to achieve. For example, to undo the effect of translation of (-2, -1), we translated by (2, 1). To undo the effect of rotation by -45°, we rotated by 45°. These cases are fairly straightforward; the undoing effect is clearly understood. If transforming a point is like asking the question, 'Where will point A end up if it's transformed in this way?', inverting this process is like trying to find the question when we know the answer. We can only be certain we've got the right question if there's one and only one question that could have given that answer. Are there affine transformations that cannot be unravelled in this way? All rotations and translations can be inverted. Most scalings can be inverted, except for the special cases when one or more of the scaling factors are zero.

In general, we can invert the effect of scaling with factors (s_x, s_y) by a scaling with factors ($1/s_x$, $1/s_y$). For example, scaling the point (3, 5) by factors (2, -1) gives the point (6, -5). We reverse this process using factors (1/2, 1/(-1)) = (1/2, -1) on this point, to give (3, 5), our original point. However, if s_x say is zero, we cannot find the factor $1/s_x$ needed to invert the scaling and the zero divide error is invoked.

In all forms of affine transformations except for scaling with factor zero, distinct points are transformed to separate distinct points. The function is 'one to one'. One point from the domain is mapped to a distinct point in the image set; indeed this has been stressed as a defining property for all functions. For a function to be 'one to one', however, any point in the image set can only have been created from one point in the domain. Separate points must map to separate points, so they can be distinguished. If, for example, two points P_1, P_2 in the domain gave the same image point y under some affine transformation f, so

        $Q = f(P_1)$
and $Q = f(P_2)$,

how can we answer the question, 'What position of P gives Q as the answer when we calculate f(P)'? 'Perhaps it's P_1', or, 'It may be P_2' are not satisfactory as mathematical answers; we need certainty for functions. Take the example of transforming the points P_1 (2, 7) and P_2 (4, 7) under a scaling with factors (0, 2). The image points of these are (0, 14) and (0, 14). In fact, any point of form (x, 2) would by transformed to (0, 14). So if someone challenges, 'I've just transformed a point by scaling (0, 2) and I've got the answer (0, 14), I bet you can't give me the answer', you'd be reduced to guessing. All you know is that the answer is something like (x, 2), where x could be any value, so you'd be very lucky to pick the right one by guesswork.

Inversion of Functions

Affine transformations are important to computer graphics as transformations, and the inversion of these, undoing their effects, has clear explanation. The inversion of functions in general is an important topic, particularly if we are to use functions for representation of shapes as graphs (this will be discussed further in the next chapter). We have seen in the case of affine transformations that if a function has more than one value from its domain (inputs) that gives the same value (output) in its codomain, then it cannot be inverted, as there would be ambiguity. One of the defining properties of a function is that it must be 'single valued': for any input, there is only one output. The inverse of a function is itself a function, so this rule must also be applied in this case. Thus, for a function to be invertible, this rule must operate in both directions. There is only one output for any input, and there is only one input that could have created a particular output. Functions with this property are called 'one-to-one' functions.

Figure 4.19 shows the graphs of three functions. In box (a), $y = f_1(x)$ cannot be inverted as it has several possible values of x for one value of y. If you draw a horizontal line through this graph, it could cut the curve at several points, giving more than one value of x for the same value of y represented by the height of the line. However, at least within the domain illustrated in fig 4.19, a horizontal line will cut the curves in boxes (b) and (c) at one point only; the curves illustrate 'one-to-one' functions. This is because they are monotonic, $f_2(x)$ in box (b) is monotonic increasing, $f_3(x)$ in box (c) is monotonic decreasing. This means that as x increases from left to right of the graph, the value of y increases (box b) or decreases (box c) consistently for each of these graphs. The graph of $f_1(x)$ is seen to rise, then fall, then rise again. Any graph that rises for part of its domain and falls for another part is not monotonic, and cannot therefore be inverted. Tests to determine whether a function is monotonic depend on the gradient evaluated using calculus, to be discussed in chapter 5.

The inverse of a function was referred to above as 'finding the question given the answer'. The normal expression for a function is '$y = f(x)$', enabling us to find y for a given value of x. Reversing this process means we need to find 'the value of x that, when entered into the function f, will give us the result y'. This is usually expressed in a confusing notation as

$x = f^{-1}(y).$

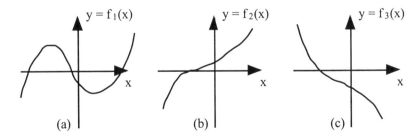

Figure 4.19 Three functions, only two of which can be inverted

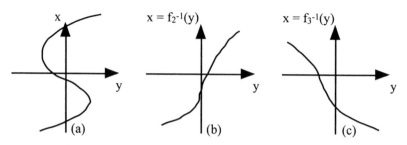

Figure 4.20 Reflections of the graphs of fig 4.19 in a line at 45° to the x-axis through the origin

the superfix '-1' indicating that the process of finding a value using function f is to be reversed. Normally, a superfix '-1' attached to a number indicates raising to the power -1, as in

$$a^{-1} = 1/a.$$

The superfix '-1' is not attached to a number this time, it is attached to a function rule, and in this context it has the special meaning of 'undoing' the function rule.

The formulae defining some functions are easily inverted. This is particularly the case with linear functions, whose equations can be manipulated to place x rather than y as the 'focus' of the equation. For example, subtracting 5 from each side of the equation

$$y = 2x + 5$$

gives $\quad y - 5 = 2x.$

Now dividing by 2, and placing x at the head of the equation, we get the functional form for x in terms of y

$$x = (y - 5)/2.$$

We now have x as the dependent variable and y as the independent variable, contrary to the usual form, as we have swapped the normal axis labels in fig 4.20. These are merely dummy names; the concept of x as a function of y is as equally tenable as the more usual y as a function of x.

In other cases, the functional form of an inverse usually requires human ingenuity to extract; it may not be easily derived or may not be directly expressed as a simple mathematical function. However, graphs of inverse functions are easy to create from the graphs of the original form by reflecting in a line at 45° to the x-axis through the origin (fig 4.20). In an inverse, the same x and y values are connected by the formula, but now the x-axis and the y-axis are interchanged; x takes the place of y and vice versa, as arranged by the reflection. Figure 4.20 shows the reflections of the three functions of fig 4.19. It is clear that the image of box (a) cannot be a function as a vertical line can cross the graph in more than one point, thus breaking a fundamental rule. The other two examples in boxes (b) and (c) are both monotonic so do not have this fault; they are valid inverse functions $f_2^{-1}(y)$ and $f_3^{-1}(y)$ of $f_2(x)$ and $f_3(x)$ respectively.

4. Functions and Transformations: Ways of Manipulating Space

In chapter 3, we used examples of inverse functions involving the trigonometric functions to convert from Cartesian to polar coordinates. Their graphs (figs 4.21–4.23) show that sine, cosine and tangent are not monotonic functions. Both sine and cosine rise and fall indefinitely in cycles of 360° or $2\pi^c$ (the symbol ^{c}

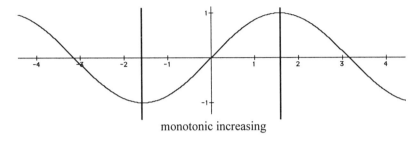

Figure 4.21 A monotonic increasing section of the sine function from $-\pi/2$ to $\pi/2$

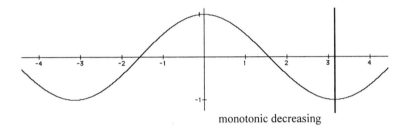

Figure 4.22 A monotonic decreasing section of the cosine function from 0 to π

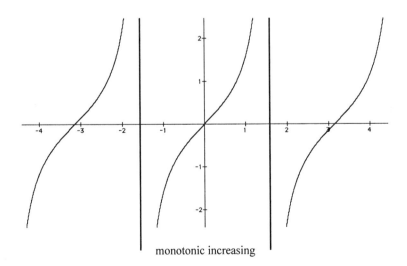

Figure 4.23 A monotonic increasing section of the tangent function from 0 to π

stands for radians as in chapter 2). Tangent (fig 4.23) always rises at points for which it is defined, but it has discontinuities or jumps that drastically diminish its value every 180° or π^c, when the value 'skips' from plus infinity to minus infinity. To find any form of inverse we must restrict the domain of the functions to sections that are monotonic increasing or monotonic decreasing, and invert those parts of the function alone. Convenient sections to choose are those including the origin. Thus, sine has a monotonic section from -90° to 90° ($-\pi/2$ to $\pi/2$), and it is this domain alone that can be inverted (fig 4.21). For simplicity, the symbol for radians is left out; the presence of π implies that this is the correct measure. This gives an inverse function with image set from -90° to 90°, or $-\pi/2$ to $\pi/2$, as the 'principal values' of the sine function (fig 4.24). We have $\sin(-\pi/2) = -1$, and $\sin(\pi/2) = 1$, so the inverse of sine, often written 'arcsin' as well as $\sin^{-1}$, therefore has domain [-1, 1], and its image set is contained in [$-\pi/2$, $\pi/2$] (fig 4.24). The graph of fig 4.24 is created by reflection of the monotonic increasing section of fig 4.21 in the line through the origin at 45° to the x-axis.

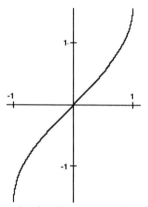

Figure 4.24 The inverse sine function, domain [-1, 1], image set [$-\pi/2$, $\pi/2$]

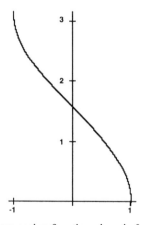

Figure 4.25 The inverse cosine function, domain [-1, 1], image set [0, π]

4. Functions and Transformations: Ways of Manipulating Space

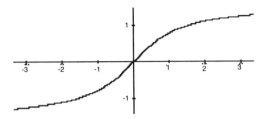

Figure 4.26 The inverse tangent function, with range [-∞, ∞], image set [-π/2, π/2]

Figure 4.25 is similarly created from the monotonic decreasing part of the cosine function in fig 4.22. This gives the inverse cosine, known as arccos or $\cos^{-1}$, a domain [-1, 1] and an image set [0, π].

The tangent function 'disappears to infinity' every 180° or π (fig 4.23). The largest monotonic increasing section containing the origin lies between -90° and 90°, or -π to π, and it is this portion that is inverted (fig 4.26). This has the unlimited domain [-∞, ∞]. In each of the three inverse functions shown here (figs 4.24–4.26), the images were created by rotating the relevant curve sections by 45°, reflecting in the y-axis, and rotating back by -45°, putting our knowledge of affine transformations to practical use. Some graininess in the graph shapes is due to pixellation during this process.

All this does is to illustrate the shapes of these curves; we have not established a formula that can be easily evaluated. This cannot be done; the meaning of arcsin(y) is simply the angle whose sine takes value y. Just as we do not have an easy formula for sine (see chapter 3), the same is true for arcsin (also arccos and arctan). Previously, laboriously calculated tables were used for evaluation; now computers and calculators have built-in routines that find these values to a given level of precision. These methods use techniques from 'numerical analysis': the study of methods for finding accurate numerical solutions of problems that are analytically intractable. Computer solutions, for example, usually involve evaluating repeated terms of a series until its sum stabilizes to a solution. We will not involve ourselves here with the development of numerical analysis; it is enough for our purposes to understand that such methods exist, and that the solutions we require are at hand from built-in computer routines.

Each of the inverse sine, cosine and tangent functions has image set or range of principal values less than 360°, so in solving real problems other clues are needed to establish the angle involved. These inverses only give the principal value, which is only one possible value. This has already been discussed in identifying how to find the correct value of θ when converting from Cartesian (x, y) to polar coordinates (r, θ). We first used

$$\theta = \arctan(y/x)$$

then added 180° or π^c whenever x < 0. Similar consideration of the quadrant (chapter 3) in which the angle lies is needed to give a full potential range of inverse sine and cosine values.

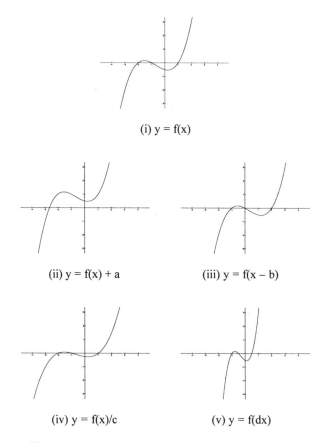

Figure 4.27 Formula changes translate and scale graphs

Shape Transformation by Function Change

The most frequently used transformations in computer graphics are the affine transformations of translation, scaling and rotation, whose formulae were given earlier in this chapter. These are applied as functions directly to the points that define computer graphics objects (chapters 6, 8 and 10 describe these uses). We have also looked at some graphs of functions in this chapter. These are also useful in describing shapes; further examples will be seen in chapter 5. From the assumption that we can describe a shape as a function graph, $y = f(x)$, this section considers how the affine transformations of translation and scaling can be performed by simple amendments to this graph equation. The example graph of the cubic polynomial $y = (x^3 + 2x^2 - x - 2)/4$ (fig 4.27(i)) is used to illustrate these effects; polynomial curves and their properties will be considered in the next chapter.

To translate the graph upwards a distance a, the value a is added to f(x) to increase y values by a (fig 4.27(ii)). This is formally stated as

$$y = f(x) + a,$$

which can be rewritten as

$$y - a = f(x).$$

This has exactly the same meaning, but associates the agent of change, the constant a, with y, the variable in its direction of change. Subtracting a from y has the effect of translating the graph a distance a in the y direction. What is the effect of subtracting the constant b from the x variable? This rearranges the function (fig 4.27 (iii)) as

$$y = f(x - b)$$

When $x = 0$, $y = f(-b)$. This means that the y value for $x = -b$ in the graph of $y = f(x)$ appears at position $x = 0$ in the graph of $y = f(x - b)$, a shift of b units to the right. This is true for any value of x; the y value at position x in the graph of $y = f(x - b)$ has moved from position $x - b$ in the graph of $y = f(x)$. That is the meaning of replacing x by $(x - b)$ in the functional form. Thus, the graph of $y = f(x - b)$ is equivalent to the graph of $y = f(x)$ moved b units in the x direction, as shown in fig 4.27(iii).

Now consider scaling in the y and x directions. Scaling of factor 1/c in y is achieved by dividing each value of y by 1/c. This functional change gives

$$y = f(x)/c.$$

Again, rearranging this to associate the agent of change with its changed variable,

$$cy = f(x).$$

This is no longer a formula giving y in terms of x, but it shows that if we multiply y by a constant c, the effect is to scale the graph in y by a factor 1/c (fig 4.27(iv)). Again, we ask what happens if we give x the same treatment? Suppose we multiply x by another constant d, to give

$$y = f(dx).$$

Comparing with $y = f(x)$, the y value is unchanged at $x = 0$; both forms have value f(0). Consider the effect on $y = f(dx)$ when $x = -1$ and $x = 1$. The graph of $y = f(dx)$ when $x = -1$ has height f(-d), the same height as the original graph of $y = f(x)$ when $x = -d$. Similarly, the graph of $y = f(dx)$ when $x = 1$ has height f(d), the same height as the graph of $y = f(x)$ when $x = d$. What was previously d units from the origin in the x direction is now one unit away. This rule is repeated in general, meaning the graph is scaled by a factor of 1/d in the x direction (fig 4.27(v)). Multiplying x by d has created a scaling in the x direction by 1/d; this is consistent with multiplying y by c, which scaled by 1/c in the y direction.

To summarize, we have established that we can change graphs in controlled ways by amending their functional rules. Subtracting a constant k from one or other of the variables moves the graph a distance k in the direction of that variable. Multiplying one or other of the variables by a constant k scales the graph by a factor 1/k in the direction of that variable. The functional inversion (subtraction of a value 'adds' to the graph, multiplication 'divides' it) may seem counter-intuitive, but some thought on the nature of what is going on should convince readers that these are correct interpretations. For the sake of completeness, the constants used in fig 4.27 are a = 1, b = 1, c = 2, d = 2. These transform the equation y = f(x) from its original form (as on page 100)

(i) $y = (x^3 + 2x^2 - x - 2)/4$

into

(ii) $y = (x^3 + 2x^2 - x - 2)/4 + 1$,
(iii) $y = \{(x - 1)^3 + 2(x - 1)^2 - (x - 1) - 2\}/4$,
(iv) $y = (x^3 + 2x^2 - x - 2)/8$,
(v) $y = \{(2x)^3 + 2(2x)^2 - 2x - 2\}/4$.

No attempt is made to clean these up; they relate to the equivalent functional forms in each of the parts of fig 4.27.

If several of the rules of fig 4.27 are imposed simultaneously, the ordering of activity is important. Consider

$y = f(dx - b)$.

When x = 0, y takes the value f(-b), indicating that the translation of the original graph of f(x) by b in the x direction fixes the value of the new curve as it crosses the y-axis to be (0, f(-b)). The scaling of factor 1/d in the x direction then takes place about this fixed point. Thus (and this is perhaps counter-intuitive), the translation takes place first followed by the scaling. A similar effect is noticed for changes in the y direction. If any variable t, say, is changed to (dt - b) in a functional formula, this represents a translation of b in the t direction followed by a scaling of 1/d in the t direction. The change of t could be represented by the equivalent d(t - b/d), producing an identical effect by scaling of factor d in the t direction followed by a translation of b/d in the t direction. The way in which the transformation is expressed affects the way in which it is interpreted. There is no easy way out of this. As in the performance of the standard computer graphic transformations to be discussed later, vigilance is the only way to ensure that operations are performed in the correct orders.

We now have a powerful set of rules for manipulating the shapes of graphs, but we cannot allow use of rotation without risking breaking a fundamental rule of functions. If the graph of fig 4.27(i) were rotated through an angle of 90°, say, we would not have a valid function graph as there would be more than one value of y for several single values of x. Any of the above effects, as well as rotation, could be performed by drawing the standard graph y = f(x) and performing the required standard computer graphic transformation to all points in the graph. It is sometimes more useful to have direct functional control over the process, which means we have to draw a new graph rather than to apply the transformations directly to *all* created points of the original drawing.

These operations can be used to animate graphs by introducing a time factor into the manipulating constants. For example, if we wish to track the graph of f(x) towards the right at a constant speed of v units per second, this involves a translation of vt in the x direction after time t. This effect can be produced by repeatedly clearing and redrawing the function

$$y = f(x - vt),$$

in which the constant b of fig 4.27(iii) is replaced by the time-related variable vt. All the constants a, b, c and d of fig 4.27, either singly or together, can be related to time to produce animations. We have seen that the sine function represents a wave form. Using the functional rules given here, we see that repeatedly clearing and redrawing

$$y = A\sin(x - vt)$$

gives a wave with height A travelling at speed v from left to right.

Conclusions

This chapter has discussed the pure mathematics concept of function as a mapping from a domain to an image set contained within an appropriate range. We have also looked at the meaning of inversion of a function: the 'undoing' of a functional evaluation. The most important functions for computer graphics are the transformations that transform from one description of 2D or 3D space to another. Affine transformations are a centrally important case; translations, scalings and rotations are at the core of computer graphic manipulation of objects and images. Readers should take away with them the concept of function, and the ways in which the affine transformations can be invoked serially to create required manipulations of shapes and objects. This theme will be looked at again in chapter 6, when we consider matrices as special tools to implement such transformations. Examples of specific functions, and the use of calculus to interpret them, is the subject of the next chapter.

5. Form from Function: Analysis of Shapes

Introduction

Some examples of functions and their graphs were given in chapter 4. In this chapter we develop this theme further, looking at ways that some important 2D shapes can be represented as the loci of functions, and further how these may be extended as the bases of standard forms in 3D. The subject of calculus is introduced as a way of analyzing graph structure. We start with the mathematical properties of the straight line. Bresenham's algorithm, the standard method for drawing straight lines on a raster computer display, is explained in most computer graphics texts, such as those listed in the section on further reading.

The Straight Line

The equation for the standard form of a straight line in 2D Cartesian coordinates is

$y = mx + c.$

All points (x, y) that satisfy this equation lie on a straight line. In effect, this is the graph (see chapter 4) of the function f(x) with rule

$f(x) = mx + c.$

When x = 0, this gives y = c; c is called the intercept on the y-axis (remember the y-axis is defined by x = 0) (fig 5.1). Adding 1 to x increases y by m, where m is called the slope or gradient of the line; it is the 'rate of change of y with respect to x'. If θ is the angle between the line and the x-axis, we have

$m = \tan(\theta).$

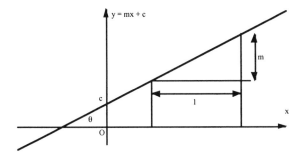

Figure 5.1 The standard form of a straight line

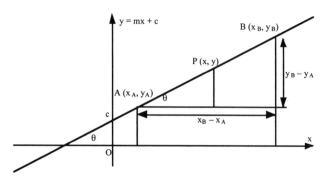

Figure 5.2 The line joining two points A (x_A, x_A) and B (x_B, y_B)

The line joining points, A and B, with coordinates (x_A, y_A) and (x_B, y_B) has slope

$$m = (y_B - y_A)/(x_B - x_A) \quad \text{provided } x_A \neq x_B,$$

the change in the y coordinate divided by the change in the x coordinate. This is seen by finding $\tan(\theta)$ in fig 5.2. Using the same form of calculation with a general point P (x, y) on the line and A, the slope is

$$m = (y - y_A)/(x - x_A),$$

so $\quad (y - y_A)/(x - x_A) = (y_B - y_A)/(x_B - x_A).$

A little algebraic manipulation reduces this to

$$y = \{(y_B - y_A)/(x_B - x_A)\}x + \{y_A - x_A(y_B - y_A)/(x_B - x_A)\},$$

or $\quad y = mx + (y_A - mx_A)$

so $\quad c = y_A - mx_A.$

Constants m and c for the 'slope-intercept' form of the line can be found in this way. If the slope is zero, the line is horizontal; if it is negative, the line slopes down to the right. A vertically upward line cannot be expressed in this way; this is the exceptional case indicated in the definition of m when $x_A = x_B$. This would give a 'zero divide' error producing an infinite slope. y does not appear in the equation of a vertical line, which is simply

$$x = x_A.$$

We assume for the purposes of this book that a method for drawing such lines between given points A, B on a raster display is available, such as the well-established Bresenham's algorithm.

Drawing General Function Graphs

Most of computer graphics is a 'fudge', producing effects that fool the human perception system into believing it is seeing what is not there. The film and television industries are similarly set up; they make us believe we are seeing continuous activity through presentation of a rapidly changing succession of static images. In drawing lines by computer, we must accept at the basic level that we see only lit-up dots on a screen that we then interpret as being joined up in a smooth way. In many cases, curves are further approximated by drawing linked sets of short straight lines. Suppose we want to display the graph of a function $y = f(x)$ for values of x from x_0 to x_1 on a pixel display. In this section, we ignore the problem of conversion from the real-valued Cartesian space (x, y) to the display screen with discrete-valued coordinates (h, v) representing horizontal and vertical pixel locations. This is the 'window-to-viewport mapping' that is part of the 'computer graphics pipeline' discussed later. At this stage, we assume we have a working command for drawing lines between two Cartesian points that will transfer this line successfully to the display screen. With this assumption, we can divide the curve of $y = f(x)$ into n straight line stages, using the following algorithm.

```
a₀ = x₀, b₀ = f(x₀), xstep = (x₁ - x₀)/n
Loop for n = 1 to 100
    a₁ = a₀ + xstep
    b₁ = f(a₁)
    Draw line from (a₀, b₀) to (a₁, b₁)
    a₀ = a₁, b₀ = b₁
End n loop
```

If n is large enough and the conversion from Cartesian (x, y) to display (h, v) coordinates is aptly defined, this produces a convincing display of a continuous curve. There is a danger of missing small variations of f(x) if the sampling step is small. The illustrations in this chapter show some pixellization, but the general shapes of curves should be considered as if smoothed by the eye.

Graphs of Polynomials

A polynomial function of power n has form

$$f(x) = a_n x^n + a_{n-1} x^{n-1} + \ldots a_2 x^2 + a_1 x + a_0,$$

where $a_0, a_1, a_2, \ldots a_n$ are constants. Here we consider the forms of general polynomial graphs $y = f(x)$. The simplest polynomials have powers zero and 1,

$\quad\quad y = a_0,$
and $\quad y = a_1 x + a_0.$

These have already been considered in the section on straight lines; the first is a horizontal line, the second case is a general line with slope $m = a_1$, intercept $c = a_0$.

Polynomials of power two are known as quadratic functions. The simplest order 2 polynomial has graph $y = x^2 = f(x)$ shown in the top left of fig 5.3 (axis markings are at one unit separation), together with variations on this (clockwise), $y = 0.5f(x)$, $y = f(x) + 1$ and $y = f(x - 1)$. These display the properties discussed in chapter 4 of, in turn, y scaling by 0.5, translation in y by 1 and translation in x by 1.

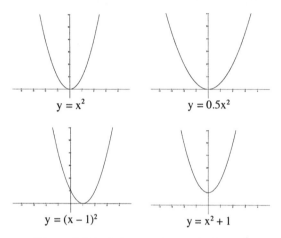

Figure 5.3 Variations on the graph of $y = x^2$

These graphs are shown in small scale; we are less interested here in individual function values than in the general shapes displayed. The main point of interest is that the graph of $y = x^2$ has a general 'U' shape, with y increasing rapidly upwards as x moves away from the value 0. It has a minimum value of $y = 0$ at $x = 0$; this is the only point at which it touches the x-axis, the function is nowhere else zero.

Now we compare this with a more general polynomial of power 2, for example, $y = x^2 - 2x - 3$ (fig 5.4, left). We see the same general shape, x^2 is the dominant term, but it is shifted both right and down so its minimum function value -4 occurs when $x = 1$. It also crosses the x-axis at two points: $x = -1$ and $x = 3$.

$$y = x^2 - 2x - 3$$

can be written in the form

$$y = (x + 1)(x - 3)$$

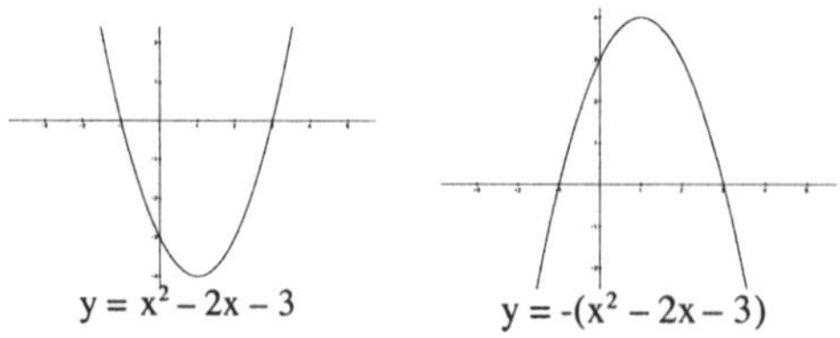

Figure 5.4 Graphs of polynomials of power 2

5. Form from Function: Analysis of Shapes

(those unfamiliar with the concept of factorization can verify the result by multiplying out the brackets of this second form). This makes it easy to identify the two 'roots' of the equation as $x = -1$ and $x = 3$, the values that make the function zero and hence the points at which the graph crosses the x-axis. If $x = -1$, $(x + 1)$ must be zero, so the full expression for y must be zero. Similarly, $x = 3$ makes $(x - 3)$ zero, so this form of factorization enables us to find roots.

Finding roots of a quadratic function is equivalent to solving a quadratic equation, often written as

$$ax^2 + bx + c = 0.$$

The formula for finding the two roots is

$$x = \frac{-b \pm \sqrt{b^2 - 4ac}}{2a},$$

perhaps a familiar memory to some readers. This can be taken on trust by those not wishing to follow the slightly tortuous derivation given on the next page, at least on first reading. They should seek the reappearance of this formula as the hint they can start reading again. The trick is to re-express

$$f(x) = ax^2 + bx + c$$

in terms of sums of squares. First, taking a factor of a, we have

$$ax^2 + bx + c = a[x^2 + (b/a)x + c/a].$$

Now we note that

$$\begin{aligned}\{x + (b/2a)\}^2 &= x\{x + (b/2a)\} + (b/2a)\{x + (b/2a)\} \\ &= x^2 + 2(b/2a)x + (b/2a)^2 \\ &= x^2 + (b/a)x + (b/2a)^2\end{aligned}$$

so $\quad x^2 + (b/a)x = \{x + (b/2a)\}^2 - (b/2a)^2.$

We have snatched an apparently arbitrary expression from the air and performed some apparently pointless manipulations on it, but we find a part of the expression for $ax^2 + bx + c$ included in this form. Readers should not be worried if they think, 'I'd never come up with that sort of reasoning'. Not many of us would; it is the great mathematical innovators who come up with such imaginative leaps. The rest of us should look on in amazement and content ourselves with being able to use their results. After taking pause for breath at this audacious policy, we can now return to our expression, to see

$$\begin{aligned}ax^2 + bx + c &= a\{x^2 + (b/a)x + c/a\} \\ &= a[\{x + (b/2a)\}^2 - (b/2a)^2 + c/a] \\ &= a[\{x + (b/2a)\}^2 - (b^2 - 4ac)/(4a^2)].\end{aligned}$$

The final term comes from

$$-(b/2a)^2 + c/a = -\{b^2/(4a^2)\} - c/a\}$$
$$= -\{b^2/(4a^2) - 4ac/(4a^2)\}.$$

Thus, $ax^2 + bx + c = 0$

when $a[\{x + (b/2a)\}^2 - (b^2 - 4ac)/(4a^2)] = 0.$

We know $a \neq 0$, or we wouldn't have a quadratic expression, so

$$\{x + (b/2a)\}^2 - (b^2 - 4ac)/(4a^2) = 0$$

or $\{x + (b/2a)\}^2 = (b^2 - 4ac)/(4a^2).$

Taking square roots of both sides (this assumes it is possible, the right-hand side of the equation is positive)

$$x + (b/2a) = \pm\sqrt{\{(b^2 - 4ac)/(4a^2)\}}$$
$$= \pm\sqrt{\{(b^2 - 4ac)\}}/(2a),$$

where '$\pm$' indicates we have two possibilities in evaluating the square root. Finally, isolating x on the left-hand side,

$$x = -(b/2a) + \sqrt{\{(b^2 - 4ac)\}}/(2a)$$

which is re-expressed as a fraction with denominator 2a as

$$x = \frac{-b \pm \sqrt{(b^2 - 4ac)}}{2a}.$$

This is where those who left out the proof should start reading once more.

From this formula, we can note three cases for solutions of a quadratic equation dependent on the sign of ($b^2 - 4ac$), giving two distinct roots, one root or no real-valued roots. Each of these is identifiable by the form of the function graph $y = f(x)$. If $b^2 - 4ac > 0$, there is a real-valued positive solution to its square root, so there are two real-valued roots of the equation: one using the positive sign, the other the negative. This is the situation of fig 5.4, where the graph crosses the x-axis at two distinct points. If $b^2 - 4ac = 0$, the square root in the formula takes the value zero, so there is only one root at $-b/(2a)$. This is the case in all but the lower right image of fig 5.3, where the graph touches the x-axis at a single point, but does not cross the axis. In the third case, when $b^2 - 4ac < 0$, there is no real-valued square root of $b^2 - 4ac$, so no real-valued solutions exist. This is the case in the lower right image of fig 5.3, when the graph neither crosses nor touches the x-axis. There are two complex numbered roots in this case, at

$$x = \frac{-b \pm i\sqrt{(4ac - b^2)}}{2a}.$$

5. Form from Function: Analysis of Shapes

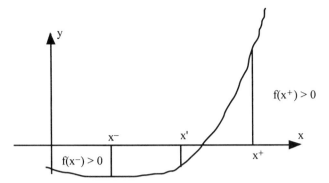

Figure 5.5 Closing in on the root of a function, interval (x^-, x^+) is replaced by (x', x^+)

This formula gives a ready resolution of any quadratic equation, whether there is a solution or not. For higher-power polynomial equations, there is no such formula: finding roots of such equations usually depends on ad hoc solutions, or using numerical methods to approach a solution.

We outline one numerical method for finding a solution of an equation $f(x) = 0$, which may or may not be a polynomial equation, to a specified level of precision when the graph of $y = f(x)$ crosses the x-axis. The solution depends on the function being continuous; its graph can be drawn without raising the pencil from the paper. When $y = f(x)$ crosses the x-axis, $f(x)$ must take some negative values and some positive values. By a stepwise search of x at a given step length, it should be possible to identify two values of x, x^+ and x^-, such that

$$f(x^+) > 0 \quad \text{and} \quad f(x^-) < 0.$$

Somewhere between x^+ and x^-, we should have a value of x such that $f(x) = 0$. Take the value midway between x^+ and x^-, $x' = (x^+ + x^-)/2$, and consider its function value $f(x')$. If $f(x') > 0$, there must be a root between x' and x^-; if $f(x') < 0$, there must be a root between x' and x^+. The latter case is shown in fig 5.5. Thus, we have found an interval containing a root that is half the length of the previous one. This process is repeated until the interval is narrow enough to achieve the desired level of precision as in the following routine.

```
Loop until |x⁺ - x⁻| < h
    x' = (x⁺ + x⁻)/2
    if f(x') < 0, x⁻ = x'
        else if f(x') > 0, x⁺ = x'
        else x⁻ = x', x⁺ = x'
    end if
end Loop
root = (x⁻ + x⁺)/2
```

Repeated halving closes in on a root relatively quickly, but this depends on finding the two starting values, one each side of the root. Although fig 5.5 depicts x^- to be less than x^+, the process is not dependent on this; it will work whether the curve rises through zero from left to right as in fig 5.5, or falls through zero. This only finds one root; the process will have to be repeated for other values of x^+ and x^- if other roots exist. There are more sophisticated numerical methods that can close in more rapidly on roots of equations; such routines are available from standard numerical analysis packages.

Returning to fig 5.4 on page 108, the second graph is the first one turned upside down; that is exactly what is achieved by multiplying the graph function by -1. We can write this graph's function,

as
$$y = -(x^2 - 2x - 3),$$
$$y = -x^2 + 2x + 3,$$

showing the highest power term to be $-x^2$. When x is far from zero, this is the dominant value, so the general shape is now an inverted 'U'. If x is large and positive, $-x^2$ is even larger but negative. Also, if x is large and negative, $-x^2$ is even larger and negative – the product is of form 'minus (minus times minus)' which is the same as 'minus (plus)', or just 'minus'. This kind of analysis can be used to determine the overall shapes of polynomial graphs.

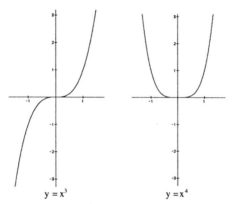

Figure 5.6 Graphs of $y = x^3$ (left) and $y = x^4$ (right)

As examples, we look at the graphs of $y = x^3$ and $y = x^4$ (fig 5.6). Note that $y = x^3$ generally moves from low left to high right. For large negative values of x, x^3 takes an even larger negative value, as the evaluation involves multiplication of three (an odd number of) minuses. For $y = x^4$, we have another U-shaped curve; for large negative values of x, we multiply an even number of minuses, so the result is positive. This is a general rule. If the highest power term is a positive multiple of an odd power, the graph goes from low at the left to high at the right as on the left of fig 5.6. If the highest power term is a negative multiple of an odd power, the graph is the reverse of this, going from high at the left to low at the right (like the left of fig 5.6 turned upside down). If the highest power term is a positive multiple of an even power, the graph has high extremities to left and right (as the right image of fig 5.6). If the highest power term is a negative multiple of an even power, the graph has low extremities to left and right (like the right image of fig 5.6 turned upside down).

5. Form from Function: Analysis of Shapes

It is useful to know what is happening at the extreme ends of the graph; now we consider how to understand what is going on in the central region.

Consider first a specific example,

$$y = (x + 1)(x - 1)(x - 2).$$

This may not look like a general polynomial, but it can be multiplied out into the standard form. Multiplying the final two brackets (remember each term of the first bracket must multiply each term of the second), we get

$$\begin{aligned}(x - 1)(x - 2) &= x(x - 2) - 1(x - 2) \\ &= x^2 - 2x - x + 2 \\ &= x^2 - 3x + 2.\end{aligned}$$

We now have to multiply this by $(x + 1)$ to give

$$\begin{aligned}(x + 1)(x - 1)(x - 2) &= (x + 1)(x^2 - 3x + 2) \\ &= x(x^2 - 3x + 2) + 1(x^2 - 3x + 2) \\ &= x^3 - 3x^2 + 2x + x^2 - 3x + 2 \\ &= x^3 - 2x^2 - x + 2,\end{aligned}$$

which is the standard form. We will not use this level of detail in future evaluations; it is worth showing how the expansion is done for those not familiar with the method.

Now we have established this as a proper polynomial of power 3 with positive multiple of its highest power, we will find the original form more convenient for analysis. First, though, we note that our rule identifies the curve as moving from low left to high right.

The original form has three multiplied brackets. When the whole form is equal to zero, at least one of these brackets must be zero. The points at which the graph of $y = f(x)$ crosses the x-axis (this is when $y = 0$) are the roots of the equation $f(x) = 0$. Thus the graph of

$$y = (x + 1)(x - 1)(x - 2)$$

crosses the x-axis when

$$(x + 1)(x - 1)(x - 2) = 0,$$

which can only occur when one of

so $x + 1 = 0$ or $x - 1 = 0$ or $x - 2 = 0$,
    $x = -1$ or $x = 1$ or $x = 2$.

This gives a lot of information about the general shape of the graph. Assuming it to be continuous and smooth, we can sketch its general shape as rising through (-1, 0), falling through (1, 0) then rising again through (2, 0) (fig 5.7).

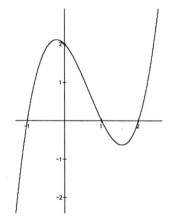

Figure 5.7 A portion of the graph of $y = (x + 1)(x - 1)(x - 2)$

The sense of 'rising' and 'falling' through zero is shown in fig 5.7. This concept has arisen in looking at straight-line graphs, where the gradient m determines if a line rises from left to right (m > 0) or falls (m < 0). This suggests that a gradient defined for curves would be useful, but what does 'gradient' mean for a curve?

Calculus: Differentiation

The slope or gradient (the terms are interchangeable) of a line joining points A (x_A, y_A) and B (x_B, y_B) was defined (fig 5.2) as

$$m = (y_B - y_A)/(x_B - x_A).$$

If the line is written as

$$y = f(x) = mx + c,$$

we can write the gradient m as

$$m = \{f(x_B) - f(x_A)\}/(x_B - x_A),$$

as $\quad y_A = f(x_A) \quad$ and $\quad y_B = f(x_B).$

The value of m remains the same regardless of which two points A and B are chosen on the line. This is the problem with curves: it does not remain constant. If m is constant, the join of any two points on the line remains at a set angle with the x-axis. This is clearly not the case with a curve representation. The slope appears to change at different points on the curve. How can we determine this local slope?

Figure 5.8 shows part of the graph of $y = x^2$, with lines joining points A (0.5, 0.25) to B_1 (2, 4) and B_2 (1, 1) on the graph. Suppose we want to know what

5. Form from Function: Analysis of Shapes

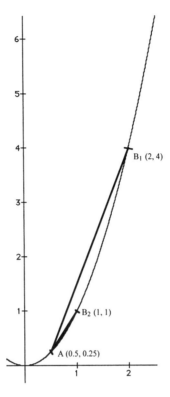

Figure 5.8 Part of the graph of $y = x^2$ showing lines joining three points

the slope should be at point A. It is clear that AB_1 and AB_2 are not at the same angle to the x-axis. The slopes of the two lines are calculated as

AB_1: $m_1 = (4 - 0.25)/(2 - 0.5) = 2.5$
AB_2: $m_2 = (1 - 0.25)/(1 - 0.5) = 1.5$.

We know how to find the slope for two distinct points A, B on the curve. To find the slope at point A, it would be easy to make B equal to A in the formula, but this gives us our zero divide error and we would end up with the apparently insoluble problem of finding 0/0. What if we make B very close to A? How close does it need

Table 5.1 Slopes of line segments AB, as B approaches A from above and below

x_B	Slope	x_B	Slope
2	2.5	-1	-0.5
1	1.5	0	0.5
0.7	1.2	0.3	0.8
0.6	1.1	0.4	0.9
0.55	1.05	0.45	0.95
0.51	1.01	0.49	0.99
0.501	1.001	0.499	0.999
0.5001	1.0001	0.4999	0.9999

to be? Table 5.1 gives some values of slopes for fixed A at (0.5, 0.25) and positions of B at values of x as they get closer to A from above (left columns) and below (right columns). In both cases, the slope appears, in some sense, to be approaching the value 1. This is adopted as the slope or gradient of $y = x^2$ at A.

The gradient is the slope of a tangent to the curve at this point. The tangent is a line that touches the curve at point A generally without crossing it; it is the limiting form of the line AB as B approaches A. The concept of touching without crossing is sound in most cases; the tangent can cross the curve, although it is still the limiting form of AB when the curve has an 'S' bend or point of inflection at A. The tangent still defines the 'direction' of the curve.

The slope of a general curve $y = f(x)$ at any point with value x can be defined by a similar theoretical method. Suppose A and B are separated horizontally by a small distance δx (where $\delta x \neq 0$). If the x coordinate of A has value x, we have $x_A = x$, $x_B = x + \delta x$, $y_A = f(x)$, $y_B = f(x + \delta x)$. For a change in x values of δx, the y coordinate changes by an amount

$$\delta y = f(x + \delta x) - f(x).$$

Thus, the gradient of line AB is

$$\delta y / \delta x = (y_B - y_A)/(x_B - x_A) = \{f(x + \delta x) - f(x)\}/\delta x.$$

The slope of the curve of $y = f(x)$ at the point A is then defined as the limiting value of this ratio as δx approaches zero, written as

$$\frac{dy}{dx} = \lim_{\delta x \to 0} \left\{ \frac{\delta y}{\delta x} \right\} = \lim_{\delta x \to 0} \left\{ \frac{f(x + \delta x) - f(x)}{\delta x} \right\}.$$

'The slope at point A' is equivalent to the 'slope of the tangent to the curve at A'. This tangent is not the trigonometric ratio, but a line that *touches* the curve at A. It is the limiting form of AB as A approaches B. The middle of these expressions is stated as 'the limit, as delta x approaches zero, of delta y over delta x'. This is known as the 'derivative of y with respect to x' or the 'derivative of f(x) with respect to x'. Alternatively, this is called the 'differential of y with respect of x', and it has alternative notation, $y' = f'(x)$. The two separate notations may stem from the two separate originators of the method, Newton (1642–1727) used a dash or dot notation to describe 'fluxions' and Leibniz (1646–1716) used the ratio form, usually stated as 'dy by dx'. There was a long and acrimonious dispute carried on mainly by Newton against Leibniz as to who had devised the method – they both had.

Strictly, we should insist that such a limit exists and that the limit from above is equal to the limit from below, as in the two parts of table 5.1. This is the case for most functions when $y = f(x)$ is a 'well-behaved' smooth curve having a continuous tangent. There is a further discussion lurking here on continuity; the function f(x) is continuous at a point if f(x), and its limit from below and its limit from above are equal. In other words, the graph of $y = f(x)$ can be drawn without raising the pencil from the paper. We side-step this potentially long diversion, and assume the phrase 'if it exists' is readily interpreted in most cases. Let us consider some special cases of slopes, or derivatives by looking at how particular functions can be manipulated.

5. Form from Function: Analysis of Shapes

We used $y = x^2$ as an example, which is a good case to consider first from a formula point of view. We need to evaluate

$$\frac{dy}{dx} = \lim_{\delta x \to 0}\left\{\frac{\delta y}{\delta x}\right\} = \lim_{\delta x \to 0}\left\{\frac{(x+\delta x)^2 - x^2}{\delta x}\right\}.$$

The numerator within the limit can be expanded as

$$(x+\delta x)^2 - x^2 = x^2 + 2\delta x + \delta x^2 - x^2 = 2\delta x + \delta x^2,$$

as $(x+\delta x)^2 = x(x+\delta x) + \delta x(x+\delta x) = x^2 + x\delta x + x\delta x + \delta x^2$. So

$$\lim_{\delta x \to 0}\left\{\frac{(x+\delta x)^2 - x^2}{\delta x}\right\} = \lim_{\delta x \to 0}\left\{\frac{2x\delta x - \delta x^2}{\delta x}\right\} = \lim_{\delta x \to 0}\left\{2x - \delta x\right\}.$$

We have been able to divide by δx to produce the right-hand form, which is no longer a fraction. We can now set δx to 0 in this expression without any zero divide error, reducing it to $2x$, so

$$\frac{dy}{dx} = 2x.$$

Firstly, we note that this gives the correct result for our example of table 5.1; when $x = 0.5$, the gradient is 1. This method can be extended to several special cases by dealing with particular formulae. We consider just two more: the first is the general power case, $f(x) = x^n$. Now we have

$$\frac{dy}{dx} = \lim_{\delta x \to 0}\left\{\frac{\delta y}{\delta x}\right\} = \lim_{\delta x \to 0}\left\{\frac{(x+\delta x)^n - x^n}{\delta x}\right\}.$$

We need to expand the bracket in the numerator,

$$(x+\delta x)^n = (x+\delta x)(x+\delta x) \ldots (x+\delta x),$$

where there are n brackets. All possible combinations of first and second terms in brackets must be multiplied. There is only one way of getting a term x^n by taking the first term of each bracket. I could get a term $x^{n-1}\delta x$ by taking the δx from the last bracket and x from the remaining $n - 1$ brackets. But I could get the same form by taking δx from the first, or second, or ... brackets. Thus, there are n ways of getting a term of form $x^{n-1}\delta x$. Similarly, we can find how many terms of form $x^{n-2}\delta x^2$ can be found, by considering how many pairs of brackets we can taken a dx from, and so on. We content ourselves by stating that

$$(x+\delta x)^n = x^n + nx^{n-1}\delta x + o(\delta x^2).$$

The mysterious '$o(\delta x^2)$' indicates that all terms remaining are 'of the order of δx^2;'

they contain δx^2, δx^3, ..., with δx having a power of at least 2. This expansion rule is covered fully by the binomial expansion, but we avoid this digression.

$$\frac{dy}{dx} = \lim_{\delta x \to 0} \left\{ \frac{x^n + nx^{n-1}\delta x + o(\delta x^2) - x^n}{\delta x} \right\}$$

$$= \lim_{\delta x \to 0} \left\{ \frac{nx^{n-1}\delta x + o(\delta x^2)}{\delta x} \right\} = \lim_{\delta x \to 0} \left\{ nx^{n-1} + o(\delta x) \right\}.$$

Again, we have divided by δx to eliminate the fractional form. Dividing the $o(\delta x^2)$ by δx reduces the smallest power of δx involved to one; each term of $o(\delta x)$ contains at least a single δx factor. Thus, if we make $\delta x = 0$, $o(\delta x)$ will also be 0, and we can invoke the limit without having a zero divide problem. When $y = x^n$,

$$\frac{dy}{dx} = nx^{n-1} \quad \text{when } n \geq 1.$$

Note that this is consistent with our special case for $n = 2$; the derivative of x^2 is $2x^{2-1} = 2x$. When $n = 1$ we have the function rule $y = x$. This has derivative

$$\frac{dy}{dx} = 1.x^0 = 1,$$

using the index rule for power zero, consistent with our straight line slope. The derivative of a constant, $y = k$, is not covered by this rule as $n = 0$. The graph of $y = k$ is horizontal, so its slope is everywhere zero.

A number of general formulae regarding differentiation can be developed from the definition of a derivative; they are given below without proof. Careful consideration of the definition of a derivative shows the validity of most. If $y = f(x)$, we can write $\frac{d}{dx}f(x)$ or $\frac{dy}{dx}$ or $f'(x)$; the most convenient form is used appropriately.

$$\frac{d}{dx}\{f(x) + g(x)\} = \frac{d}{dx}f(x) + \frac{d}{dx}g(x), \qquad \frac{d}{dx}\{kf(x)\} = k\frac{d}{dx}f(x).$$

The first rule gives the derivative of the sum of two functions $f(x)$ and $g(x)$; the second gives the derivative of a function multiplied by a constant k. With $k = 2$, the second is consistent with the first when $f(x) = g(x)$. These two rules also show how a general polynomial can be differentiated. For example, if

$$f(x) = a_n x^n + a_{n-1} x^{n-1} + \ldots a_2 x^2 + a_1 x + a_0,$$

$$f'(x) = na_n x^{n-1} + (n-1)a_{n-1} x^{n-2} + \ldots 2a_2 x + a_1.$$

Each term is multiplied by its power, and that power is reduced by 1, with the exception of the constant term a_0, which disappears completely. The derivative is reduced in power by 1 and has one fewer term, a_0 being eliminated.

5. Form from Function: Analysis of Shapes

Two important rules are those for the derivative of a product of two functions, and the 'function of a function' rule, when the argument of a function is a function itself.

$$\frac{d}{dx}\{f(x)g(x)\} = g(x)\frac{d}{dx}f(x) + f(x)\frac{d}{dx}g(x) = g(x)f'(x) + f(x)g'(x),$$

$$\frac{d}{dx}f\{g(x)\} = g'(x)f'\{g(x)\}.$$

These are stated for completeness; they are very useful for those who wish to differentiate complicated formulae, but are not of great use in the following discussion. They can be derived from the definition of differentiation of a function.

Special forms have to be developed for many other functions; these include the sine, cosine and exponential functions,

$$\frac{d}{dx}\sin(x) = \cos(x), \qquad \frac{d}{dx}\cos(x) = -\sin(x) \qquad \frac{d}{dx}e^x = e^x.$$

We will return to these after we have considered the relationships between a derivative and graphs of functions, as illustrated in fig 5.9, which shows the graphs of a function f(x), its derivative f'(x) and the derivative of its derivative, f''(x). The last expression is known as the second derivative; this can be taken as $\frac{d}{dx}\left\{\frac{d}{dx}f(x)\right\}$ and is usually written as $\frac{d^2}{dx^2}f(x)$. The functional form is irrelevant; we are concerned with identifying features of the graph shapes, and how they interrelate. The derivative of a function represents the gradient of its graph, so we note firstly from fig 5.9 that when f(x) is increasing (has an upward slope from left to right), f'(x) > 0. This occurs when x < 3 and x > 0.5. Also, the graph of f(x) is decreasing (has a downward slope) when f'(x) < 0, for -3 < x < 0.5.

There are a number of features of note in the graph of f(x). There is a local maximum at x = -3. As its name suggests, this is a locally highest point on the graph; it is not an absolute maximum as there are higher points to the far right of the graph. Comparing it with the gradient graph, this is the point when the gradient decreases through zero. There is a local minimum at x = 0.5, where the gradient graph is increasing through zero. Maxima and minima occur when the gradient is zero (the tangent to the original curve is horizontal). They can usually be distinguished by considering the second derivative; does it rise (minimum) or fall (maximum) through zero?

Similarly, the second derivative curve can identify maxima and minima in the gradient curve. It rises through zero at about x = -1.5, which is where the minimum of the gradient curve occurs. As the second derivative is negative for x < -1.5, the gradient is decreasing for this range. Consider the meaning of a gradient that decreases as you move from from left to right on the original graph. This can only occur if the graph is curved like an 'inverted U' shape, like the intersection symbol ∩. To the right of x = -1.5, the second derivative is positive, so the

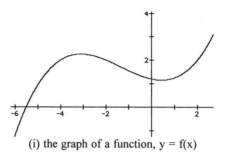

(i) the graph of a function, y = f(x)

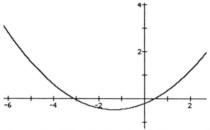

(ii) the graph of the function's gradient, y = f'(x)

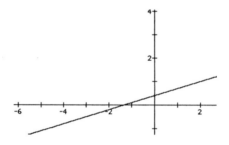

(iii) the graph of the second derivative, y = f''(x)

Figure 5.9 The graph of a function and its first two derivatives

gradient increases for this range, meaning the original graph is curved in a 'U' shape, like the union symbol ∪. What is going on at the point between these two, at x = -1.5? In terms of curvature, the answer is precisely nothing. Cyclists are familiar with the point at which they turn out of a right-hand bend and go directly into a left-hand bend. It is the point when they instantaneously steer straight. This is the curve's point of inflection (or inflexion), at which the curve does not bend or flex itself. A point of inflection occurs when the second derivative passes through zero. Now imagine yourself riding along the curve of fig 5.9(i), and sense when that instant occurs; it is at the x value when the gradient curve has its minimum, when the second derivative curve passes through zero.

We discussed the nature of polynomial curves, and opened the question earlier of how we know a curve is 'well behaved'. The answer is through the calculus. By finding roots of the gradient curve, we know where all maxima and minima occur. Similarly, the roots of the second derivative curve tell us where points of inflection occur, and thus where changes of the kind of curvature occur. Knowing these, we

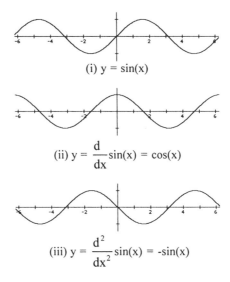

Figure 5.10 The graph of y = sin(x) and its first two derivatives

can be confident of the curve's general behaviour, that it does not throw in extra curves or wobbles at intermediate values.

Earlier, we gave the derivatives of sin(x) and cos(x),

$$\frac{d}{dx}\sin(x) = \cos(x), \qquad \frac{d}{dx}\cos(x) = -\sin(x).$$

Thus, the second derivative of sin(x) is -cos(x). Looking at the curves of fig 5.10, we see once more the links between curves and their derivatives. The graph of sin(x) (fig 5.10(i)) has maxima when its derivative curve (fig 5.10(ii)) falls through zero and minima when its derivative curve rises through zero. Its points of inflection (the border points between 'cup shape' ∪ and 'cap shape' ∩) occur when the second derivative curve (fig 5.10(iii)) crosses zero. By understanding these properties on a now familiar set of shapes, their use on new forms may become clearer.

We have discussed differentiation of a simple function of one variable with respect to that variable, but other forms are available. For example, suppose we have a curve defined parametrically, with x and y linked by an intermediate parameter, t say. We will find examples of this form later in the chapter. A point on the curve is defined at (x(t), y(t)). How do we find the gradient of this curve, when the functional form is not given in y = f(x) form? We can find the gradients or rates of change of x and y with respect to t by differentiation with respect to t,

$$\frac{dx}{dt} \quad \text{and} \quad \frac{dy}{dt}.$$

The ratio of these gives the derivative of y with respect to x, the gradient of the parametric curve, as

$$\frac{dy}{dx} = \frac{dy}{dt} \Big/ \frac{dx}{dt}.$$

Considering the definitions of the derivatives involved as limiting ratios of small quantities, this seems to make 'cancelling' sense as a fraction. It is very tempting to follow the rules for manipulation of fractions and write

$$\frac{dy}{dt} \Big/ \frac{dx}{dt} = \frac{dy}{dt}\frac{dt}{dx} = \frac{dy}{dx}.$$

However, the derivative is not a true fraction; we should not separate the 'dx' and the 'dy' parts. Remember that

$$\frac{dx}{dt} = \lim_{\delta t \to 0}\left\{\frac{\delta x}{\delta t}\right\} \quad \text{and} \quad \frac{dy}{dx} = \lim_{\delta x \to 0}\left\{\frac{\delta y}{\delta x}\right\}.$$

In this expression we have switched from a limit involving δt to a limit involving δx; this is the real theoretical difficulty. In most 'well-behaved' cases, as δt approaches 0, so does δx, and the 'cancellation' works. However, it is not really fractional cancellation, its proof is part of the mathematical topic of analysis, and readers should beware arbitrarily splitting the upper and lower parts of derivatives.

To take one simple example, a circle with radius 1 centred at the origin can be defined in terms of a parameter θ as

$$x = \cos(\theta), \qquad y = \sin(\theta).$$

This follows from the definitions of sine and cosine in terms of polar coordinates (chapter 3). To find the gradient of the circle at a point with parameter θ, we have

$$\frac{dx}{d\theta} = -\sin(\theta), \qquad \frac{dy}{d\theta} = \cos(\theta),$$

so

$$\frac{dy}{dx} = \frac{dy}{d\theta} \Big/ \frac{dx}{d\theta} = -\cos(\theta)/\sin(\theta) = -x/y.$$

Another aspect of differentiation that can be of use in computer graphics is partial differentiation. In finding isosurfaces (chapter 8), a function $\phi(x, y, z)$ of position is used. This means that for any point (x, y, z) in the function domain (a region of 3D space), a real value $\phi(x, y, z)$ can be calculated from the function rule. It is sometimes useful to know the rate of change of the function value in a particular direction, for example, in the x direction. This can be defined in limiting form as

$$\frac{\partial \phi}{\partial x} = \lim_{\delta x \to 0}\left\{\frac{\phi(x+\delta x,\ y,\ z) - \phi(x,\ y,\ z)}{\delta x}\right\},$$

5. Form from Function: Analysis of Shapes

the symbol '∂' or 'curly d' showing this is different from the normal form of differentiation (and note that it is different from the Greek 'δ'). In finding the derivative, the rules for normal differentiation are followed, but values of y and z are treated as constants. So, for example, our rules for differentiation of powers of x,

$$\frac{dy}{dx} = nx^{n-1},$$

gives $\quad \dfrac{d}{dx}(x^3) = 3x^2, \quad \dfrac{d}{dy}(y^2) = 2y, \quad \dfrac{d}{dz}(z) = 1.$

Using these on the separate parts of the function

$$\phi(x, y, z) = x^3 y^2 z$$

we find the partial derivatives of $\phi(x, y, z)$ with respect to each of x, y and z as

$$\frac{\partial \phi}{\partial x} = 3x^2 y^2 z, \quad \frac{\partial \phi}{\partial y} = 2x^3 y z, \quad \frac{\partial \phi}{\partial z} = x^3 y^2.$$

For the partial derivative with respect to x, the terms $y^2 z$ are carried through as multiplying constants, just as if they were numbers; the x^3 alone is differentiated.

Isosurfaces (chapter 8) are defined by the equation

$$\phi(x, y, z) = \phi_0,$$

where ϕ_0 is some constant. A normal to this surface (a direction orthogonal, or at right angles, to the surface), is given by the vector function grad,

$$\mathrm{grad}(\phi) = \left(\frac{\partial \phi}{\partial x}, \frac{\partial \phi}{\partial y}, \frac{\partial \phi}{\partial z} \right).$$

In this section, we have assumed that functions to be differentiated are 'well-behaved'; this means they are continuous and have continuous tangents. Discontinuous functions have gaps; discontinuous tangents mean instant direction changes in the function graph known as cusps. We have already seen in the case of the tan function it has discontinuities. If $\delta x > 0$, a function f(x) is continuous at a domain value x if and only if

$$f(x) = \lim_{\delta x \to 0} f(x + \delta x) = \lim_{\delta x \to 0} f(x - \delta x).$$

Calculus: Integration

Integration is anti-differentiation; it undoes the effect of differentiation, as far as is possible. We have discussed how in figs 5.9 and 5.10 we can move from an original curve definition through its derivative or gradient to its second derivative, and how these forms are related. Now, can we reverse that process?

Taking a physical example may be useful, and it also indicates how calculus is of use in computer animation. Speed is the rate of change of distance, the derivative of distance with respect to time. The units of speed, miles per hour, metres per second, indicate that relationship. If you have travelled 120 kilometres in two hours, your average speed is (from distance divided by time) 60 kilometres per hour. Suppose you travelled steadily at that rate. Then you know that you had travelled 30 kilometres after 30 minutes, 10 kilometres after 10 minutes, and so on. We have calculated a distance travelled for a particular time of travel from our knowledge of speed; we have anti-differentiated.

If the speed is changing, speed at a specific time is found by differentiating position with respect to time. (Now we see the need for differentiating 'with respect to' something: here we are considering time as the independent variable, rather than a position.) The reverse problem is that of finding the position of a particle given its speed – that is done by integration. Suppose we have a sophisticated tracking device that can tell us precisely the position of a car throughout a journey on a known route. We can find its speed from the slope of the position/time graph, differentiating the position with respect to time. Now suppose that, instead of the tracking device, we have a tachograph in the car that gives us a graph of the car's speed at all times during the journey. In order to find the car's position at any time, we need to reverse the result of the previous problem: we must anti-differentiate or integrate the speed with respect to time. Position will still not be uniquely determined. Suppose two cars give us identical tachograph graphs. Does this mean that both cars travelled alongside each other? Not necessarily, as the two cars could have started simultaneously from considerably different positions. As well as having complete information on the speed, we need to know the starting position (or initial position) in order to find a unique solution. This is typical of the solution of dynamical problems. (It is practically impossible for two vehicles to have such identical tachograph readings; you are asked to accept the above description as a 'mind experiment'.)

The upper part of fig 5.11 gives two examples of distance/time graphs. The straight-line graph of $s_1(t)$ indicates a steady speed as the graph has constant slope, as shown by the speed/time graph of $v_1(t)$ below it. The speed/time graph $v_2(t)$ for the variable speed model to the right is found from the slope or gradient of its distance/time graph $s_2(t)$; this is the derivative of the distance/time graph. Note how the speed is greatest when the distance/time graph has highest slope. If all we have is the speed/time graph, we can re-create the *shape* of the distance/time graph, but we do not know from where to start drawing this shape. Knowledge of any single (time, distance) point identifies the particular distance/time graph for the particular application. This is equivalent to selecting the *particular solution* of the problem from a set of *general solutions*. Objects that travel at the same speeds will not end up at the same point if they started from different points. We can tell, from a speed

5. Form from Function: Analysis of Shapes

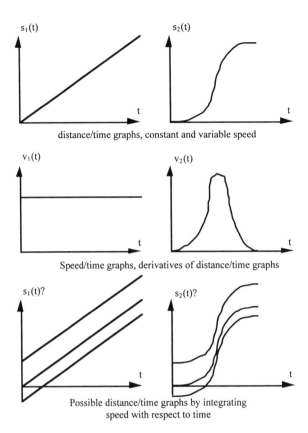

Figure 5.11 The relationship between integration and differentiation illustrated using speed and distance travelled

function, how far each object has travelled relative to some starting point, but absolute positioning needs a reference position.

The process of integration or anti-differentiation is equivalent to the 'area so far' concept. For constant speed v, it is fairly easy to understand that distance s travelled in time t is equal to v times t. If we travel at 60 kilometres per hour for two hours, we have travelled 120 kilometres. The distance $s_1(t)$ travelled for a journey at constant speed will give a straight-line graph when plotted against time, as on the left of fig 5.12. We see that the height of the distance/time graph $s_1(t)$ is related to the shaded 'area so far' under the speed/time graph $v_1(t)$. When the speed varies, the graph of distance travelled will be a curve, as on the right of fig 5.12, $s_2(t)$, but the relationship between the two curves is equivalent; the height of the distance graph $s_2(t)$ at a particular time t is equivalent to the shaded area below the speed/time graph $v_2(t)$ from the start of the journey to the specified time t. The finding of this curve for specific functions v(t) is the process of integration. A general area of this type from start time t_0 to end time t is symbolized as the integral

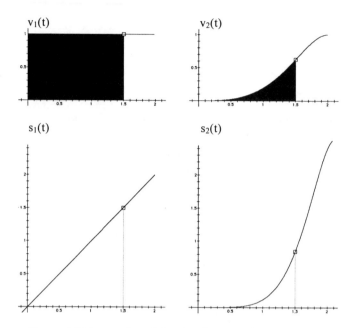

Figure 5.12 Integration of distance from speed as 'area so far'

$$\int_{t_0}^{t} v(t) dt$$

The simple interpretation of this is 'the area below the graph of v(t) against time between start time t_0 and end time t'. In our example of distance being the integral of speed, the interpretation is the distance travelled from the time t_0 to the time t. As indicated above, to find an absolute position, we would need to add this to a known starting point at $t = t_0$. Using limits makes this a 'definite integral'.

We concentrate solely on understanding of the concept here, so do not follow the huge range of skills exercises needed to be able to find integrals for specific functions. For most computer purposes, numerical methods are available to do this, either as specifically written or package functions. Specification of the function to be integrated and the limits of the integral (v(t), t_0 and t in our example) as parameters of such functions enable the correct integral value to be found. Analytic solution for specific examples is often difficult, sometimes impossible. Numerical methods such as addition of small rectangular areas, as shown in the next paragraph, can be resorted to in all cases, and these are most suitable for computer implementation. For the purposes of this book, we leave the process of integration as a black box art, as our main purpose is to understand the concept rather than master the method.

Approximations to integrals can be found based on the theoretical definition of an integral as the limiting value of a sum of small areas. Figure 5.13 shows how small strip rectangle areas can be overlaid on the area between the graph of the function f(x) and the x-axis between two limits. In this case, the sum of these strip areas is an underestimate of the required integral (as seen from the shaded box which lies below the curve), but the limiting form of this sum gives the correct value as

5. Form from Function: Analysis of Shapes   127

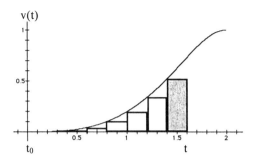

Figure 5.13 An area estimate is found by summing strip areas between t_0 and t

the strip widths approach zero. This limiting argument is similar to that used in differentiation. In this case we are adding together an increasingly large number of ever smaller areas to approach the correct value. The theoretical argument is based on defining a similar slight overestimate by basing strips on the maximum function value above the strip base (the minimum is used in fig 5.13). As the strip widths become smaller, the overestimate and underestimate can be shown to get closer together. Thus they are both, in the limit, equivalent to the true area, which must lie between them. If the height adopted for a strip is v(t) and its width is δt, the area of all strips is

$$\sum_{t_0}^{t} v(t)\delta t,$$

the symbol Σ (sigma is the Greek 'S' for sum) denoting the sum of all values following it between the limits below and above it. This is the sum of all strip areas v(t)δt from t = t_0 to t = t (fig 5.13). The limiting form of this is

$$s(t) = \int_{t_0}^{t} v(t)dt = \lim_{dt \to 0} \left\{ \sum_{t_0}^{t} v(t)\delta t \right\},$$

the stylized 'S' shape integral sign ∫ and 'dt' replacing Σ' and 'δt' to indicate that the limiting form has been taken. A straightforward practical way of estimating an integral between known limits is to evaluate the sum of strips for a reasonably small strip width δt. More sophisticated forms are used to give better approximations, but this simple method usually suffices if strip width δt is taken small enough. Note that this result, shown as s(t) to indicate the distance travelled, is a function of its upper limit t.

Our example of distance and time has led us to use symbols s, v and t, but these are arbitrary in the context of other problems. A general integral of function f(x) with respect to x between limits x_0 and x_1 is written

$$\int_{x_0}^{x_1} f(x)dx.$$

The symbols are arbitrary, the process remains the same.

One further extension is mentioned: the double integral. As an integral finds the area under a curve, a double integral can find the volume under a surface. Suppose f(x, z) is a function of the two real values x and z, so that the equation

$$y = f(x, z)$$

represents a surface. The volume under the surface can be approximated by slicing it into narrow strips and dividing these strips into small box-shaped pieces. As with approximation of curve area, the boxes are approximations to the actual volume, as a flat top is substituted for a portion of the curved surface, but if boxes are small enough these inaccuracies will also be relatively small. Figure 5.14 shows a volume from such a function above the ranges $[x_0, x_1]$ in x and $[z_0, z_1]$ in z. The z range is divided into narrow intervals of width δz, say, one of which is shown as a thin cross-sectional strip. This strip is itself divided into small boxes by cutting the x range into lengths of size δx. If the height of this particular box is $f(x, z)$, then its volume (length x width x height) is $f(x, z)\delta x \delta z$. We can approximate the total volume under the surface within the given x and z ranges as

$$\sum_{z=z_0}^{z_1} \sum_{x=x_0}^{x_1} \left(f(x, z) \delta x \delta z \right),$$

the sum of all the individual small boxes. The precise volume is found as the limiting value of this as both δx and δz approach zero, adding an increasingly large number of smaller volumes,

$$\int_{z=z_0}^{z_1} \int_{x=x_0}^{x_1} f(x, z) dx dz = \lim_{\delta z \to 0} \lim_{\delta x \to 0} \sum_{z=z_0}^{z_1} \sum_{x=x_0}^{x_1} \left(f(x, z) \delta x \delta z \right).$$

Again, as the double limit is invoked, the sigma 'Σ' signs have become stylized 'S' integrals '$\int$', and δx and δz have become dx and dz. From the symmetry of the problem, we could have summed strips in the z direction rather than the x direction

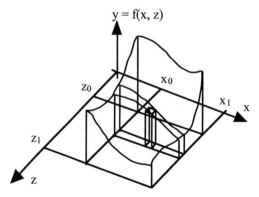

Figure 5.14 Finding volume by summing small volume elements

to achieve the same result, so the ordering of the x/z integral signs is immaterial. Higher-order integrals also have their uses; for example, triple integrals can be used to find the mass of an object by integrating its density in x, y and z. In computer graphics rendering, the method of radiosity implicitly uses a quadruple integral over two surfaces to estimate the amount of light radiation exchanged between the two surfaces. This may be approximated by a quadruple summation. For computing purposes, approximations from the summation formulae with relatively small values of δx and δz (or the appropriate variable increment) are usually satisfactory.

Series Expansions

The computer generation of sine, cosine, exponential and other functions is through series expansions, evaluated term by term. The method of differentiation gives us a way of generating power series expansions for these forms, provided we know the values of their derivatives at a set point. Suppose we have a function f(x), and we wish to find a power series expansion for it:

$$f(x) = a_0 + a_1 x + a_2 x^2 + a_3 x^3 + a_4 x^4 + \ldots,$$

the '…' indicating that this continues indefinitely. The power series is like a polynomial (explained above) without an upper limiting power. If the constants a_0, a_1, … are appropriately chosen to diminish, successive terms can get smaller and smaller in such a way that the series approaches a limiting value. We ask the reader to accept this, as we did the concept of integration as the sum of an increasingly large number of ever-decreasing values. The pure mathematical proof of the concept would take us on too long a diversion. So, to return to our supposition, we first set x = 0, to give

$$f(0) = a_0 \qquad \text{so,} \qquad a_0 = f(0).$$

That's a good start; now can we isolate other constants? Try differentiating the original form; we know how to differentiate a polynomial now, so we get

$$f'(x) = a_1 + 2a_2 x + 3a_3 x^2 + 4a_4 x^3 + \ldots,$$

the term in a_0 has disappeared, and the other terms have their powers diminished by one. Setting x = 0 gives

$$f'(0) = a_1, \qquad \text{so} \qquad a_1 = f'(0).$$

This is still promising, so we continue the process (sneaking in an extra term from the … range occasionally), repeatedly differentiating and setting x = 0.

$$f''(x) = 2a_2 + 3.2a_3 x + 4.3a_4 x^2 + 5.4a_5 x^3 + \ldots,$$
gives $\quad f''(0) = 2a_2 \qquad$ so $\qquad a_2 = f''(0)/2.$

gives
$$f'''(x) = 3.2a_3 + 4.3.2a_4x + 5.4.3a_5x^2 + + 6.5.4a_6x^3 + ...,$$
$$f'''(0) = 3.2a_3 \quad \text{so} \quad a_3 = f'''(0)/(3.2).$$

gives
$$f^{(4)}(x) = 4.3.2a_4 + 5.4.3.2a_5x + + 6.5.4.3a_6x^2 + ...,$$
$$f^{(4)}(0) = 4.3.2a_4 \quad \text{so} \quad a_4 = f^{(4)}(0)/(4.3.2),$$

(the superscript (4) indicates that there should be four dashes; the fourth derivative is taken). The pattern should by now be clear, in general

$$a_n = f^{(n)}(0)/n!,$$

where the function $n! = n(n-1) ... 3.2.1$. This is the factorial function, defined in this way for integers $n > 0$ and specially for $n = 0$ as $0! = 1$. This case is counter-intuitive, but it makes many formulae 'work' in extreme cases. So, for example,

$4! = 4.3.2.1 = 24$, (the same as 4.3.2 in our form above)
$5! = 5.4.3.2.1 = 5.4! = 120$,
$6! = 6.5.4.3.2.1 = 6.5! = 720$,
$7! = 7.6.5.4.3.2.1 = 7.6! = 5040$,

so the factorial function gets extremely large as n increases. Putting values of a_0, a_1, ... into the original power series form gives the Maclaurin expansion,

$$f(x) = f(0) + f'(0)x + f''(0)x^2/2! + f'''(0)x^3/3! + ... f^{(n)}(0)x^n/n! + ...$$

As n! gets very large for relatively small n, dividing by n! makes the terms get very small as n increases, so good approximations can usually be found in a few terms. The usual procedure is to accumulate the series term by term, stopping when the current term is numerically smaller than some precision threshold. A more general form, known as the Taylor series, is found by substituting $(a + x)$ for x,

$$f(a + x) = f(a) + f'(a)x + f''(a)x^2/2! + f'''(a)x^3/3! + ... f^{(n)}(a)x^n/n! + ...$$

We show three examples of how Maclaurin series can be found, for $\sin(x)$, $\cos(x)$ and e^x. The easiest expansion is $f(x) = e^x$. We gave, without proof, the derivatives of this as

$$\frac{d}{dx}e^x = e^x,$$

so repeated differentiation continues to return the same form. The value of e, about 2.71828, is defined in this way. The rate of change of e^x is equal to the value of e^x; this is the basis of 'exponential growth'. Thus,

$$f'(x) = f''(x) = ... e^x, \quad \text{so} \quad f'(0) = f''(0) = ... e^0 = 1.$$

Remember the fundamental power law of chapter 2, $x^0 = 1$, for any non-zero value of x. Thus, in the Maclaurin series, the constant $a_n = f^{(n)}(0)/n! = 1/n!$, so the series can be written

5. Form from Function: Analysis of Shapes

$$e^x = 1 + x + x^2/2! + x^3/3! + x^4/4! + \ldots .$$

There is an interesting correspondence between this expansion and that for sin and cos. Let us represent $\sin(x) = s(x)$ and $\cos(x) = c(x)$. Using (from chapter 2)

$$\sin(0) = 0 \qquad \cos(0) = 1,$$

and the unproven derivatives stated above,

$$\frac{d}{dx}\sin(x) = \cos(x), \qquad \frac{d}{dx}\cos(x) = -\sin(x).$$

$$\begin{array}{ll}
s(0) = 0, & c(0) = 1; \\
s'(0) = c(0), \text{ so } s'(0) = 1, & c'(0) = -s(0), \text{ so } c'(0) = 0; \\
s''(0) = -s(0), \text{ so } s''(0) = 0, & c''(0) = -c(0), \text{ so } c''(0) = -1; \\
s'''(0) = -c(0), \text{ so } s'''(0) = -1, & c'''(0) = s(0), \text{ so } c'''(0) = 0; \\
s^{(4)}(0) = s(0), \text{ so } s^{(4)}(0) = 0, & c^{(4)}(0) = c(0), \text{ so } c^{(4)}(0) = 1.
\end{array}$$

The change of sign should be tracked carefully through these from the definitions of derivatives of sin and cos. The original state has been reached again. This sequence of derivatives will cycle through the same sequence of four values; the sin sequences through 0, 1, 0, -1 repeatedly; the cos sequence through 1, 0, -1, 0 repeatedly. Substituting these into the Maclaurin series,

$$\sin(x) = x - x^3/3! + x^5/5! - x^7/7! + \ldots$$
$$\cos(x) = 1 - x^2/2! + x^4/4! - x^6/6! + \ldots .$$

The sin series consists of odd powers, the cos of even powers, with alternate terms missed out and remaining terms alternating in sign. These series work only when x is measured in radians; this is the real reason for mathematical use of radians rather than degrees. Consider a very simple numerical example to illustrate how this is evaluated. Suppose we need the value of sin(1) correct to five decimal places:

$$\sin(1) = 1 - 1/3! + 1/5! - 1/7! + \ldots .$$

This is approached by evaluating successive terms and accumulating them into the sum until a term is small enough not to affect the fifth decimal place, working is shown to six decimal places.

term(1) = 1	sum(1) = 1;
term(2) = -1/3! = -0.166667	sum(2) = 0.833333
term(3) = 1/5! = 0.008333	sum(3) = 0.841667
term(4) = -1/7! = -0.000198	sum(4) = 0.841468
term(5) = 1/9! = 0.000003	sum(5) = 0.841471
term(6) = -1/11! = -0.000000	sum(6) = 0.841471.

One more term than is necessary has been evaluated here. It should be clear after term 5 that no more terms will affect the fifth decimal place, so we can state confidently $\sin(1) = 0.84147$ (correct to five decimal places). Checking with a calculator confirms this value.

The sin and cos expansions extract alternate exponential series terms, with appropriate sign changes. This leads to a surprising result involving complex numbers. Consider the series for e^{ix}, where i is the imaginary number such that $i^2 = -1$. This also means $i^3 = i.i^2 = -i$, $i^4 = i^2.i^2 = 1$, $i^5 = i.i^4 = i$, and so on. There is no apparent meaning for e^{ix} until we substitute ix instead of x in the expansion of e^x. Splitting the series separates real and imaginary parts.

$$e^{ix} = 1 + (ix) + (ix)^2/2! + (ix)^3/3! + (ix)^4/4! + (ix)^5/5! \ldots$$

$$= 1 + i^2x^2/2! + i^4x^4/4! \ldots$$
$$+ ix + i^3x^3/3! + i^5x^5/5! \ldots$$

$$= 1 - x^2/2! + x^4/4! + \ldots$$
$$+ ix - ix^3/3! + ix^5/5! \ldots$$

$$= 1 + (-x^2/2! + x^4/4! + \ldots$$
$$+ i(x - x^3/3! + x^5/5! \ldots)$$

so, $\quad e^{ix} = \cos(x) + i \sin(x)$.

This astounding result links trigonometry, exponential and imaginary numbers. Setting x to π gives

$$e^{i\pi} = -1.$$

The great American mathematician and physicist Richard Feynman found this formula shortly before his fifteenth birthday. He wrote jubilantly in large letters in his notebook, 'The most remarkable formula in Math, $e^{i\pi} + 1 = 0$'.[1] This links five of the most important symbols in mathematics in a remarkably simple equation.

The real and imaginary parts of complex numbers are mapped onto x and y coordinates to represent complex numbers as points in a plane, called the Argand plane (after its deviser). The form $e^{i\pi}$ links Cartesian and polar representations. From the triangle of fig 5.15

$$z = x + iy = r \cos(\theta) + ir \sin(\theta)$$
so $\quad z = r\{\cos(\theta) + i \sin(\theta)\} = re^{i\theta}$.

r is the modulus and θ is the argument of the complex number z. We do not attempt to justify that power laws can be extended to complex numbers, but this can be proved acceptable. Thus, two complex numbers z_1, z_2 such that

$$z_1 = r_1 e^{i\theta_1} \quad \text{and} \quad z_2 = r_2 e^{i\theta_2},$$

---

[1] This is recorded in James Gleick's excellent 1992 biography of Feynman, *Genius*, Little Brown, pages 35–36.

5. Form from Function: Analysis of Shapes   133

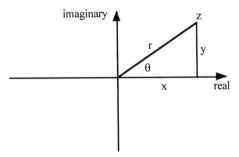

Figure 5.15 The representation of a complex number in the Argand plane

have product

$$z_1 z_2 = r_1 e^{i\theta_1} r_2 e^{i\theta_2} = r_1 r_2 e^{i(\theta_1 + \theta_2)}.$$

Multiplication of two complex numbers is achieved by multiplying their moduli and adding their arguments. We can show similarly that division is achieved by dividing the moduli and subtracting the arguments. The square of z is

$$z^2 = (re^{i\theta})^2 = r^2 e^{i2\theta},$$

the modulus is squared and the argument is doubled. How can we reverse this to find a square root? It seems natural to find the square root of the argument and to halve the argument. This is correct, but we should have two square roots; the fundamental theorem of algebra tells us this. The angle θ is the 'same' as $2\pi + \theta$ (one full revolution more). Halving this gives the other root with argument $\pi + \theta/2$. The equivalent method is used to find three cube roots by dividing θ, $2\pi + \theta$ and $4\pi + \theta$ by 3, and so on for higher powers. We are concerned with square roots, as they will prove useful later in a method for generating fractal figures. The two square roots of a complex number are

$$\sqrt{r}\, e^{i(\theta/2)} \qquad \text{and} \qquad \sqrt{r}\, e^{i(\pi + \theta/2)}.$$

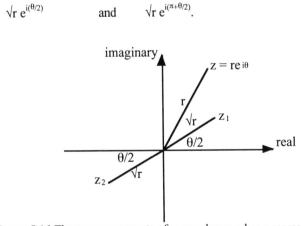

Figure 5.16 The two square roots of a complex number z represented as z_1 and z_2

From their positions on an Argand plane (fig 5.16), these have 'x + iy' forms

$$\sqrt{r}\cos(\theta/2) + i\sqrt{r}\sin(\theta/2) \quad \text{and} \quad \sqrt{r}\cos(\pi + \theta/2) + i\sqrt{r}\sin(\pi + \theta/2).$$

The formulae for sin and cos of added angles were established in chapter 4,

$$\cos(\alpha + \beta) = \cos(\alpha)\cos(\beta) - \sin(\alpha)\cos(\beta),$$
$$\sin(\alpha + \beta) = \sin(\alpha)\cos(\beta) + \cos(\alpha)\cos(\beta),$$

so
$$\cos(\pi + \theta/2) = \cos(\pi)\cos(\theta/2) - \sin(\pi)\sin(\theta/2) = -\cos(\theta/2),$$
$$\sin(\pi + \theta/2) = \sin(\pi)\cos(\theta/2) + \cos(\pi)\sin(\theta/2) = -\sin(\theta/2),$$

using $\cos(\pi) = -1$ and $\sin(\pi) = 0$. The two roots of z can be written as

$$\sqrt{r}\cos(\theta/2) + i\sqrt{r}\sin(\theta/2) \quad \text{and} \quad -\sqrt{r}\cos(\theta/2) - i\sqrt{r}\sin(\theta/2).$$

This is equivalent to setting the second root of r to be $-\sqrt{r}$, which appears to make sense, but the rules of complex numbers ban us from using a negative modulus. The second root is half a revolution away from the first (fig 5.16). The principle can be extended to finding the nth roots of a complex number $re^{i\theta}$. The modulus is the nth root of r, the arguments are θ/n, $(2\pi + \theta)/n$, $(4\pi + \theta)/n$, ... continued until n roots are found; the $(n + 1)^{th}$ will be coincident with the first. The first root has argument θ/n; others are equally spaced out in argument steps of $2\pi/n$ around a circle with radius $\sqrt[n]{r}$. This result is stated for completeness. The square root method is used in chapter 8 when considering fractal objects in the complex plane.

Calculus and Animation

In animation, there is a conflict between the amount of work to be done and the control of the animator over the process. The simulation of movement is produced by repeatedly drawing slightly different images fast enough to convince the human perception system that there is continuous activity. There is considerable work involved if all images are directly under human control, as in early film animation. If control, in some sense, is passed over to a computer system, the animator is more remote from the end product. A compromise is needed. The most common solution is 'key framing', where the animator sets a few important animation frames, and the system fills in the gaps in some sense smoothly in a process of interpolation often known as 'inbetweening'.

Physically based animation is a popular concept, with motion created to satisfy physical or pseudo-physical laws. It is sometimes easier to create rules that mimic the effects of physics rather than to follow the accepted physical laws, as these are often extremely difficult to solve. Newton's laws are fairly straightforward, and are still very useful. Newton's second law (of three) directly relates to animation:

> A particle of constant mass m subject to a resultant force **F** moves in accordance with **F** = m**a**, where **a** is the acceleration of the particle.

5. Form from Function: Analysis of Shapes

If the force is known, the acceleration can be found. Acceleration is the rate of change of velocity, where velocity is the vector equivalent of speed (see chapter 7). Thus, acceleration is the derivative of velocity with respect to time. Similarly, velocity is the rate of change or the derivative of position, where position is measured as a vector location from a set location, such as the origin of a Cartesian axis system. We can use force **F** to find the acceleration **a**. This can be integrated to find a velocity **v**, which in turn can be integrated again to find a location P. This is a two-stage integral problem. Given starting values for position and velocity, in theory, the absolute position can be found at any subsequent time. In practice, the process of integration is analytically intractable, so position is usually found using approximate methods for a sequence of short forward time steps, with each of these being used to give an animation frame. After each forward step, the process is recalculated as if starting from the current position.

Suppose we are given a starting time t, an object with original position P(t), velocity **v**(t) and acceleration **a**(t) and a known controlling force **F**(t). If δt is a short time step, a subsequent position $P(t + \delta t)$ can be estimated from the formula

$$P(t + \delta t) = P(t) + \mathbf{v}(t)\delta t.$$

This is equivalent to using the first two terms of a Taylor series for P, or to locating the new point a short forward step along the tangent of its motion curve. For animation, the object is drawn in this new position. Now the same method is used to update the velocity from the acceleration at time t,

$$\mathbf{v}(t + \delta t) = \mathbf{v}(t) + \mathbf{a}(t)\delta t.$$

The process is now restarted from position $P(t + \delta t)$ with velocity $\mathbf{v}(t + \delta t)$ to find the new location at $P(t + 2\delta t)$, and so on. At each stage, the known force **F**(t) will enable us to find the correct acceleration **a**(t), so **v** and hence P can be updated repeatedly. This is an approximation; the sequence of values of P will gradually drift away from the physically correct location as the number of time steps increases. If exact physical interpretation is important, more precise integration methods can be adopted, for example by using more Taylor series terms or other methods such as Runge-Kutta approximation in which the error values are more tightly controlled. We do not explore these; 'off the shelf' routines are available to solve such problems using standard numerical analysis techniques. If the requirement is to produce an interesting animation effect without the need for physical veracity, the less precise method can give perfectly acceptable results with less computational complexity.

This is an extremely brief mention of an important topic. Physically based animation, where actual or pseudo-physical laws are used to control motion, is used in many up-to-date applications. The topic is included merely to point out the direct connection between calculus and animation, an important branch of computer graphics. We touch on the topic briefly, once more, when considering motion control using spline curves (chapter 9). There are, however, other things that we need to pursue, so we must move on.[2]

---

[2] A. Watt and M. Watt (1992) *Advanced Animation and Rendering Techniques*, Addison-Wesley, is a good source of information on a range of animation methods.

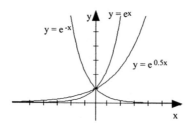

Figure 5.17 Some exponential curves

The Exponential Function

An exponential function exp(x) or e^x was discussed in terms of its Maclaurin expansion above, found from its defining property

$$\frac{d}{dx}e^x = e^x.$$

The slope or gradient of the graph is equal to the graph height. The value of e, approximately 2.71828, is defined to make this property 'work'. We have seen how amending the function definition can change the graph shape. Figure 5.17 shows three versions of $y = e^{kx}$, with k = -1, 0.5 and 1. For $y = e^x$ (k = 1), the graph hugs the x-axis very close without meeting it for large negative x (the x-axis is known as an asymptote), rises to pass through y = 1 for x = 0 and continues to rise 'exponentially' (this is the literally correct form of an often abused term). $y = e^{0.5x}$ has a similar shape but with a less steep rise; the curve is stretched by factor two in the x direction. $y = e^{-x}$ is the mirror image of $y = e^x$ in the y-axis, showing exponential decay towards, but never reaching, y = 0. The function is useful in modelling the effect of friction or 'damping' in animation. The position of a pendulum swinging in air is controlled as a sin or cos function of time. If this is multiplied by e^{kt}, a small negative value of k simulates the effect of a slightly rusty hinge, with oscillations gradually fading. Larger negative values of k give more severe damping, as though the pendulum is swinging in water or even syrup.

The Conic Sections

Several graphs of functions y = f(x) were given earlier in this chapter. This explicit form gives a formula for calculation of dependent variable y from the independent variable x. Some relationships must be expressed implicitly as f(x, y) = 0; the variables cannot be separated. Although not clearly expressed, such forms can also define shapes through their loci, the set of all values (x, y) that satisfy the equation.

5. Form from Function: Analysis of Shapes 137

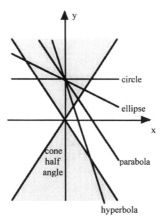

Figure 5.18 Cutting plane angles for generating conic sections

A particularly fruitful set of equations is found from polynomial functions of x and y with maximum power two. These are the conic sections; their shapes can be created by slicing through a cone with a plane. A mathematical cone has two unbounded cone shapes, each known as a nappe, with a common vertex and axis of symmetry (plate 5.1). By slicing through the cone with planes at different angles, the family of conic sections circle, ellipse, parabola and hyperbola is created (fig 5.18 shows a cross-section of cutting plane angles; the results are shown in plate 5.2). The circle has cutting plane orthogonal to the cone axis; as the angle between the cutting plane and axis decreases, first an ellipse is formed. When the angle is equal to the cone half angle, a parabola is formed and a hyperbola results when the angle is smaller than this. The hyperbola has two parts as the plane cuts both nappes. A limiting form of the hyperbola is a pair of straight lines. This occurs when the cutting plane passes through the cone vertex at (0, 0, 0).

Two of these forms are important in computer graphics: the circle and the ellipse (fig 5.19). When set in the most convenient position in a 2D Cartesian system, a circle with radius 1, centre (0, 0), has equation

$$x^2 + y^2 = 1.$$

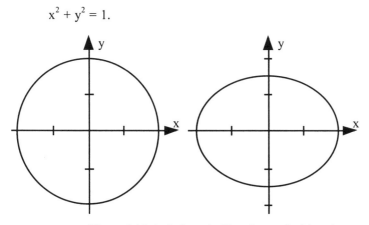

Figure 5.19 A circle and ellipse in standard locations

Using the method for functional scaling in equations, we scale by factor a in x and by b in y through substituting x/a for x and y/b for y. This gives equation

$$(x/a)^2 + (y/b)^2 = 1,$$

or $\quad \dfrac{x^2}{a^2} + \dfrac{y^2}{b^2} = 1$

in standard form. Figure 5.19 shows a circle with r = 2 and an ellipse with a = 2, b = 1.5. Axis ticks are at one unit separation, so the effects of these values can be seen. In the ellipse, when x = 0, $(y/b)^2 = 1$, so $y^2 = b^2$ and y = ±b (this means 'plus or minus b', showing there are two possible values). As there are two values of y for a single value of x, this shows that the full graph cannot be expressed as an explicit formula y = f(x). For this reason, neither circles nor ellipses can be easily drawn by tracking in x. A parametric representation makes depiction easier. From our definitions of polar coordinates (chapter 3), we have

$$x = r\cos(\theta) \quad \text{and} \quad y = r\sin(\theta).$$

When r is constant and θ varies from 0° to 360°, plotting all values of x and y linked by the same values of θ across these formula generates a circle of radius r, centre (0, 0); in other words, all points at distance r from the origin. Applying our scaling method to a circle with radius 1 (r = 1), we get

or $\quad\quad \begin{array}{lll} x/a = \cos(\theta) & \text{and} & y/b = \sin(\theta), \\ x = a\cos(\theta) & \text{and} & y = b\sin(\theta). \end{array}$

Figure 5.20 shows the relationships between parametric forms for a circle and an ellipse. θ in the ellipse form is equivalent to the angle on a point projected outwards

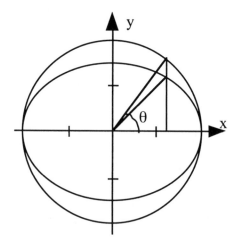

Figure 5.20 An ellipse with a surrounding circle showing the parameter θ

5. Form from Function: Analysis of Shapes

in the y direction onto the smallest surrounding circle radius a (here we assume a > b; an equivalent formulation is available for b > a). This is equivalent to scaling a circle, radius a, by b/a in the y direction.

The circle can be drawn by fixing r and plotting points $(x, y) = (r \cos(\theta), r \sin(\theta))$ for short steps of θ from 0° to 360°. The ellipse plots $(x, y) = (a \cos(\theta), b \sin(\theta))$.

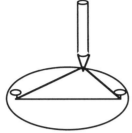

Figure 5.21 Drawing an ellipse using a loop of string and two pins

We can create a circle by holding a loop of string between a fixed pin in a page and a pencil point rotating around the pin with the loop help taut. Similarly an ellipse can be created by keeping a loop of string taut between a pencil and two pins (fig 5.21). The positions of the two pins are the foci (singular focus) of the ellipse. Given the values of a and b for the standard Cartesian form of an ellipse when a > b, the eccentricity of the ellipse is defined as

$$e = \sqrt{\{1 - b^2/a^2\}}.$$

a and b are swapped in this formula if b > a. A circle has b = a; its eccentricity is zero. Eccentricity is a measure of non-circularity. The two foci of an ellipse are placed at (ae, 0) and (-ae, 0); these are the pin points of fig 5.21. As e = 0 for the circle, it is clear that the circle is a limiting form of the ellipse with both pin points coinciding at (0, 0). Again, we should be clear that there is an equivalent formulation for cases when b > a.

We will not need the other conic sections for our purposes, but for the sake of completeness we give the standard forms of their equations as

parabola: $\quad y^2 = 4ax.$

hyperbola: $\quad \dfrac{x^2}{a^2} - \dfrac{y^2}{b^2} = 1$

The parabola is the familiar quadratic curve (fig 5.22 left shows $y^2 = 4x$, with a = 1). In this form, y is squared rather than x, so it is shown lying on its side compared to fig 5.3. The hyperbola (fig 5.22 right) has two separate branches that approach two 'asymptotes' (not shown). An asymptote is some boundary beyond which a curve will not pass. Far from the origin, x and y are very large compared to the 1 on the right-hand side of the hyperbola equation, so we can approximate

$$\dfrac{x^2}{a^2} - \dfrac{y^2}{b^2} \approx 0.$$

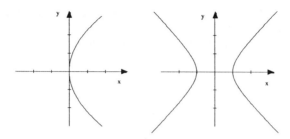

Figure 5.22 A parabola (left) and hyperbola (right) in standard orientation

The symbol ≈ means 'is approximately equal to'. This can be factorised or separated into two parts (we do not go into the detail of how this is done, but the result can be verified by multiplying out the brackets):

$$\left(\frac{x}{a} - \frac{y}{b}\right)\left(\frac{x}{a} + \frac{y}{b}\right) = 0.$$

If two items multiplied together are equal to zero, this means that one or other of them must be zero. Thus, either

$$\left(\frac{x}{a} - \frac{y}{b}\right) \approx 0 \quad \text{or} \quad \left(\frac{x}{a} + \frac{y}{b}\right) \approx 0.$$

These both resolve to straight-line equations,

$$y = (b/a)x \quad \text{and} \quad y = -(b/a)x,$$

representing a pair of straight lines passing through the origin, as c = 0 in the standard form y = mx + c. As the slopes are b/a and -b/a, the lines are at angles of arctan(b/a) and -arctan(b/a) to the x-axis. The hyperbola of fig 5.22 (right) has

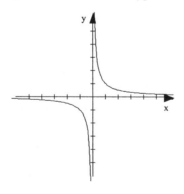

Figure 5.23 The rectangular hyperbola xy = 1

5. Form from Function: Analysis of Shapes 141

$a = b = 1$, so its asymptotes are at ±45° to the x-axis, as is seen in the figure from the near straight-line forms of the curve sections when x is far from zero. This forms a 'rectangular hyperbola' as the asymptotic lines are at right angles to each other. The rectangular hyperbola form can be created in another orientation with the x- and y-axes as asymptotes using the equation

$$xy = c^2.$$

This is the original rotated by 45° to align the asymptotic lines with the axes. Figure 5.23 shows the rectangular hyperbola with constant $c = 1$, plotted using the form $y = 1/x$. Using two loops for values of x from slightly above zero upwards, and from slightly below zero downwards, the problem of zero divide when $x = 0$, a discontinuity in the curve, is avoided.

Some Standard 3D Forms and their Equations

The circle is the basis of several 3D structures frequently encountered in computer graphics. These can be represented in standard form as equations involving Cartesian coordinates in 3D. The simplest is the sphere, with equation

$$x^2 + y^2 + z^2 = r^2.$$

This was developed as an offshoot of the spherical polar coordinate representation in chapter 3 (figs 3.28, 5.24). A point on the surface of a sphere is at a fixed radial distance r from its centre. The distance of a point (x, y, z) from the origin is $\sqrt{\{x^2 + y^2 + z^2\}}$, so the equation

$$\sqrt{\{x^2 + y^2 + z^2\}} = r$$

must be satisfied by all points on the surface of a sphere with centre at the origin and radius r. This equation is normally represented with the square root eliminated by squaring both sides, as above.

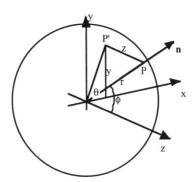

Figure 5.24 The point P on the surface of a sphere with its outward normal **n**

The parametric relationship between points on the surface of the sphere and the spherical polar coordinate angles θ and φ can be useful in computer graphics. The surface of a globe can be scanned as 360° of longitude change (-180° to 180° from the Greenwich meridian) and 180° of latitude change (from -90° at the South Pole to 90° at the North Pole). The surface of a sphere with radius r can similarly be covered by varying the angles of a spherical polar coordinate point (r, θ, φ), keeping the radius r constant. Changing θ from 0° to 360° and φ from 0° to 180° (fig 5.24 shows a point P on the surface of a sphere) covers all surface points. By subdividing these ranges into small enough sections, sufficient points on the surface of the sphere can be generated to form the vertices of a polyhedral approximation to the surface, replacing the curved surface with small plane faces (as explained in chapter 8). Surface points can be converted into Cartesian coordinates for the transformations needed for display,

$$x = r\cos(\theta)\sin(\phi); \qquad y = r\sin(\theta)\sin(\phi); \qquad z = r\cos(\phi),$$

using the spherical polar to Cartesian conversion formulae given in chapter 3. Alternatively, individual points can be created at a dense enough cover to ensure at least one appears in each pixel of the display surface when transformed for display (chapter 10). The beach ball in plate 10.4 was created using this method, generating enough points by dense filling of the θ and φ ranges. By avoiding the polygon surfaces of a polyhedral approximation, an exact sphere surface is generated, although this process is slower than the approximate version. Plate 10.4 also shows the effect of the light on the surface of the ball. This is done (as explained in chapter 10) using the surface normal, a vector pointing orthogonally out of the surface. Figure 5.24 shows how by extending the radius OP through the surface point P a normal is created. Thus, the surface normal (also to be explained later in describing vectors, chapter 7) can be readily found from this direction. We will also need to locate spheres at suitable positions for scene descriptions. This can be done using the routines for translation, scaling and rotation (chapter 4), usually implemented using the matrix methods developed in chapter 6.

This sketchy description of the properties of a sphere points the way ahead to several uses in computer graphics. In describing two other standard objects, the cylinder and the cone, we will omit all the forward references.

As we used spherical polar coordinates to help with definition of a sphere, cylindrical polar coordinates can help with definition of a cylinder. From the definition of chapter 3, all points with r = constant in a spherical polar coordinate definition (r, φ, z) lie on the curved surface of a cylinder. Figure 5.25 shows a point P on the curved surface of a cylinder with axis of symmetry along the z-axis. The curved surface is bounded by the planes z = 0 and z = c. By generating angles θ from 0° to 360° and values of z from 0 to c while holding r at the required constant value, the full curved surface of the cylinder can be found. Conversion formulae for cylindrical polar coordinates give the equivalent Cartesian coordinates,

$$x = r\cos(\theta); \qquad y = r\sin(\theta); \qquad z = z.$$

These are parametric equations for the curved surface of the cylinder. The surface normal marked **n** is parallel to the line OP' in fig 5.25, where P' has coordinates (r cos(θ), r sin(θ), 0). The upper part of the plant pot in plate 10.4 was created as a

5. Form from Function: Analysis of Shapes

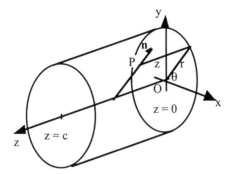

Figure 5.25 A point P on the curved surface of a cylinder, and its surface normal **n**

cylinder by scattering z within the required limits and θ between 0° and 360° densely enough to fill the pixel representation of the surface. The outward normals for the circular ends of the cylinder of fig 5.25 are in the direction of the z-axis for plane z = c and opposite the direction of the z-axis for z = 0. These pieces can be created as filled-in circles such that

$$x^2 + y^2 \leq r,$$

in the planes z = 0 and z = c. Again, the circles can be created by taking sufficient subdivisions of θ from 0° to 360° and of the radius from 0 to r to fill the disc shapes in the two z planes using parametric equations

$$x = r \cos(\theta), \qquad y = r \sin(\theta).$$

To draw an annulus, or a flat ring shape, like the upper surface of the plant pot's rim in plate 10.4, the lower limit of the range for the radius is made greater than zero in this method.

The lower section of the plant pot in plate 10.4 was created as part of a cone. Figure 5.26 shows how a parametric representation of a cone can be generated with major axis along the z-axis and vertex at the origin. The curved surface lies between the two z planes at z = 0 and z = c. If z = 0 is not specified, the mathematical formulation would produce another nappe of the cone as in plate 5.1. Let the radius of the end circle at z = c be r_0. To reach P, we move from O a distance z to Q and then to P a distance r at angle θ to the x-axis direction. We need to determine a formula for r; this varies as we change the value of z. Figure 5.27 shows the triangle OQP embedded in a cross-section of the cone along its axis. With φ as the cone half angle, we have tan(φ) = QP/OQ = r/z. But from the cone itself, it is clear that tan(φ) = r_0/c. Putting these two results together, we obtain

$$r = z\, r_0/c = z \tan(\phi).$$

The normal **n** is shown in fig 5.27. We do not derive its direction here, but note that it is at an angle φ to the z–axis, within a plane at θ to the x direction.

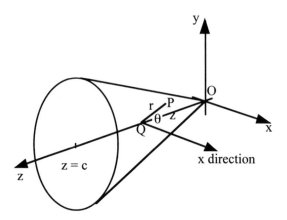

Figure 5.26 The point P is on the curved surface of a cone

The lower part of the plant pot in plate 10.4 is a frustum of a cone: a slice through the cone that in this case does not include its vertex. This is easily created by setting limits for the extent of z so as not to include the vertex.

With all these objects, placement in the necessary part of a scene is achieved by the standard transformations. Placing an object in a mathematically neat position does not lead to particularly interesting images, so each point generated by the methods described above can be subjected to translation and rotation as needed. When transforming surface normals, only the rotations should be used; normals remain the same regardless of their locations. We have not mentioned scaling here, as the objects described above have size values included within their definitions. Variable scaling would be needed to generate ellipsoid shapes, as squashed spheres, but the transformation of normals after scaling needs some care. If **n** is a normal that can be separated into x, y and z components (more of this in chapter 7), each component should be *divided* by the scale factor in its component direction to maintain the normal's orthogonality to the surface. We are looking too far ahead now; we merely state this as a potentially useful result.

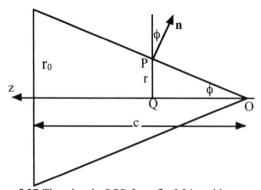

Figure 5.27 The triangle OQP from fig 5.24 enables us to find r

Summary

This chapter has looked at several relationships between shape and function, some of them quite difficult. A number of important functional forms have been introduced, the polynomial, conic sections and the exponential functions, and the trigonometric forms of sine and cosine revisited in a 2D domain. In 3D, the sphere, cylinder and cone structures, often used as primitive forms in computer graphics applications, are described. Both 2D and 3D forms use parametric representations, by which Cartesian coordinates are linked by one or more intermediate parameters. Most importantly, differential calculus has been identified as a means of obtaining shape information, and of identifying major features, such as maxima, minima and points of inflection of curves. The linking topic of integral calculus has been seen to undo the effects of differentiation, so that physical laws involving differentials, such as acceleration and velocity, can be used to control animation. Calculus has also been used to devise series expansions of functions, leading to a result in complex numbers that will prove useful.

This portmanteau chapter has clustered together a number of topics for convenience; they all express relationships between shape, location and functions. Although some of the detail may be hazy on initial reading, readers should carry away with them the concepts of how 2D forms and 3D shapes can be described as functional forms, and how the calculus can be useful in analyzing such shapes.

6. Matrices: Tools for Manipulating Space

Matrices in Computer Graphics

Matrices are used in 2D and 3D computer graphic systems to perform the standard affine transformations of translation, scaling and rotation that form part of the modelling, viewing and coordinate changing processes. Perspective viewing can also be described by a matrix method ('matrix' is the singular of 'matrices'). These different operations can be dealt with in a unified way through their representation as matrix operations, simplifying the code used for these purposes. Matrices, used in conjunction with homogeneous coordinate methods, also enable sequences of transformations to be 'concatenated' into a single operation, thus offering considerable time saving. A matrix maps directly into the two-dimensional array data structure that is readily available in most computer languages. It is a compact method of storing several numbers. These are all reasons for matrix methods to lie at the heart of most computer graphics systems.

The following sections define and develop laws controlling matrices and explain their properties before describing explicitly their main applications to computer graphics. It is hoped that this justification will give readers enough motivation to pursue the chapter through the properties and definitions until the importance of the use of matrices for computer graphics is made clear. Only those properties that are useful in the context of computer graphics are explained.

Definition and Notation

A matrix is a rectangular array of elements. For our purposes, the elements are real numbers, although there is no reason why they cannot be other entities, even matrices themselves. The array is a structure that holds elements of interest. The word 'matrix' can mean the womb or a cavity in which things are embedded. These include the structure of claws which hold jewels in a ring or brooch, the base material within which nuggets of ore or precious materials are located in geology or the mould from which type is cast. The rectangular form of a mathematical matrix means that elements appear in complete rows and columns.

Most of the notation concerning matrices is made clear by a single statement.

$$A = \begin{bmatrix} a_{11} & a_{12} & a_{13} & \dots & a_{1n} \\ a_{21} & a_{22} & a_{23} & \dots & a_{2n} \\ \dots & \dots & \dots & \dots & \dots \\ a_{m1} & a_{m2} & a_{m3} & \dots & a_{mn} \end{bmatrix}.$$

The matrix itself is denoted by an upper case (capital) character, with its elements, enclosed within square brackets, carrying the same lower case character. Some texts use round brackets '(' and ')'. This is not an important issue; curly braces '{' and '}' or plain vertical lines should, however, not be used as they can have different meanings. Each element has a pair of suffixes, indicating its row and column position. The (horizontal) row is always denoted before the (vertical) column. The general example given above has m rows and n columns and is said to have order (m x n) – pronounced 'm by n'. Again, rows before columns is the rule. For example, if we have

$$B = \begin{bmatrix} 3 & 8 & -4 & 0 \\ 2 & -8 & 8 & 5 \\ 1 & 1 & 4 & 11 \end{bmatrix} \quad \text{and} \quad C = \begin{bmatrix} 4 & 1 & 7 \\ -3 & 1 & -5 \\ 2 & -4 & 1 \end{bmatrix},$$

matrix B has order (3 x 4), with $b_{13} = -4$ and $b_{31} = 1$, and C has order (3 x 3), with $c_{13} = 7$ and $c_{31} = 2$.

The simplest matrix operation is to find a transpose. The transpose A^T of a matrix A is found by writing the rows of A as the columns of A^T. Some texts use $\overline{A}$ or A' for transpose. If matrix P has order (m x n) and $Q = P^T$, then Q has order (n x m) and $q_{ij} = p_{ji}$ for all i and j (that is, for i = 1 to n and j = 1 to m). For example, the transposes of the example matrices B and C given above are

$$B^T = \begin{bmatrix} 3 & 2 & 1 \\ 8 & -8 & 1 \\ -4 & 8 & 4 \\ 0 & 5 & 11 \end{bmatrix} \quad \text{and} \quad C^T = \begin{bmatrix} 4 & -3 & 2 \\ 1 & 1 & -4 \\ 7 & -5 & 1 \end{bmatrix}.$$

Forms of Matrices

Many of the matrices used in computer graphics have the same number of rows as columns. Such matrices are known as square matrices. C above is an example of a 'square matrix of order 3', which means the same as a 'matrix of order (3 x 3)'.

Particularly for a square matrix, the set of elements whose row and column numbers are equal is known as the leading diagonal. For our example matrix C, the elements c_{11}, c_{22} and c_{33} are emboldened to show the leading diagonal below.

$$C = \begin{bmatrix} \mathbf{4} & 1 & 7 \\ -3 & \mathbf{1} & -5 \\ 2 & -4 & \mathbf{1} \end{bmatrix}.$$

For reasons that should be obvious on inspection, the matrices D, E, F and G below are known as diagonal, upper triangular, lower triangular and symmetric respectively.

$$D = \begin{bmatrix} 4 & 0 & 0 \\ 0 & 8 & 0 \\ 0 & 0 & 5 \end{bmatrix}, \quad E = \begin{bmatrix} -2 & 2 & -1 \\ 0 & -4 & 3 \\ 0 & 0 & 5 \end{bmatrix},$$

6. Matrices: Tools for Manipulating Space   149

$$F = \begin{bmatrix} 4 & 0 & 0 \\ -3 & -3 & 0 \\ 2 & -4 & -1 \end{bmatrix}, \qquad G = \begin{bmatrix} 3 & 1 & -2 \\ 1 & -5 & -5 \\ -2 & -5 & 1 \end{bmatrix}.$$

More formally stated, the rules for such square matrices are
- D is diagonal if and only if: $d_{ij} = 0$ whenever $i \neq j$;
- E is upper triangular if and only if: $e_{ij} = 0$ whenever $i > j$;
- F is lower triangular if and only if: $f_{ij} = 0$ whenever $i < j$;
- G is symmetric if and only if: $g_{ij} = g_{ji}$ for all i and j.

Operations on Matrices: Addition

There is no point in merely defining and describing several forms of matrices and looking at them in wonder, as one might admire the jewels held in a brooch. Matrices were developed in the mid to late nineteenth century, particularly by Arthur Cayley and James Joseph Sylvester, because they were useful in a number of topical problems. Werner Heisenberg developed some methods in the mid 1920s in advancing the topic of quantum mechanics; he was advised by Max Born that he had redeveloped some results from matrix theory. We now look at laws for comparing, manipulating and combining matrices that are useful in the context of computer graphics.

For two matrices to be equal, they have to be identical in all aspects. Equality of two matrices implies that they have the same orders and all pairs of corresponding elements are equal. Matrices also have to be of the same form before they can be added. Two matrices are conformable for addition if and only if they have the same orders. If this is the case, their sum has the same order and is formed by filling its elements with the sums of corresponding elements from the matrices to be added. Addition of matrices is a binary operator, as it takes in *two* matrices and blends them to give one as a result. Both addition and equality are illustrated in the statement

$$\begin{bmatrix} 4 & -2 & 8 & 1 \\ -8 & 0 & -4 & -5 \\ 5 & 3 & 0 & 3 \end{bmatrix} + \begin{bmatrix} -2 & 3 & -1 & 0 \\ -4 & -4 & 4 & 4 \\ 5 & 1 & 3 & -3 \end{bmatrix} = \begin{bmatrix} 2 & 1 & 7 & 1 \\ -12 & -4 & 0 & -1 \\ 10 & 4 & 3 & 0 \end{bmatrix},$$

but no sum can be formed from two matrices that are not compatible for addition, such as

$$\begin{bmatrix} 4 & -2 & 8 & 1 \\ -8 & 0 & -4 & -5 \\ 5 & 3 & 0 & 3 \end{bmatrix} \text{ and } \begin{bmatrix} -2 & 3 & -1 \\ -4 & -4 & 4 \\ 5 & 1 & 3 \end{bmatrix}.$$

If a matrix is added to itself (this is always possible as the order must be the same), the effect is to multiply each element by 2. Similar repeated addition gives multiples by 3, 4, and so on. This suggests the rule for multiplying a matrix by a scalar (a scalar is simply a single real valued number; this term is used to distinguish it from other mathematical 'objects' such as matrices and vectors). To multiply a matrix A by a scalar value λ, simply multiply all elements of A by λ. For example,

$$-4 \begin{bmatrix} -5 & -2 \\ 3 & 1 \\ -4 & 3 \end{bmatrix} = \begin{bmatrix} 20 & 8 \\ -12 & -4 \\ 16 & -12 \end{bmatrix}.$$

Operations on Matrices: Multiplication

We have seen how to multiply a matrix by a scalar; now we consider the binary operation that multiplies a matrix by another matrix. This is not as straightforward as the operations described above. It may seem bewildering why a relatively complex operation is performed, when an easier one could have been defined. The reason is its usefulness. Defining an easy operation that has no obvious uses serves no purpose. A relatively complicated definition is used because it is useful across a wide range of potential applications. It is worth persevering to master the skill of multiplying matrices.

As in the case of addition, it is not always possible to multiply a pair of matrices. Matrices A and B can be multiplied only if they are conformable for multiplication, which occurs when the number of columns of A is equal to the number of rows of B. Thus, if A has order (m x p) and B has order (p x n), multiplication is allowed. Note that the row length of A is the same as the column length of B; they are both equal to p. This is essential for the multiplication process described below. For example, when

$$A = \begin{bmatrix} 2 & -2 & 1 \\ 4 & 4 & -1 \\ -2 & 0 & 5 \end{bmatrix} \text{ and } B = \begin{bmatrix} -3 & 3 \\ -1 & -5 \\ 0 & 4 \end{bmatrix},$$

the order of A is (3 x 3) and that of B is (3 x 2), so A and B are compatible for multiplication. The rows of A and the columns of B contain three elements. To decide on compatibility for multiplication, write down the orders of both matrices next to each other, as '(3 x 3) (3 x 2)'. If the central numbers can be equally 'linked', as in

(3 x $\boxed{3) (3}$ x 2)

then the matrices are conformable for multiplication. The numbers outside the linking box are also significant. They give the order of the matrix that results from the multiplication, (3 x 2) in this example. Thus, in general, if A has order (m x p) and B has order (p x n), multiplication is allowed and the resulting product matrix, C say, has order (m x n).

6. Matrices: Tools for Manipulating Space

Now we need to find the elements of C. The ith row of A and j^{th} column of B have elements

$$a_{i1} \quad a_{i2} \quad ... \quad a_{ip} \quad \text{and} \quad \begin{array}{c} b_{1j} \\ b_{2j} \\ ... \\ b_{pj} \end{array}$$

respectively (remember that the first suffix indicates the row position and the second indicates the column position). To find the element c_{ij} of C, these elements are multiplied together pair by pair, the resulting values being summed to give

$$c_{ij} = a_{i1}b_{1j} + a_{i2}b_{2j} + ... + a_{ip}b_{pj}.$$

The '...' can be avoided using more formal mathematical notation as

$$c_{ij} = \sum_{k=1}^{p} a_{ik} b_{kj}.$$

This is be interpreted as 'the sum of all values of $a_{ik}b_{kj}$ when k changes from 1 to p'. The Greek equivalent of capital S, 'Σ', is used to denote that a sum is being found. This form of sum of repeated products is sometimes called an 'inner product' in mathematical texts.

For example, to find element c_{32} of the product C of our example matrices A and B,

$$A = \begin{bmatrix} 2 & -2 & 1 \\ 4 & 4 & -1 \\ -2 & 0 & 5 \end{bmatrix} \text{ and } B = \begin{bmatrix} -3 & 3 \\ -1 & -5 \\ 0 & 4 \end{bmatrix},$$

we take the third row of A and the second column of B,

$$-2 \quad 0 \quad 5 \quad \text{and} \quad \begin{array}{c} 3 \\ -5 \\ 4, \end{array}$$

multiply pair-wise as

$$-2 \times 3 = -6$$
$$0 \times -5 = 0$$
$$5 \times 4 = 20$$

and add to give

$$c_{32} = -6 + 0 + 20 = 14.$$

This is done for all six elements of C to give

$$\begin{bmatrix} 2 & -2 & 1 \\ 4 & 4 & -1 \\ -2 & 0 & 5 \end{bmatrix} \begin{bmatrix} -3 & 3 \\ -1 & -5 \\ 0 & 4 \end{bmatrix} = \begin{bmatrix} -4 & 20 \\ -16 & -12 \\ 6 & 14 \end{bmatrix}.$$

This rather unpleasant, but useful, process has a number of consequences. Matrix multiplication is not commutative; in general AB ≠ BA. Even if AB can be formed, it may not be possible to find the product BA. This is the case in the example above, where A has order (3 x 3) and B has order (3 x 2). A and B are compatible for multiplication, but B and A are not. In a product AB, A is said to pre-multiply B and B is said to post-multiply A.

Given matrices A order (m x p), B order (p x q) and C order (q x n), the product AB can be formed, having order (m x q). This is conformable for multiplication with C, and the permitted product (AB)C has order (m x n). In (AB)C, the brackets indicate that the product AB is the first to be performed in time. Alternatively, we could form the product BC with order (p x n) and then pre-multiply by A to give A(BC) with order (m x n). Both (AB)C and A(BC) are valid sequences of multiplication; both give matrices of the same order (m x n) and both end up giving exactly the same result. Thus, we can write both forms as ABC, the order of performing the multiplication being irrelevant. No proof is offered, but readers can justify this by trying a few examples. This appears to imply that the order of multiplication does not matter, whereas the last paragraph stressed that it does. The apparent paradox is due to different interpretations of the word 'order'. It is important to maintain *positional* order when multiplying matrices, but change of *time* order does not affect the outcome. This result will be useful later when we consider matrix operations to perform sequences of transformations on points.

The product AB involves the rows of A associated with the columns of B. In forming a transpose as defined above, rows and columns are exchanged, leading to the following rule for transpose of a product:

$$(AB)^T = B^T A^T.$$

This, and the multiple product rule described above, can be verified by trying out a few examples.

The Identity Matrix

In number theory, the identity element for any binary operation leaves the other value concerned unchanged. For example, zero has this role for addition (adding zero to any value leaves it unchanged) and 1 has this role for multiplication. An identity element for matrix multiplication is known as the identity matrix or unit matrix, designated by the symbol I. There is more than one form of identity matrix, depending on the order of the matrix needed to perform the multiplication. I is always a square diagonal matrix. If required, it can be designated as I_n, where n indicates its order. It is easy to verify that

6. Matrices: Tools for Manipulating Space

$$I = \begin{bmatrix} 1 & 0 & \dots & 0 \\ 0 & 1 & \dots & 0 \\ \dots & \dots & \dots & \dots \\ 0 & 0 & \dots & 1 \end{bmatrix},$$

with all elements zero except the unit values in the leading diagonal, satisfies these requirements. If I pre-multiplies or post-multiplies a matrix, the product leaves the other matrix unchanged.

The identity matrix is closely tied to the idea of an inverse matrix. In number theory, the inverse of the number x for addition is -x, as

$$x + (-x) = 0,$$

the identity for addition. The inverse of the number x for multiplication is 1/x, as

$$x*(1/x) = 1,$$

the identity for multiplication. Note that we must exclude $x = 0$ from this definition, it has no inverse as we are not allowed to divide by zero. In terms of function theory, we must exclude the value zero from the domain of the multiplicative inverse function. In general, an inverse 'undoes' the effect of an operation. If A is a square matrix, it *may* have an inverse – some matrices are excluded, as for multiplication of numbers. If such an inverse exists, it is designated as A^{-1} and has the property

$$AA^{-1} = A^{-1}A = I.$$

For square matrices of order two, the inverse can be easily stated. If

$$A = \begin{bmatrix} a_{11} & a_{12} \\ a_{21} & a_{22} \end{bmatrix},$$

then

$$A^{-1} = \frac{1}{(a_{11}a_{22} - a_{12}a_{21})} \begin{bmatrix} a_{22} & -a_{12} \\ -a_{21} & a_{11} \end{bmatrix},$$

provided $(a_{11}a_{22} - a_{12}a_{21})$, known as the determinant of A, is not zero. This can be checked by multiplying A and A^{-1}. All square matrices have a determinant, a single value calculated from the elements of the matrix. For square matrices of order 2, the determinant is formed as shown here, multiplying the elements on the leading diagonal and subtracting the product of the two remaining elements. For square matrices of order 3 or more, the calculation of determinant and inverse is more complicated. This will be revisited later. The value of the determinant indicates whether a square matrix has an inverse. If the determinant is non-zero, an inverse exists and the matrix is said to be non-singular. A square matrix with no inverse (in other words with zero determinant) is singular, it has no 'partner'.

Matrices and Equations

One of the major uses of matrices in general applications is in the solution of systems of equations. This is not the main use of matrices in computer graphics, but can be useful in some circumstances, for example in solving the systems of equations produced in the radiosity shading method (chapter 10) or solving some intersection problems in vector work (chapter 7). This section and the section following on general methods for calculating matrix inverses can therefore be considered as optional – they can be returned to if needed but may be skipped at first reading. For those who take this option, it would be sensible to restart, without penalty, at the section on 'Matrices, transformations and homogeneous coordinates' (page 160).

The ability to solve simple systems of linear equations (containing only single powers and no products of unknown quantities) is a skill learned at secondary school. For example, consider the two-variable system

$$2x + 5y = -1 \qquad (1)$$
$$3x - 4y = 10 \qquad (2).$$

The system contains two equations with two unknowns. Generally, if we have the same number of distinct equations as unknown values, the system can be solved. If plotted as graphs in a Cartesian plane, equations (1) and (2) represent straight lines (hence the description as linear equations). Many points separately satisfy each of the equations, for example equation (1) is satisfied by (x, y) pairs (-0.5, 0), (2, -1) and (-8, 3). A similar list of values can be generated to satisfy (2). The problem is to find the one point that satisfies both equations, if such a value exists. Interpreting the problem graphically, this involves finding the point at which the lines intersect. Some pairs of lines may have no intersection (they may be parallel), others may have an infinite number of common points (the two equations represent the same line). *One* way (not the only one) of approaching this problem is as follows.

Multiply both sides of equation (1) by 3 and both sides of equation (2) by 2 (these are the 'coefficients' of x in the alternate equation). Such multiplication involves multiplying all parts of the equation, both left-hand and right-hand sides, by the required constant. If the same operation is performed on both sides of an equation, the balance of the equation is unchanged. This gives

$$6x + 15y = -3 \qquad (3)$$
$$6x - 8y = 20 \qquad (4).$$

Note that (3) and (4) are simply disguised versions of (1) and (2), with the coefficients of x made the same in both equations; the multiplying values for the original equations were chosen so that this should occur. The x terms can now be removed by subtracting equation (4) from equation (3). This involves subtracting left-hand side from left-hand side and right-hand side from right-hand side to maintain the balance, to give

$$23y = -23 \qquad (5).$$

6. Matrices: Tools for Manipulating Space

The number of unknown values and the number of equations is reduced by one, leaving one equation for one unknown, y. Note that if the equations represented parallel or coincident lines, the y term would also have disappeared, indicating that no single solution is possible. Equation (5) is solved by dividing both sides by 23, the coefficient of y, to give

$$y = -1.$$

We can substitute this value – this means replacing y wherever it occurs by the particular value -1, into either of the original equations to find x. Using (1) for this purpose gives

$$2x - 5 = -1,$$

which can be rearranged by adding 5 to both sides, to give

$$2x = 4.$$

Dividing both sides by 2 (equivalent to the operation used to solve equation (5)),

$$x = 2.$$

Thus our final solution is x = 2, y = -1. This is the only pair of values of x and y that simultaneously satisfy both equations (1) and (2) and represent the intersection point of the two lines represented by these equations.

This relatively simple problem has been dealt with in some detail to illustrate some of the processes used in solving systems of equations. The validity of an equation is maintained if the same operation is performed on both sides of the equation. An exception to this rule is division by zero, which must be checked for in any computer routine. Useful operations on a system of equations, used in the general algorithm described later, are:

- multiplication of any equation by a constant;
- addition of a multiple of any equation to any other equation;
- swapping the positions of any two equations within the system.

The second of these could be replaced by the two operations of multiplying one equation by a constant and then adding it to another equation, which is effectively what was done in the above example.

Matrices have not yet been mentioned in this section. Systems of linear equations can be represented in matrix form as

$$AX = B.$$

For equations (1) and (2), A, X and B take the form

$$A = \begin{bmatrix} 2 & 5 \\ 3 & -4 \end{bmatrix}, \quad X = \begin{bmatrix} x \\ y \end{bmatrix} \text{ and } B = \begin{bmatrix} -1 \\ 10 \end{bmatrix}.$$

Multiplying out AX, this gives

$$\begin{bmatrix} 2x + 5y \\ 3x - 4y \end{bmatrix} = \begin{bmatrix} -1 \\ 10 \end{bmatrix}.$$

Equating individual elements of these two matrices gives us the original equations (1) and (2). The structure AX = B suggests a solution using the inverse of matrix A. If that inverse exists, we can pre-multiply both sides of this by A^{-1} to give

$$A^{-1}AX = A^{-1}B.$$

$A^{-1}AX$ can be 'grouped' as $(A^{-1}A)X = IX = X$, so the left-hand side is simply the matrix of unknowns. Thus, a complete solution is given as

$$X = A^{-1}B.$$

If the inverse A^{-1} does not exist (the determinant of A is zero), then there is no single solution of the system of equations. This method seems attractive, but the cost in evaluating A^{-1} can be prohibitively expensive, particularly for large systems of equations with several unknowns, so we investigate another method below.

All values involved in the problem can be placed in a single matrix M, combining the elements of A and B,

$$M = \begin{bmatrix} 2 & 5 & -1 \\ 3 & -4 & 10 \end{bmatrix},$$

where each row of M contains all information from one of the original equations. The process of solving the system can be performed using the three equation operations defined above on rows of the matrix. The method described here, known as 'Gaussian elimination with partial pivoting', was developed by Gauss[1] and uses 'partial pivoting' to ensure rounding errors do not make computer-generated solutions unreliable. The procedure, which gives a standard method for dealing with any set of equations, is demonstrated on this limited system before a general algorithm is stated. It is important to keep in mind that we have a system of two equations for two unknown quantities (there are two rows of M). Pivoting is the first stage of this process, which starts by searching column 1 of M for its numerically largest element. This is 3, in row 2. (Note that in finding a 'numerically largest' element, the sign is ignored. For this purpose, -3 would be selected before 2, although 2 is strictly larger than -3.) The 'pivot' for column 1 is found in row 2, so we swap rows 1 and 2 to place the pivot in position m_{11}.

$R_1 \leftrightarrow R_2$: $\qquad\qquad M' = \begin{bmatrix} 3 & -4 & 10 \\ 2 & 5 & -1 \end{bmatrix}.$

---

[1] Carl Friedrich Gauss (1777–1855) was a prolific German mathematician, astronomer and physicist.

6. Matrices: Tools for Manipulating Space   157

Now divide row 1 by m_{11} (the 'pivot'). This is later used to clear zeros in the remainder of its column (only one element in this case). Working to the very limited precision of one decimal place, we get the matrix,

$R_1 \leftarrow R_1/m_{11}:$ $M'' = \begin{bmatrix} 1 & -1.3 & 3.3 \\ 2 & 5 & -1 \end{bmatrix}.$

We continue to work with limited precision for this section as an extreme example of dealing with rounding errors. Now produce zeros in column 1 below row 1 by subtracting relevant multiples of row 1 from the remaining rows (this statement may sound artificial as there is only one such other row in this case, but this more general form is valid for any size of problem). The correct multiples for each row are found by taking all elements m_{i1}, where i >1. Note that these are all numerically less than the original pivot value of m_{11}, so the net effect of the last two operations is to multiply elements of column 1 by a value numerically less than one, thus reducing any rounding error. For this example, we only have $m_{21} = 2$ to deal with (so the net effect of the last two operations is to multiply by 2/3, which is less than one). $m_{21}R_1$ is subtracted from R_2.

$R_2 \leftarrow R_2 - m_{21}R_1:$ $M''' = \begin{bmatrix} 1 & -1.3 & 3.3 \\ 0 & 7.6 & -7.6 \end{bmatrix}.$

We have now reduced column 1 to a simple form. We should now search through column 2 from m_{22} through m_{32} and so on for its pivot element and exchange rows where necessary, but only one such element remains, m_{22} with value 7.6 (this would be $7^2/_3$ working precisely). We divide row 2 by this (effectively a default pivot).

$R_2 \leftarrow R_2/m_{22}:$ $M'''' = \begin{bmatrix} 1 & -1.3 & 3.3 \\ 0 & 1 & -1 \end{bmatrix}.$

As we have reached the bottom row, the elimination phase is over.

Remember that the three terms in a row of the equation in this case represent the multiple of x, multiple of y and the constant on the right-hand side of the equation. Thus, a row with 0, 1 and -1 in these positions represents equation

   0x + 1y = -1,

or   y = -1.

The first row of M'''' can also be represented as an equation,

   1x − 1.3y = 3.3.

The value of y can be substituted into this, to give the result x = 2, y = -1, as before. Although there were intermediate rounding errors, the final solution turns out to be exactly correct – this does not always happen, but the pivoting does give some protection against rounding errors when many variables are involved. In small-scale problems, human ingenuity can often produce short-cut solutions. Computers can be programmed more easily to follow a standard method such as that described here.

Suppose we have a general system of n simultaneous equations for n unknowns

$$a_{11}x_1 + a_{12}x_2 + ... + a_{1n}x_n = b_n$$
$$a_{21}x_2 + a_{22}x_2 + ... + a_{2n}x_n = b_2$$
$$...\quad ...\quad ...\quad ...\quad ...$$
$$a_{n1}x_n + a_{n2}x_2 + ... + a_{nn}x_n = b_n.$$

The unknowns are designated as $x_1, x_2, ... x_n$ rather than the x, y used above. This is generally enough information to solve for the n unknowns, provided the equations are linearly independent. This means that none of the equations can be derived from the others using the standard equation manipulation rules. If the equations should be linearly dependent, the fact that there is no single solution will become clear during the process, as at some stage there will be no available non-zero value to act as pivot. An intuitive interpretation of this is that in the n-dimensional space of the variables $(x_1, x_2, ... x_n)$, some of the 'surfaces' represented by the equations are parallel. This concept relates to the 'degrees of freedom' concept introduced earlier. If one or more of the equations can be derived from others, it acts as a constraint as its constants cannot be freely determined. Thus the dimensionality of the system of equations is reduced accordingly, and less information is carried.

We start by setting up a single matrix from the coefficients of the equations,

$$M = \begin{bmatrix} a_{11} & a_{12} & ... & a_{1n} & b_1 \\ a_{21} & a_{22} & ... & a_{2n} & b_2 \\ ... & ... & ... & ... & ... \\ a_{n1} & a_{n2} & ... & a_{nn} & b_n \end{bmatrix}.$$

Within a loop, from i = 1 to n, we start by searching for the numerically largest of values in or below the leading diagonal in column i, a_{ii} to a_{in}. 'Numerically largest' means the element with largest modulus; the negative signs are ignored. Suppose this is in row k. If k and i are not equal, row k is exchanged with row i to place the pivotal element in position a_{ii}. Row i is then divided by the current pivot value in a_{ii}. Looping from j = i + 1 to n when i < n (this stage is not needed for the final row when i = n), values in column i below a_{ii} are then reduced to zero by subtracting $a_{ji}R_i$ from R_j. When this loop is exhausted, M will usually have the form

$$M' = \begin{bmatrix} 1 & a_{12} & ... & a_{1n} & b_1 \\ 0 & 1 & ... & a_{2n} & b_2 \\ ... & ... & ... & ... & ... \\ 0 & 0 & ... & 1 & b_n \end{bmatrix}.$$

The values of $a_{12}, ... b_n$ in M' are not the same as those in M above. If at any stage during this loop, the search for the pivot (the numerically largest of the values a_{ii} to a_{in}) shows all these values to be zero, it is clear that there is no distinct solution to the set of equations. The search for a solution can be abandoned at this stage.

6. Matrices: Tools for Manipulating Space

Now the process of 'back substitution' starts. $y_n = b_n$ can be immediately allocated. Looping backwards for i from n − 1 to 1, we can define

$$x_i = b_i - \sum_{j=i+1}^{n} a_{ij} x_j ,$$

to give the full set of values of $x_1, x_2, \ldots x_n$.

A routine based on this method can be built into a suite of graphics functions. Alternatively, a similar routine can be used from a commercially marketed numerical analysis system; this kind of numerical equation solver is properly a part of the subject known as 'numerical analysis'.

The Inverse of a Square Matrix

The method given above for solving equations can be adapted to find the inverse of a square matrix. This is only one of several possible methods that could be used, but is relatively efficient and safe in its avoidance of rounding error. The square matrix A to be inverted is augmented by placing an identity (or unit) matrix to its right, as

$$M = \begin{bmatrix} a_{11} & a_{12} & \ldots & a_{1n} & 1 & 0 & \ldots & 0 \\ a_{21} & a_{22} & \ldots & a_{2n} & 0 & 1 & \ldots & 0 \\ \ldots & \ldots & \ldots & \ldots & \ldots & \ldots & \ldots & \ldots \\ a_{n1} & a_{n2} & \ldots & a_{nn} & 0 & 0 & \ldots & 1 \end{bmatrix}.$$

M is operated on very much as in the Gaussian elimination method with partial pivoting, except that, in reducing elements in a particular column to zero, this is performed for *all* elements in the column (those above the pivot as well as those below) except for the pivot a_{ii} itself. We should replace the statement 'values in column i below a_{ii} ...' by 'all values in column i except for a_{ii} itself are then reduced to zero by looping from j = 1 to n (excluding the case j = i), subtracting $a_{ji}R_i$ from R_j'. If at any stage it is impossible to find a non-zero pivot, this indicates that the determinant of the original matrix A is zero, so no inverse exists. Otherwise, M will reduce to the form

$$M' = \begin{bmatrix} 1 & 0 & \ldots & 0 & b_{11} & b_{12} & \ldots & b_{1n} \\ 0 & 1 & \ldots & 0 & b_{21} & b_{22} & \ldots & b_{2n} \\ \ldots & \ldots & \ldots & \ldots & \ldots & \ldots & \ldots & \ldots \\ 0 & 0 & \ldots & 1 & b_{n1} & b_{n2} & \ldots & b_{nn} \end{bmatrix}.$$

Now the unit matrix is to the left, with a new square matrix to its right. This new square matrix is extracted as A^{-1}, the inverse of A,

$$A^{-1} = B = \begin{bmatrix} b_{11} & b_{12} & \ldots & b_{1n} \\ b_{21} & b_{22} & \ldots & b_{2n} \\ \ldots & \ldots & \ldots & \ldots \\ b_{n1} & b_{n2} & \ldots & b_{nn} \end{bmatrix}.$$

No attempt is made to justify this method, other than to state that it works. The matrices produced form true inverses.

As a side issue, the determinant of matrix A can be evaluated from this process. During each stage of the original i loop, 'row i is divided by the current pivot value in a_{ii}'. As the loops proceed, these pivot values should be multiplied together and the number of row swaps performed, m, counted (if a row is swapped with itself, this does not count). The determinant of A is the product of the pivot multiplied by -1^m. It is sometimes of interest in computer graphics systems to find if the determinant of a square matrix is numerically less than one. This can be found directly from the pivotal product.

Matrices, Transformations and Homogeneous Coordinates: Two Dimensions

We now return to a major issue of importance to computer graphics. In discussing coordinate systems earlier, the transformations of translation, scaling and rotation were defined and equations for these developed. In this section, we show how matrix methods enable all these transformations to be performed by a standard method, giving a unified way of treated them in a computer graphics system. This technique has the added bonus of enabling sequences of transformations to be merged or 'concatenated' into one operation, speeding computer graphics routines. We consider first the 2D transformations (fig 6.1).

Translation $x' = x + t_x,$
$y' = y + t_y.$

Scaling $x' = s_x x,$
$y' = s_y y.$

Rotation $x' = x \cos(\theta) - y \sin(\theta),$
$y' = x \sin(\theta) + y \cos(\theta).$

Shear in x $x' = x + k_x y,$
$y' = y.$

Shear in y $x' = x,$
$y' = k_y x + y.$

The shear operations are not given explicitly in many texts as they can be performed by combinations of rotations and scalings, but they can be more efficiently

6. Matrices: Tools for Manipulating Space

performed if separately defined. The vertices of a 2D object are represented in Cartesian form as (x, y). The formulae above transform (x, y) into an 'image' point or vertex (x', y'). When applied to all vertices defining an original object, such as the unit square of fig 6.1, the method changes the object into a transformed version. The effect of the translation defined here is to move an object t_x units horizontally and t_y units vertically. Scaling expands or contracts by factors s_x horizontally and s_y vertically, the origin (0, 0) remaining 'pinned down'. A negative scale factor gives a reflection in the relevant axis. Rotation again pins down the origin (0, 0), rotating the object through an angle θ in the positive anti-clockwise sense.

These sets of equations are directly expressed as matrix operations.

Translation $\quad \begin{bmatrix} x' \\ y' \end{bmatrix} = \begin{bmatrix} x \\ y \end{bmatrix} + \begin{bmatrix} t_x \\ t_y \end{bmatrix};$

Scaling $\quad \begin{bmatrix} x' \\ y' \end{bmatrix} = \begin{bmatrix} s_x & 0 \\ 0 & s_y \end{bmatrix} \begin{bmatrix} x \\ y \end{bmatrix};$

Rotation $\quad \begin{bmatrix} x' \\ y' \end{bmatrix} = \begin{bmatrix} \cos(\theta) & -\sin(\theta) \\ \sin(\theta) & \cos(\theta) \end{bmatrix} \begin{bmatrix} x \\ y \end{bmatrix}.$

Shear in x $\quad \begin{bmatrix} x' \\ y' \end{bmatrix} = \begin{bmatrix} 1 & k_x \\ 0 & 1 \end{bmatrix} \begin{bmatrix} x \\ y \end{bmatrix};$

Shear in y $\quad \begin{bmatrix} x' \\ y' \end{bmatrix} = \begin{bmatrix} 1 & 0 \\ k_y & 1 \end{bmatrix} \begin{bmatrix} x \\ y \end{bmatrix}.$

The point (x, y) is represented in matrix form as $\begin{bmatrix} x \\ y \end{bmatrix}$. A perfectly valid alternative is to use [x y] for (x, y). In this case, the transformations would involve the transposes of the versions given above, with the order of multiplication for scaling and rotation reversed. The choice of method is a matter of personal choice – they are both equivalent in power – but once chosen, the method should not be changed as inconsistency could lead to error.

This is a matrix method, but it holds no advantage over the direct use of equations. A

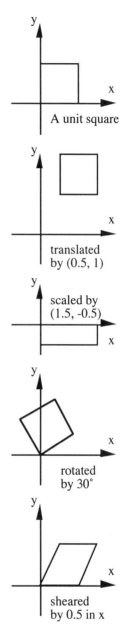

Figure 6.1 Effects of 2D transformations on a unit square

drawback is that the matrix methods used are not the same for each type of operation. Translation involves addition, whereas scaling and rotation use multiplication. Matrix methods become useful when all three basic transformations can be performed in the same way. Shear is not counted as one of these basic transformations as it can be performed by combining scaling and rotations. It is not possible to reduce all to the simpler form of matrix addition; some multiplication is involved. They can all be performed by matrix multiplication using the device of homogeneous coordinates.

The point (x, y) is represented in homogeneous coordinate form as a matrix $\begin{bmatrix} wx \\ wy \\ w \end{bmatrix}$, where w is any arbitrary non-zero constant, called the 'weight'. Thus, there are many possible interpretations for a single point. For example, (3, -4) can be represented as $\begin{bmatrix} 3 \\ -4 \\ 1 \end{bmatrix}$, $\begin{bmatrix} 6 \\ -8 \\ 2 \end{bmatrix}$ or $\begin{bmatrix} -9 \\ 12 \\ -3 \end{bmatrix}$ with weights 1, 2 and -3 respectively. Given a point in homogeneous coordinate form, its Cartesian (x, y) form can easily be extracted by dividing the first two row values by the third. In the homogeneous form $\begin{bmatrix} 11 \\ -8 \\ -2 \end{bmatrix}$ both 11 and -8 are divided by -2 to give the Cartesian point (-5.5, 4). As the choice of weight w is arbitrary, it makes sense to choose the easiest possible value to work with. Whenever possible, which is almost always in computer graphics, we choose w = 1. A general point (x, y) can be represented in the simplest homogeneous form as $\begin{bmatrix} x \\ y \\ 1 \end{bmatrix}$. The exception to this in computer graphics occurs in performing perspective projection, when use of a non-unit weight can simplify other calculations. This will be considered in a later section on 3D work.

We are now ready to devise the unified method for dealing with our three transformations. The required transformations are simply stated – their validity can be checked by multiplying out the matrices concerned and comparing results with the equations given above.

Translation $\quad \begin{bmatrix} x' \\ y' \\ 1 \end{bmatrix} = \begin{bmatrix} 1 & 0 & t_x \\ 0 & 1 & t_y \\ 0 & 0 & 1 \end{bmatrix} \begin{bmatrix} x \\ y \\ 1 \end{bmatrix} = T(t_x, t_y) \begin{bmatrix} x \\ y \\ 1 \end{bmatrix};$

Scaling $\quad \begin{bmatrix} x' \\ y' \\ 1 \end{bmatrix} = \begin{bmatrix} s_x & 0 & 0 \\ 0 & s_y & 0 \\ 0 & 0 & 1 \end{bmatrix} \begin{bmatrix} x \\ y \\ 1 \end{bmatrix} = S(s_x, s_y) \begin{bmatrix} x \\ y \\ 1 \end{bmatrix};$

6. Matrices: Tools for Manipulating Space

Rotation
$$\begin{bmatrix} x' \\ y' \\ 1 \end{bmatrix} = \begin{bmatrix} \cos(\theta) & -\sin(\theta) & 0 \\ \sin(\theta) & \cos(\theta) & 0 \\ 0 & 0 & 1 \end{bmatrix} \begin{bmatrix} x \\ y \\ 1 \end{bmatrix} = R(\theta) \begin{bmatrix} x \\ y \\ 1 \end{bmatrix};$$

Shear in x
$$\begin{bmatrix} x' \\ y' \\ 1 \end{bmatrix} = \begin{bmatrix} 1 & k_x & 0 \\ 0 & 1 & 0 \\ 0 & 0 & 1 \end{bmatrix} \begin{bmatrix} x \\ y \\ 1 \end{bmatrix} = Sh_y(k_x) \begin{bmatrix} x \\ y \\ 1 \end{bmatrix},$$
(the notation indicates that y is unchanged);

Shear in y
$$\begin{bmatrix} x' \\ y' \\ 1 \end{bmatrix} = \begin{bmatrix} 1 & 0 & 0 \\ k_y & 1 & 0 \\ 0 & 0 & 1 \end{bmatrix} \begin{bmatrix} x \\ y \\ 1 \end{bmatrix} = Sh_x(k_y) \begin{bmatrix} x \\ y \\ 1 \end{bmatrix},$$
(the notation indicates that x is unchanged).

The six matrices $T(t_x, t_y)$, $S(s_x, s_y)$, $R(\theta)$, $Sh_y(k_x)$ and $Sh_x(k_y)$ represent the operations of translation, scaling, rotation, shearing in the x direction and shearing in the y direction respectively. The clumsy notations used for shearing matrices indicate by suffixes which coordinate is unchanged by the shear. This method is used for consistency with the 3D forms introduced below. It is worth reiterating that some texts represent their homogeneous coordinates as a row matrix [x y 1] rather than the column matrices used here. If this is done, the transformation matrices are the transposes of those given above and the order of multiplication is reversed. Care should be taken when interpreting texts to understand which method is being used. In a particular work, consistency is important. It does not matter which method is used, as long as the same one is used throughout.

Now we can illustrate the benefits of this technique. In a naïve application of multiple transformations, all transformations would be applied successively to every point of an object to be transformed. Using the matrix method for homogeneous coordinates, sequences of transformations can be reduced to a single matrix. As an example, suppose we wish to rotate a figure through 90° clockwise about the point (3, -2). This can be performed by the sequence of 3 standard transformations shown in fig 6.2).

- translate the whole figure so that (3, -2) is placed at the origin using a translation of (-3, 2);
- rotate through the required angle of -90° (remember that a clockwise rotation is in the negative sense);
- return the figure to the correct location by moving its origin to (3, -2) using a translation of (3, -2).

The first translation stage is defined by matrix

$$T(-3, 2) = \begin{bmatrix} 1 & 0 & -3 \\ 0 & 1 & 2 \\ 0 & 0 & 1 \end{bmatrix}.$$

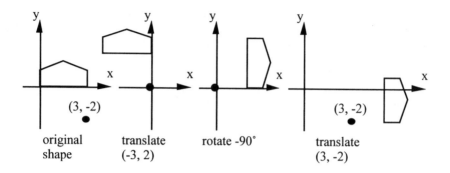

Figure 6.2 Rotation by 90° clockwise about (3, -2)

The standard form of rotation allows us to rotate by any angle about the origin. This initial translation rearranges the figure so that the required centre of rotation is now located at the origin, enabling us to rotate by -90°. From chapter 3, we use the trigonometric values sin(-90°) = -1 and cos(-90°) = 0. The required rotation matrix is

$$R(-90°) = \begin{bmatrix} 0 & 1 & 0 \\ -1 & 0 & 0 \\ 0 & 0 & 1 \end{bmatrix}.$$

The last translation, replacing the centre of rotation in its original position, has matrix

$$T(3, -2) = \begin{bmatrix} 1 & 0 & 3 \\ 0 & 1 & -2 \\ 0 & 0 & 1 \end{bmatrix}.$$

Using matrix multiplication, these three operations can be combined into one matrix before processing the figure itself, so only one operation needs to be applied to each of the object's points. Figure 6.2 shows only a simple object defined by five points. When the method is applied to objects defined by several thousands of points, considerable time saving can be achieved. It is important to combine the matrices in the correct order as matrix multiplication is not commutative.

The transformation matrices given in this text operate on points by pre-multiplication. The point representation appears to the right, with the operator matrix to its left. Thus the first operation must be placed to the left of the point to be transformed, the next operation appearing to the left of the previous one, and so on. In general if we have to apply several transformation matrices M_1, M_2, M_3, ... in that time order to a point represented in homogeneous form as matrix X, the composite operation is given by

$$X' = ... (M_3(M_2(M_1 X))),$$

where X' is the homogeneous representation of the final point reached. A discussion above showed that the time order (but not the space order) of matrix multiplication could be changed, so the above line can be restated generally as

$$X' = \ldots M_3 M_2 M_1 X,$$

and finally rearranged as

$$X' = (\ldots M_3 M_2 M_1) X,$$

where the matrix product ($\ldots M_3 M_2 M_1$) represents the complete sequence of transformations. By multiplying these together, the sequence is reduced to a single matrix that can be applied to each of the points of the figure in turn. For example, if the figure has 1000 points, and the transformation comprises five standard transformations, the naïve approach would involve 5000 multiplications of a matrix by a point. By combining the matrices into one, at an initial cost of four matrix multiplications, there would subsequently be only 1000 multiplications of a matrix by a point. The 'concatenation' of all operations into a single matrix has enabled considerable time saving. This is the real reason for the use of this technique in computer graphics.

Returning to our specific example, the sequence of individual matrices gives the single matrix

$$M = T(3, -2) R(-90°) T(-3, 2) = \begin{bmatrix} 1 & 0 & 3 \\ 0 & 1 & -2 \\ 0 & 0 & 1 \end{bmatrix} \begin{bmatrix} 0 & 1 & 0 \\ -1 & 0 & 0 \\ 0 & 0 & 1 \end{bmatrix} \begin{bmatrix} 1 & 0 & -3 \\ 0 & 1 & 2 \\ 0 & 0 & 1 \end{bmatrix}.$$

Remembering that we cannot change the positional order of these matrices, but we may multiply them in any time order; multiplying the right-hand pair first, we get

$$M = \begin{bmatrix} 1 & 0 & 3 \\ 0 & 1 & -2 \\ 0 & 0 & 1 \end{bmatrix} \begin{bmatrix} 0 & 1 & 2 \\ -1 & 0 & 3 \\ 0 & 0 & 1 \end{bmatrix} = \begin{bmatrix} 0 & 1 & 5 \\ -1 & 0 & 1 \\ 0 & 0 & 1 \end{bmatrix}.$$

This single matrix can now be applied to all points of the figure, effectively cutting out the middle two phases of fig 6.2.

The determinant of a transformation matrix (which can be found by a computer routine as discussed above) indicates the change of area effected by that transformation. If we denote the determinant of matrix A as det[A], we have

$$\det[T(t_x, t_y)] = 1,$$

$$\det[S(s_x, s_y)] = s_x s_y,$$

$$\det[R(\theta)] = \cos^2(\theta) + \sin^2(\theta) = 1.$$

The last case uses a standard trigonometric identity that can be easily established from Pythagoras' theorem. These values clearly indicate the change of area given when these transformations are applied to a two-dimensional shape. Scaling and rotation do not affect area, but the scaling factors s_x and s_y stretch or squash an object linearly, so their combined effect on area is represented by their product. A theorem on determinants of square matrices tells us that when matrices are multiplied, the determinant of the result is the product of the determinants of the original matrices. From this, it is clear that the determinant of a combination of affine transformations gives the overall area change effected by that combination.

Matrices, Transformations and Homogeneous Coordinates: Three Dimensions

Similar methods are used in the three-dimensional world. We move directly to the homogeneous representation of a three-dimensional point (x, y, z) in Cartesian coordinates as $\begin{bmatrix} wx \\ wy \\ wz \\ w \end{bmatrix}$ when the 'weight' $w \neq 0$. For purposes other than performing the perspective projection transformation (discussed later), we take w = 1 for simplicity, representing the point (x, y, z) as $\begin{bmatrix} x \\ y \\ z \\ 1 \end{bmatrix}$. Matrices for the operations of translation and scaling are obvious extensions of the 2D case. Rotation must now be defined about the three coordinate axes as shown below.

Translation

$$x' = x + t_x,$$
$$y' = y + t_y,$$
$$z' = z + t_z.$$

$$T(t_x, t_y, t_z) = \begin{bmatrix} 1 & 0 & 0 & t_x \\ 0 & 1 & 0 & t_y \\ 0 & 0 & 1 & t_z \\ 0 & 0 & 0 & 1 \end{bmatrix}.$$

Scaling

$$x' = s_x x,$$
$$y' = s_y y,$$
$$z' = s_z z.$$

$$S(s_x, s_y, s_y) = \begin{bmatrix} s_x & 0 & 0 & 0 \\ 0 & s_y & 0 & 0 \\ 0 & 0 & s_z & 0 \\ 0 & 0 & 0 & 1 \end{bmatrix}.$$

Rotation about the x-axis

$$x' = x,$$
$$y' = y \cos(\theta) - z \sin(\theta),$$
$$z' = y \sin(\theta) + z \cos(\theta).$$

$$R_x(\theta) = \begin{bmatrix} 1 & 0 & 0 & 0 \\ 0 & \cos(\theta) & -\sin(\theta) & 0 \\ 0 & \sin(\theta) & \cos(\theta) & 0 \\ 0 & 0 & 0 & 1 \end{bmatrix}.$$

6. Matrices: Tools for Manipulating Space

Rotation about the y-axis

$x' = x \cos(\theta) + z \sin(\theta),$
$y' = y,$
$z' = -x \sin(\theta) + z \cos(\theta).$

$$R_y(\theta) = \begin{bmatrix} \cos(\theta) & 0 & \sin(\theta) & 0 \\ 0 & 1 & 0 & 0 \\ -\sin(\theta) & 0 & \cos(\theta) & 0 \\ 0 & 0 & 0 & 1 \end{bmatrix}.$$

Rotation about the z-axis

$x' = x \cos(\theta) - y \sin(\theta),$
$y' = x \sin(\theta) + y \cos(\theta),$
$z' = z.$

$$R_z(\theta) = \begin{bmatrix} \cos(\theta) & -\sin(\theta) & 0 & 0 \\ \sin(\theta) & \cos(\theta) & 0 & 0 \\ 0 & 0 & 1 & 0 \\ 0 & 0 & 0 & 1 \end{bmatrix}.$$

Shear with z unchanged

$x' = x + k_x z,$
$y' = y + k_y z,$
$z' = z.$

$$Sh_z(k_x, k_y) = \begin{bmatrix} 1 & 0 & k_x & 0 \\ 0 & 1 & k_y & 0 \\ 0 & 0 & 1 & 0 \\ 0 & 0 & 0 & 1 \end{bmatrix}.$$

The first two methods, for translation and scaling, are very similar to their 2D equivalents with the addition of an extra z coordinate, which needs an extra row and column for the transformation matrix.

Rotations are almost always described in 3D as being 'about an axis', invoking the idea of a 'right-hand screw' for positive rotation, in the sense of tightening a standard screw with a screwdriver pointing in the positive direction of the axis of rotation. If the right thumb points in the positive direction of this axis, the fingers curl naturally in the sense of rotation. This form of definition is specific to three-dimensional cases. No such axis of rotation exists in 2D, unless we artificially create an axis outside the 2D universe, and in four- and higher-dimensional systems there is more than one 'axis' used for each rotation. At first reading, the concept of rotation in 4D space may seem a bit bewildering, but there is little difficulty in extending mathematical laws to cope with extra dimensions. We have just extended a 2D transformation system to one in 3D, and we discussed in chapter 3 how 'objects' in four and higher dimensions can be created. These are sometimes useful for computer graphics as well as in other disciplines. In 4D systems, it is far easier to consider rotation as 'from one axis towards another'. For example, in a 4D system with axes (x, y, z, u), rotation 'from x towards y' leaves z and u unchanged, so the rotation is conceptually 'about z and u'. The description of rotation in the sense 'from x towards y' can be used in all coordinate systems from 2D upwards. It represents the only form of rotation in 2D systems, rotation about z in 3D, rotation 'about z and u' in 4D, and so on, giving a uniform way of describing all cases.

We are particularly concerned with the 3D case here, so it is worth describing the three possible forms of rotation explicitly (fig 6.3). Rotation about the x-axis turns y towards z; rotation about the y-axis turns z towards x; rotation about the z-axis turns x towards y. Once the alphabetic x to y to z to x ... cycle is identified, this seems a natural system to adopt. As in other texts, the more convenient form of 'R_x' is used here rather that the cumbersome '$R_{y \text{ to } z}$', but knowledge of this cyclic effect may help readers to understand the process. This was discussed in chapter 4, but it does no harm to revisit the concept.

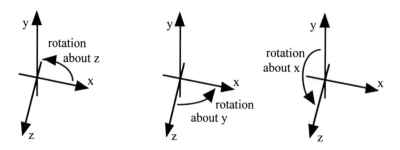

Figure 6.3 Rotations about the three coordinate axes

As we have to consider three forms of rotation in a 3D system, we must also take into account three forms of shear, each of which leaves one of the three spatial coordinates unchanged. The single example given above leaves the z coordinate unchanged, shearing in the x and y directions. The shear matrices

$$Sh_x(k_y, k_z) = \begin{bmatrix} 1 & 0 & 0 & 0 \\ k_y & 1 & 0 & 0 \\ k_z & 0 & 1 & 0 \\ 0 & 0 & 0 & 1 \end{bmatrix} \quad \text{and} \quad Sh_y(k_z, k_x) = \begin{bmatrix} 1 & k_x & 0 & 0 \\ 0 & 1 & 0 & 0 \\ 0 & k_z & 1 & 0 \\ 0 & 0 & 0 & 1 \end{bmatrix}$$

leave the x coordinate and the y coordinate unchanged respectively, shearing in the two altered coordinate directions. To visualize this, imagine having a stacked pack of cards on a table, then drag a finger across the edge of the pack to make it non-vertical. The vertical location of cards has not changed, but position in both horizontal directions can be changed. Suffixes in the above formulae are consistently given in the now familiar alphabetical cycle. Only one shear matrix was given in the main list above, for brevity. As for 2D systems, many texts do not mention shearing explicitly, as it can be performed by a combination of rotation and scaling operations, but having the shear operation directly available can add to the usability of modelling and computer graphics systems.

If we have a sequence of transformations, $M_1, M_2, M_3, \ldots$ in matrix form, to be performed on the homogeneous point $X = \begin{bmatrix} x \\ y \\ z \\ 1 \end{bmatrix}$, the complete operation can be represented as

$X' = \ldots M_3 M_2 M_1 X = MX$, say,

where the single matrix $M = \ldots M_3 M_2 M_1$ represents the complete sequence of transformations. As in the two-dimensional case, pre-computation of M can lead to considerable time savings in performing such a sequence on an object defined by many vertices in three dimensions. As in the 2D case, this is illustrated with a specific example.

6. Matrices: Tools for Manipulating Space

Suppose we want to rescale an object by a factor of 2 in the x direction about the origin, and subsequently rotate it by 45° positively about the x-axis, keeping the point (1, 1, 1) fixed. Remembering that $\cos(45°) = \sin(45°) = 1/\sqrt{2}$, the required matrices for the composite operation can be written down from the standard formulae given above. The composite matrix representing the complete set of operations required is found by repeated pre-multiplication as the process develops, rather than just identifying the matrices to be used and then multiplying the stored values at the end. In a computer context, this is more memory efficient as just the one intermediate result is carried forward each stage, and no more costly in terms of time taken.

- Scale by 2 in the x direction: $S(2, 1, 1) = \begin{bmatrix} 2 & 0 & 0 & 0 \\ 0 & 1 & 0 & 0 \\ 0 & 0 & 1 & 0 \\ 0 & 0 & 0 & 1 \end{bmatrix}$

- Place (1, 1, 1) at the origin: $T(-1,-1,-1) = \begin{bmatrix} 1 & 0 & 0 & -1 \\ 0 & 1 & 0 & -1 \\ 0 & 0 & 1 & -1 \\ 0 & 0 & 0 & 1 \end{bmatrix}$

Composite matrix:

$$\begin{bmatrix} 1 & 0 & 0 & -1 \\ 0 & 1 & 0 & -1 \\ 0 & 0 & 1 & -1 \\ 0 & 0 & 0 & 1 \end{bmatrix} \begin{bmatrix} 2 & 0 & 0 & 0 \\ 0 & 1 & 0 & 0 \\ 0 & 0 & 1 & 0 \\ 0 & 0 & 0 & 1 \end{bmatrix} = \begin{bmatrix} 2 & 0 & 0 & -1 \\ 0 & 1 & 0 & -1 \\ 0 & 0 & 1 & -1 \\ 0 & 0 & 0 & 1 \end{bmatrix}$$

- Rotate about x-axis by 45°: $R_x(45°) = \begin{bmatrix} 1 & 0 & 0 & 0 \\ 0 & 1/\sqrt{2} & -1/\sqrt{2} & 0 \\ 0 & 1/\sqrt{2} & 1/\sqrt{2} & 0 \\ 0 & 0 & 0 & 1 \end{bmatrix}$

Composite matrix:

$$\begin{bmatrix} 1 & 0 & 0 & 0 \\ 0 & 1/\sqrt{2} & -1/\sqrt{2} & 0 \\ 0 & 1/\sqrt{2} & 1/\sqrt{2} & 0 \\ 0 & 0 & 0 & 1 \end{bmatrix} \begin{bmatrix} 2 & 0 & 0 & -1 \\ 0 & 1 & 0 & -1 \\ 0 & 0 & 1 & -1 \\ 0 & 0 & 0 & 1 \end{bmatrix} = \begin{bmatrix} 2 & 0 & 0 & -1 \\ 0 & 1/\sqrt{2} & -1/\sqrt{2} & 0 \\ 0 & 1/\sqrt{2} & 1/\sqrt{2} & -\sqrt{2} \\ 0 & 0 & 0 & 1 \end{bmatrix}$$

(in this stage, we note that $2/\sqrt{2} = \sqrt{2}$, as $2 = \sqrt{2}*\sqrt{2}$).

- Replace the origin to (1, 1, 1): $T(1, 1, 1) = \begin{bmatrix} 1 & 0 & 0 & 1 \\ 0 & 1 & 0 & 1 \\ 0 & 0 & 1 & 1 \\ 0 & 0 & 0 & 1 \end{bmatrix}$

Composite matrix:

$$\begin{bmatrix} 1 & 0 & 0 & 1 \\ 0 & 1 & 0 & 1 \\ 0 & 0 & 1 & 1 \\ 0 & 0 & 0 & 1 \end{bmatrix} \begin{bmatrix} 2 & 0 & 0 & -1 \\ 0 & 1/\sqrt{2} & -1/\sqrt{2} & 0 \\ 0 & 1/\sqrt{2} & 1/\sqrt{2} & -\sqrt{2} \\ 0 & 0 & 0 & 1 \end{bmatrix} = \begin{bmatrix} 2 & 0 & 0 & 0 \\ 0 & 1/\sqrt{2} & -1/\sqrt{2} & 1 \\ 0 & 1/\sqrt{2} & 1/\sqrt{2} & 1-\sqrt{2} \\ 0 & 0 & 0 & 1 \end{bmatrix}$$

The final matrix generated,

$$M = \begin{bmatrix} 2 & 0 & 0 & 0 \\ 0 & 1/\sqrt{2} & -1/\sqrt{2} & 1 \\ 0 & 1/\sqrt{2} & 1/\sqrt{2} & 1-\sqrt{2} \\ 0 & 0 & 0 & 1 \end{bmatrix},$$

represents all operations. Applying this overall transformation to a cube, say, with eight vertices, at the initial cost of three matrix multiplications, we now only have to perform eight direct matrix by point multiplications. Had we not 'concatenated' the series of operations into one matrix in this way, four such operations would have been performed on each point – a total of 24 matrix by point multiplications. With more complex objects, the savings are more dramatic.

As a check on our calculations, it is illustrative to apply the resulting matrix to a few points. We consider the point (0.5, 1, 1). The scaling in x will place this point at (1, 1, 1), the centre of the subsequent rotation, so it will not be moved further by the rotation. Applying our transformation M to the point $X = \begin{bmatrix} 0.5 \\ 1 \\ 1 \\ 1 \end{bmatrix}$, we have

$$MX = \begin{bmatrix} 2 & 0 & 0 & 0 \\ 0 & 1/\sqrt{2} & -1/\sqrt{2} & 1 \\ 0 & 1/\sqrt{2} & 1/\sqrt{2} & 1-\sqrt{2} \\ 0 & 0 & 0 & 1 \end{bmatrix} \begin{bmatrix} 0.5 \\ 1 \\ 1 \\ 1 \end{bmatrix} = \begin{bmatrix} 1 \\ 1 \\ 1 \\ 1 \end{bmatrix}.$$

This can be interpreted as the Cartesian point (1, 1, 1) as required. As another example check, the point (1, 2, 2) will be moved to (2, 2, 2) by the x-scaling. A little thought indicates that rotation of 45° about the x-axis with centre of rotation (1, 1, 1) will place this resulting point on the plane y = 1, distant $1 + \sqrt{2}$ from the z-axis. Applying matrix M to the representation $\begin{bmatrix} 1 \\ 2 \\ 2 \\ 1 \end{bmatrix}$ of the point (1, 2, 2), we get

$$MX = \begin{bmatrix} 2 & 0 & 0 & 0 \\ 0 & 1/\sqrt{2} & -1/\sqrt{2} & 1 \\ 0 & 1/\sqrt{2} & 1/\sqrt{2} & 1-\sqrt{2} \\ 0 & 0 & 0 & 1 \end{bmatrix} \begin{bmatrix} 1 \\ 2 \\ 2 \\ 1 \end{bmatrix} = \begin{bmatrix} 2 \\ 1 \\ 1+\sqrt{2} \\ 1 \end{bmatrix}.$$

This gives the correct location for the final point as $(2, 1, 1 + \sqrt{2})$.

Inverse of a Transformation Matrix

A relatively complicated routine for finding an inverse of a general square matrix was given above. For affine transformation matrices, inverses are easy to find. An inverse matrix 'undoes' the effect of the original matrix, so we can write down

2D $T^{-1}(t_x, t_y) = T(-t_x, -t_y),$

$S^{-1}(s_x, s_y) = S(1/s_x, 1/s_y),$ provided $s_x \neq 0, s_y \neq 0,$

$R^{-1}(\theta) = R(-\theta),$

$Sh_y^{-1}(k_x) = Sh_y(-k_x),$ $Sh_x^{-1}(k_y) = Sh_x(-k_y).$

3D $T^{-1}(t_x, t_y, t_z) = T(-t_x, -t_y, -t_z),$

$S^{-1}(s_x, s_y, s_z) = S(1/s_x, 1/s_y, 1/s_z),$ provided $s_x \neq 0, s_y \neq 0, s_z \neq 0,$

$R_x^{-1}(\theta) = R_x(-\theta),$ $R_y^{-1}(\theta) = R_y(-\theta),$ $R_z^{-1}(\theta) = R_z(-\theta).$

$Sh_z^{-1}(k_x, k_y) = Sh_z(-k_x, -k_y),$ $Sh_x^{-1}(k_y, k_z) = Sh_x(-k_y, -k_z),$

$Sh_y^{-1}(k_z, k_x) = Sh_y(-k_z, -k_x).$

Rotation of -θ undoes a rotation of θ. This result can also be checked by multiplying original matrices by their inverses. Each gives the unit matrix I.

In multiplying together sequences of transformation matrices, we have

$$M^{-1} = (\ldots M_3 M_2 M_1)^{-1} = M_1^{-1} M_2^{-1} M_3^{-1} \ldots .$$

This can be justified by considering the product

$$(\ldots M_3 M_2 M_1)(M_1^{-1} M_2^{-1} M_3^{-1} \ldots) = \ldots M_3 M_2 (M_1 M_1^{-1}) M_2^{-1} M_3^{-1} \ldots$$

$$= \ldots M_3 M_2 I M_2^{-1} M_3^{-1} \ldots$$

$$= \ldots M_3 M_2 M_2^{-1} M_3^{-1} \ldots$$

$$= \ldots = I.$$

Central matrices are repeatedly paired off, until the whole sequence is reduced to I, the unit matrix. This gives an easy way of creating the inverse of a sequence of transformation matrices at the same time as creating the product of the sequence. When a new transformation is absorbed into the sequence, its matrix pre-multiplies the existing transformation matrix and its inverse post-multiplies the composite inverse. Both processes go hand in hand; there is no need to use a difficult matrix inversion routine.

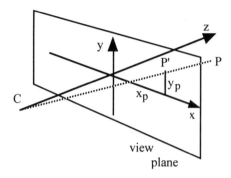

Figure 6.4 Perspective projection of a general point P onto a view plane using centre of projection C

Perspective Projection

Part of the process of viewing objects in 3D is to project them onto a 2D surface that will be mapped onto the viewing surface, such as a VDU screen or plotter. 'Realistic' effects are achieved by using perspective projection. We assume that a special view coordinate system has been set up for this purpose, with origin on this view plane and a pseudo observer (or centre of projection, C) at a point $(0, 0, -d)$ in this system, as shown in fig 6.4. These systems are often set up as left-handed sets of coordinates, as shown in this figure. The z-axis points away from the centre of projection, measuring the depth of an object into the scene. The other direction would give the more orthodox right-handed set of axes.

Given the location of a point P in space, the perspective projection calculates the position P' where the line CP intersects the view plane, the plane of $z = 0$ in this method. An alternative method places the centre of projection at the origin, with the view plane having equation $z = d$. The result for this alternative formulation is given later.

If we 'look down' upon fig 6.4 from the positive y-axis, fig 6.5 results. From this view, it is more easy to see how the x coordinate of P' can be found. The tan of

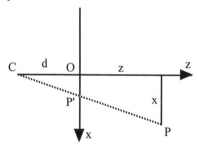

Figure 6.5 Vertical view of fig 6.4

6. Matrices: Tools for Manipulating Space

angle $\angle OCP'$ can be found from two triangles as OP'/d or as $x/(d + z)$. As $OP' = x_p$, we have

$$\frac{x_p}{d} = \frac{x}{d + z},$$

so
$$x_p = \frac{dx}{d + z} = \frac{d}{d + z}x.$$

Similarly, the y coordinate of P' is given by

$$y_p = \frac{d}{d + z}y = \frac{dy}{d + z}.$$

Trivially, P' has z coordinate $z_p = 0$. Note that this method gives a zero divide error if $z = -d$. We cannot view a point that is alongside the observer. This makes physical sense. Most viewing systems get around this problem by including a 'near plane' to exclude any point with z less than some small positive value from the potentially visible region. This also avoids the situation of points behind the observer being projected in a negative sense onto the view plane. It could be useful in real life to have 'eyes in the back of one's head' but this would be most confusing in a computer graphics system. A 'perspective transformation matrix' to perform this operation, given a homogeneous representation of P as $P = \begin{bmatrix} x \\ y \\ z \\ 1 \end{bmatrix}$, is

$$\text{Per}(d) = \begin{bmatrix} d & 0 & 0 & 0 \\ 0 & d & 0 & 0 \\ 0 & 0 & 0 & 0 \\ 0 & 0 & 1 & d \end{bmatrix}.$$

This requires some explanation. P' is given by

$$P' = \text{Per}(d)P = \begin{bmatrix} d & 0 & 0 & 0 \\ 0 & d & 0 & 0 \\ 0 & 0 & 0 & 0 \\ 0 & 0 & 1 & d \end{bmatrix}\begin{bmatrix} x \\ y \\ z \\ 1 \end{bmatrix} = \begin{bmatrix} dx \\ dy \\ 0 \\ d + z \end{bmatrix}.$$

The result is a homogeneous matrix that does not have weight $w = 1$. Above, the convention has been to hold $w = 1$ for simplicity, as in most computer graphic texts; this is the single case when the more general form is required. The Cartesian representation of P' is found by dividing its first three elements by the fourth, its 'weight' w. This gives the required result,

$$(x_p, y_p, z_p) = \left(\frac{dx}{d + z}, \frac{dy}{d + z}, 0\right).$$

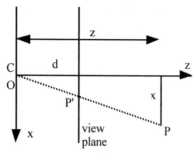

Figure 6.6 Alternative perspective set-up with origin at the centre of projection

In many algorithms, for example hidden surface algorithms that determine which of several objects is visible from the point of view of an observer, it is necessary to pass on the 'depth information' z. In such cases, the matrix above can be used to evaluate x_p and y_p, but a special addition to the algorithm can carry the value of z through unchanged as $z_p = z$.

An alternative method sets up the view coordinate system with its origin at the centre of projection C and the view plane passing through (0, 0, d) (fig 6.6 has O and C coincident). The equations for the position of P' (x'_p, y'_p, z'_p) are

$$x'_p = \frac{dx}{z}, \quad y'_p = \frac{dy}{z} \quad \text{and} \quad z'_p = d.$$

The matrix $\text{Per}'(d) = \begin{bmatrix} d & 0 & 0 & 0 \\ 0 & d & 0 & 0 \\ 0 & 0 & d & 0 \\ 0 & 0 & 1 & 0 \end{bmatrix}$ can be used to perform this version of perspective projection. As in the previous method, points alongside the centre of projection (alongside the ears of an observer, with z = 0 in this formulation) would give a 'zero divide' error and cannot be viewed by this method. Once more, use of a 'near plane' elsewhere in the system prevents this anomalous situation and avoids the projection of points from behind the centre of projection onto the view plane.

Applying $\text{Per}'(d)$ to the homogeneous representation of P as $P = \begin{bmatrix} x \\ y \\ z \\ 1 \end{bmatrix}$, using the general form of homogeneous point representation with weight w, we get

$$\begin{bmatrix} wx'_p \\ wy'_p \\ wz'_p \\ w \end{bmatrix} = \text{Per}(d)P = \begin{bmatrix} d & 0 & 0 & 0 \\ 0 & d & 0 & 0 \\ 0 & 0 & d & 0 \\ 0 & 0 & 1 & 0 \end{bmatrix}\begin{bmatrix} x \\ y \\ z \\ 1 \end{bmatrix} = \begin{bmatrix} dx \\ dy \\ dz \\ z \end{bmatrix}.$$

The result can be reduced to its Cartesian equivalent by dividing other terms by the 'weight' z, to give $x'_p = \frac{dx}{z}$, $y'_p = \frac{dy}{z}$ and $z'_p = d$, as required. As above, the value of z'_p produced may be ignored, the previous value of z being carried through for use in algorithms that need depth information.

The different structures of these matrices compared with the affine transformation matrices (those for translation, scaling, rotation and all combinations of them have bottom row [0 0 0 1]) pose difficulties with computer implementation. Space and time-saving considerations make it attractive to eliminate this bottom row in storage of standard affine transformations, giving special concatenation routines that take this into account rather than use 'off the shelf' matrix multiplication. The creation of a perspective view in computer graphics comprises a modelling stage (when objects are composed to create the scene, as a photographer would arrange objects to be photographed), a viewing stage (the view parameters are set up, equivalent to the photographer locating the tripod, choosing a lens) and a projection and rendering stage (as in the actual exposure of the film).

The use of a 'near plane' to avoid division by zero in perspective projection has been introduced. Other forms of clipping are used in computer graphics, to eliminate all objects or parts of objects outside the field of view, or to cut out objects so far away as not to contribute meaningfully to an image. This has to intrude somewhere within this three-stage image synthesis process, so the smooth use of matrix concatenation has to be interrupted into two stages. Modelling and viewing can be concatenated into one matrix operation, although it may be useful to store intermediate data on the modelling stage if several views of the same scene are to be taken. It is more difficult to include the projection transformation as part of the same process, as the abbreviated form of matrix convenient for modelling transformations is no longer suitable. These two forms of process may be treated separately if specialized matrix operators are used for affine transformations. It is worth noting that some sophisticated modellers allow non-affine forms of space warping such as tapering or twisting as modelling transformations. Thus, the smooth concatenation of matrices must be interrupted at one or more stages in the process of image synthesis, sometimes called the 'computer graphics pipeline'. Many texts omit to mention this. Although this means that two or three matrix operations per point are needed, the use of matrix concatenation can still cut down considerably the amount of calculation needed to view a complex object; it is still a time-saving method. Some implementation issues are discussed later in the chapter.

Parallel projection is a simple alternative to perspective projection. It is equivalent to viewing a point from a centre of projection placed an enormous distance away from the view plane (technically, a parallel projection is the limit of a perspective projection as d approaches infinity). When all created points have been converted to a view coordinate system (as in figs 6.4 and 6.5), the parallel projection simply takes the x and y coordinates from the view projection. The equations for the 'view plane origin' form of perspective projection are

$$x_p = \frac{dx}{d+z} \quad \text{and} \quad y_p = \frac{dy}{d+z}.$$

If we divide top and bottom of the fractions by d, this can be rearranged as

$$x_p = \frac{x}{1+z/d} \quad \text{and} \quad y_p = \frac{y}{1+z/d}.$$

It is clear from this form that as d becomes very large the term z/d gets smaller because the denominator of z/d increases. In the limit, as d approaches infinity, we take z/d = 0, so the limiting form of the perspective projection gives the expected parallel projection formulae

$$x_p = x \quad \text{and} \quad y_p = y.$$

We have here taken a 'limit as d approaches infinity'; this is similar to the 'limit as δx approaches zero' that we used in developing the differential and integral calculus in chapter 5.

Perspective projection could also be used to distort geometric models adding a non-affine option to the modelling process. It should be clear that perspective projection is a non-affine process from the presence of 'vanishing points' in classical perspective images or the visual effect of straight railway lines converging upon a distant point. A defining property of affine transformations is the preservation of parallel lines. Under perspective transformation, parallel lines are made to converge to a vanishing point, and are therefore non-parallel. The effect is equivalent to distorting an oval to an egg shape, broadening some parts and narrowing others. Distortion using centre of projection (0, 0, -d) can be applied to both x and y variables of an object's vertices, or either x or y alone. Care must be taken in carrying the z coordinate through this process – it can be retrieved through the 'weight'. Equivalent formulae can be devised for perspective transformation with the centre of projection on the x or y coordinate axes. If allowed, this process makes concatenation of modelling operations very difficult. For example, direct matrix implementation of a sequence of perspective modelling transformations, one along each coordinate axis, without extraction of the perspective axis coordinate would reduce all vertices to the origin (0, 0, 0). A way to safeguard against this in sequences of perspective modelling operations, perhaps with different axes, is to evaluate vertices after every perspective operation, losing the benefit of matrix concatenation. If required, perspective transformations could be incorporated into modelling as an initial stage, before passing object vertices on to the affine modelling and viewing stage.

Computer Implementation of Matrix Methods

Matrices are excellent conceptual devices for *devising* ways of performing affine and projective transformations. They are often described as methods for *performing* such transformations, but, once the concepts from matrix methods are developed, better time performance is achieved by specially tailored routines. A similar situation was outlined in chapter 2; ways of storing and manipulating numbers in a computer are based on the concepts of binary numbers, but the implementation uses a number of convenient divergences. Many texts hint at potential efficiencies leading from the peculiar structures of transformation matrices, but few give details of what they mean; fewer discuss limitations introduced by such efficiencies. Concatenation of two 3D transformation matrices (order 4 x 4) takes 4^3 or 64 multiplications using a

6. Matrices: Tools for Manipulating Space

standard matrix multiplication routine. Ignoring, for the time being, perspective transformations, the standard modelling and viewing matrices all have fourth rows [0 0 0 1]. If this is taken into account, concatenation can be performed using only 36 multiplications. Further savings are possible using specific routines for concatenation of different forms of transformations. As shown in the methods suggested below, in 3D translations take only 3 additions, scalings use 12 multiplications, rotations involve 16 multiplications and 8 additions or subtractions after the necessary calculations of cosine and sine terms. Such savings are desirable in the time-consuming activity of computer graphics. The 'working parts' of such transformations fit only three rows of four elements, compared to the 16 elements using direct matrix methods, so space saving is also possible, although this is less of an issue as relatively few such matrices need to be stored.

A further divergence from the mathematical matrix model gives potential memory and time benefits. Accessing an element of a two-dimensional array is more time consuming than accessing an element of a one-dimensional array, so it is more efficient to code 3D transformations as 12 elements of a one-dimensional array, ensuring through the code that elements are properly matched. This can be done by storing a general 3D transformation matrix B as a 12 element one-dimensional array A,

$$\begin{bmatrix} b_{11} & b_{12} & b_{13} & b_{14} \\ b_{21} & b_{22} & b_{23} & b_{24} \\ b_{31} & b_{32} & b_{33} & b_{34} \\ 0 & 0 & 0 & 1 \end{bmatrix} = \begin{bmatrix} a_0 & a_1 & a_2 & a_3 \\ a_4 & a_5 & a_6 & a_7 \\ a_8 & a_9 & a_{10} & a_{11} \\ 0 & 0 & 0 & 1 \end{bmatrix},$$

where a_i is equivalent to the C language array element A[i]. A similar device can hold a general 2D transformation in a six-element array. Whilst this method does give some memory and time savings, some developers may not find these factors important and may prefer to implement more standard direct matrix methods requiring fewer functions to be developed. Some sample manipulations of the 'pared down' 3D version are given below, on the understanding that this is just *a* method, not the only one or necessarily the 'best' for all circumstances. Similar methods are available for 2D manipulation; these will not suffer the problems caused by the special case of perspective transformations.

The fourth row of a perspective transformation is not of the form [0 0 0 1]. Ingenuity in interpretation of the array structure that holds transformations can allow this to be accommodated without defining a different data type. Direct application of the perspective transformation converts the 'z' depth coordinate uniformly to the constant z value of the view plane for all vertices, so this can be ignored. However, the value of z used in hidden-surface or line calculations can be recovered from the non-zero weight (directly when the origin is at the centre of projection, or by subtracting d if the origin is on the view plane), which is calculated from the fourth row of the overall matrix. This row can be held in place of the unneeded values in locations 8 to 11 of the transformation array. Care must then be taken in the correct interpretation of these array elements. This form of operation is shown below.

Following is a typical computer graphics sequence of transformations used in rendering a 3D scene, with the type of coordinate generated by each shown:

- modelling transformations used to define objects (3D world coordinates);
- viewing transformations (3D view coordinates);
- perspective transformation on all vertices (2D window coordinates and z);
- window to viewport transformation (2D device coordinates).

These could all be concatenated to one operation, preferably using 'pseudo-matrix' methods. Many texts state this blandly, but do not indicate the special treatment that must be given to the perspective transformation. This process has to be broken into at some stage in classical image synthesis algorithms to clip vertices, edges, faces and parts of edges and faces that lie outside the visible region defined in the viewing transformations. Clipping after perspective transformation is an easier process, involving simple comparisons of coordinates against fixed limits, compared to the more difficult testing of points against general planes used in clipping before perspective projection. The disadvantage of the 'project first, clip last' strategy is that all vertices in the data set must be projected; this is unsuitable when a large number of objects lie outside the potentially visible region. Sophisticated pre-selection strategies are often used in such cases to eliminate obviously external objects before all vertices that define objects (as described in chapter 8) pass through the modelling and viewing transformation stages, so the extra cost of performing perspective projection as part of this process is relatively small. Systems designers must make a strategic decision as to the positioning of their clipping routine in the general 3D viewing pipeline. The ordering adopted below is to clip after the perspective transformations has been performed, allowing concatenation of all 3D transformations into one operation. This neatly separates the 2D operations of window to viewport mapping from the 3D routines.

Examples are given below to show how such a matrix-operated affine transformation system could operate in a 3D computer graphics system. Routines set up initial matrices of the three standard transformations; these are concatenated with an existing transformation matrix using a multiplication routine to build a composite transformation matrix. (Routines for concatenating rotation about x and y and shear are not shown for brevity; the concept of how to develop these routines should be clear.) Initialization using a 'near unit' matrix (an identity matrix of order 4 with its bottom row sliced off) would avoid the need for separate translation, scaling and rotation matrix creation routines, but having such routines does eliminate one matrix multiplication. Suffixes indicating the content of a one-dimensional array are used, so that they can be directly coded, but operations are shown below in matrix formation to indicate their relationships with their matrix derivations. If speed is the major consideration in generating routines based on these formulae, inelegant but fast direct allocation of all elements is the best method. This can increase speed by avoiding the index checks used in neater looping methods.

3D Initial Matrices

With $A = \begin{bmatrix} a_0 & a_1 & a_2 & a_3 \\ a_4 & a_5 & a_6 & a_7 \\ a_8 & a_9 & a_{10} & a_{11} \\ 0 & 0 & 0 & 1 \end{bmatrix}$,

Plate 3.1 A (r, g, b) cube (top left) and, clockwise, slices through it at 0.75, 0.50 and 0.25 respectively of the green axis

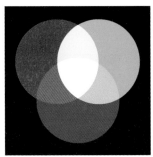

Plate 3.2 Additive colour – combinations of red, green and blue – produce cyan, magenta, yellow and white

Plate 3.3 The (r, g, b) colour (0.2, 0.8, 0.6); its colour may be distorted due to its printed representation in (c, m, y)

Plate 3.4 A (h, s, v) cone, lines (anticlockwise) show the hues of red, magenta, blue, cyan, green, and yellow

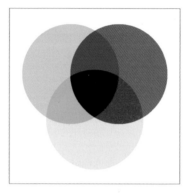

Plate 3.5 Subtractive colour – combinations of cyan, magenta and yellow – produce red, green, blue and black

Plate 5.1 A mathematical cone with two nappes; it extends to infinity beyond the bounds of the image

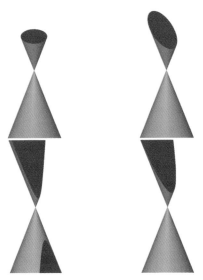

Plate 5.2 Slices at different angles through the cone revealing (clockwise from top left) a circle, ellipse, parabola and hyperbola

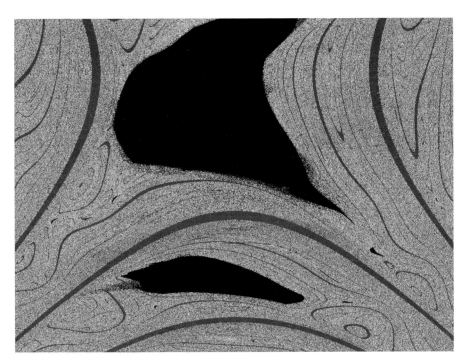

Plate 8.1 A chaotic pattern produced using a Mandelbrot type iteration in $-2 < x < 2$, $-1.5 < y < -1.5$ for the function $x' = \tan(x^2 - y^2) + c_x$, $y' = \cos(x^2 + y^2) - c_y$. Many functions produce interesting patterns

Plate 8.2 A tree generated by Paul Briggs using parametric L-systems

Plate 8.3 A Sierpinski tetrahedron generated by Aurelio Campa using 3D IFS

Plate 8.4 A fractal copse of 10 IFS trees using some non-affine transformations

Plate 8.5 A terrain model by Denis Crampton using functional variation of height to produce flatter valley features without making mountain peaks over-pointed

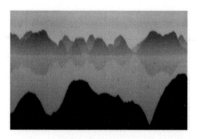

Plate 8.6 A landscape created by Semannia Luk Cheung with the support of John Vince

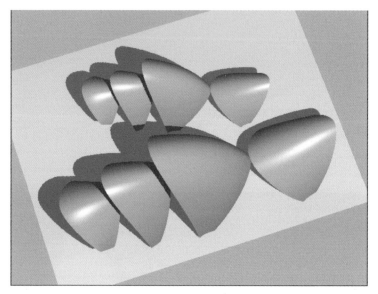

Plate 9.1 Eight 'shells', each created as a pair of cubic Bézier spline patches, joined at the central 'ridge' of symmetry. They are depicted using the exact object method (chapter 10) with Phong shading, highlights and shadows. Some pixellation is seen due to the random nature of the image generation, in which points are scattered across the two Bézier patch parameters

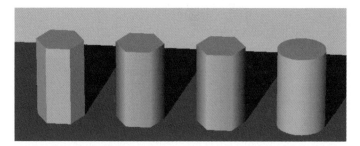

Plate 10.1 Four 'cylinders' displayed using, from the left, Lambert (flat) shading, Gouraud shading, Phong shading and exact object rendering. The first three examples are created as octagonal based prisms; there is little observed difference between Gouraud and Phong examples, as no specular highlights are seen

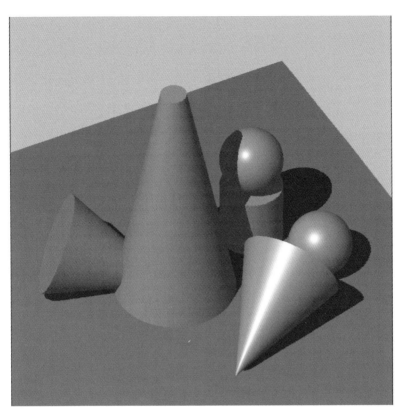

Plate 10.2 An abstract construction of cones, a cylinder and spheres. This shows how the z-buffer copes with interpenetration, the shadow buffer copes with complex cast shadows and how gloss factors may vary (higher for the rear sphere than the other two glossy objects)

Plate 10.3 Part of a maple tree modelled by John Thum. Texture mapping is used to create surface detail

Plate 10.4 This plate shows many of the effects described in the book. Curved surfaces use exact object rendering; the walls, ball and floor are procedurally texture mapped and stochastically anti-aliased although some 'jaggies' still appear; the plants are generated by 3D IFS; a z-buffer and shadow buffer was used and the ball has a specular highlight

6. Matrices: Tools for Manipulating Space

- translation: $a_i = 0$ for $i = 0$ to 11, except for $a_0 = a_5 = a_{10} = 1$, $a_3 = t_x$, $a_7 = t_y$, $a_{11} = t_z$;

- scaling: $a_i = 0$ for $i = 0$ to 11, except for $a_0 = s_x$, $a_5 = s_y$, $a_{10} = s_z$;

- rotation(x): $a_i = 0$ for $i = 0$ to 11, except for $a_5 = \cos(\theta)$, $a_9 = \sin(\theta)$, $a_6 = -a_9$, $a_{10} = a_5$, $a_0 = 1$;

- rotation(y): $a_i = 0$ for $i = 0$ to 11, except for $a_{10} = \cos(\theta)$, $a_2 = \sin(\theta)$, $a_8 = -a_2$, $a_0 = a_{10}$, $a_5 = 1$;

- rotation(z): $a_i = 0$ for $i = 0$ to 11, except for $a_0 = \cos(\theta)$, $a_4 = \sin(\theta)$, $a_1 = -a_4$, $a_5 = a_0$, $a_{10} = 1$;

- shear(x same): $a_i = 0$ for $i = 0$ to 11 except for $a_0 = a_5 = a_{10} = 1$, $a_4 = k_y$, $a_8 = k_z$;

- shear(y same): $a_i = 0$ for $i = 0$ to 11 except for $a_0 = a_5 = a_{10} = 1$, $a_9 = k_z$, $a_1 = k_x$;

- shear(z same): $a_i = 0$ for $i = 0$ to 11 except for $a_0 = a_5 = a_{10} = 1$, $a_2 = k_x$, $a_6 = k_y$.

3D Concatenation Routines

The following assume that the specified transformation is performed after a general

transformation $A = \begin{bmatrix} a_0 & a_1 & a_2 & a_3 \\ a_4 & a_5 & a_6 & a_7 \\ a_8 & a_9 & a_{10} & a_{11} \\ 0 & 0 & 0 & 1 \end{bmatrix}$ and use $c = \cos(\theta)$ and $s = \sin(\theta)$ in the

rotation matrices:

- translation:

$$\begin{bmatrix} 1 & 0 & 0 & t_x \\ 0 & 1 & 0 & t_y \\ 0 & 0 & 1 & t_z \\ 0 & 0 & 0 & 1 \end{bmatrix} \begin{bmatrix} a_0 & a_1 & a_2 & a_3 \\ a_4 & a_5 & a_6 & a_7 \\ a_8 & a_9 & a_{10} & a_{11} \\ 0 & 0 & 0 & 1 \end{bmatrix} = \begin{bmatrix} a_0 & a_1 & a_2 & (a_3 + t_x) \\ a_4 & a_5 & a_6 & (a_7 + t_y) \\ a_8 & a_9 & a_{10} & (a_{11} + t_z) \\ 0 & 0 & 0 & 1 \end{bmatrix};$$

- scaling:

$$\begin{bmatrix} s_x & 0 & 0 & 0 \\ 0 & s_y & 0 & 0 \\ 0 & 0 & s_z & 0 \\ 0 & 0 & 0 & 1 \end{bmatrix} \begin{bmatrix} a_0 & a_1 & a_2 & a_3 \\ a_4 & a_5 & a_6 & a_7 \\ a_8 & a_9 & a_{10} & a_{11} \\ 0 & 0 & 0 & 1 \end{bmatrix} =$$

$$\begin{bmatrix} s_x a_0 & s_x a_1 & s_x a_2 & s_x a_3 \\ s_y a_4 & s_y a_5 & s_y a_6 & s_y a_7 \\ s_z a_8 & s_z a_9 & s_z a_{10} & s_z a_{11} \\ 0 & 0 & 0 & 1 \end{bmatrix};$$

- rotation(z):

$$\begin{bmatrix} c & -s & 0 & 0 \\ s & c & 0 & 0 \\ 0 & 0 & 1 & 0 \\ 0 & 0 & 0 & 1 \end{bmatrix} \begin{bmatrix} a_0 & a_1 & a_2 & a_3 \\ a_4 & a_5 & a_6 & a_7 \\ a_8 & a_9 & a_{10} & a_{11} \\ 0 & 0 & 0 & 1 \end{bmatrix} =$$

$$\begin{bmatrix} (ca_0 - sa_4) & (ca_1 - sa_5) & (ca_2 - sa_6) & (ca_3 - sa_7) \\ (sa_0 + ca_4) & (sa_1 + ca_5) & (sa_2 + ca_6) & (sa_3 + ca_7) \\ a_8 & a_9 & a_{10} & a_{11} \\ 0 & 0 & 0 & 1 \end{bmatrix};$$

3D Transformation Applied to a Vertex

- $$\begin{bmatrix} a_0 & a_1 & a_2 & a_3 \\ a_4 & a_5 & a_6 & a_7 \\ a_8 & a_9 & a_{10} & a_{11} \\ 0 & 0 & 0 & 1 \end{bmatrix} \begin{bmatrix} x_0 \\ x_1 \\ x_2 \\ 1 \end{bmatrix} = \begin{bmatrix} (a_0 x_0 + a_1 x_1 + a_2 x_2 + a_3) \\ (a_4 x_0 + a_5 x_1 + a_6 x_2 + a_7) \\ (a_8 x_0 + a_9 x_1 + a_{10} x_2 + a_{11}) \\ 1 \end{bmatrix}.$$

Concatenation of perspective transformation

If the **origin is on the view plane**, with centre of projection at (0, 0, 0, -d),

- $$\begin{bmatrix} d & 0 & 0 & 0 \\ 0 & d & 0 & 0 \\ 0 & 0 & 0 & 0 \\ 0 & 0 & 1 & d \end{bmatrix} \begin{bmatrix} a_0 & a_1 & a_2 & a_3 \\ a_4 & a_5 & a_6 & a_7 \\ a_8 & a_9 & a_{10} & a_{11} \\ 0 & 0 & 0 & 1 \end{bmatrix} =$$

$$\begin{bmatrix} da_0 & da_1 & da_2 & da_3 \\ da_4 & da_5 & da_6 & da_7 \\ 0 & 0 & 0 & 0 \\ a_8 & a_9 & a_{10} & (d + a_{11}) \end{bmatrix}.$$

All useful information in the latter matrix can be held efficiently by reinterpreting the 12-element array transformation data structure as its first, second and *fourth* rows. If array B represents the result of applying this perspective transformation to a previous transformation (possibly composite) represented by array A, we can interpret B as

- $$\begin{bmatrix} b_0 & b_1 & b_2 & b_3 \\ b_4 & b_5 & b_6 & b_7 \\ * & * & * & * \\ b_8 & b_9 & b_{10} & b_{11} \end{bmatrix} = \begin{bmatrix} da_0 & da_1 & da_2 & da_3 \\ da_4 & da_5 & da_6 & da_7 \\ 0 & 0 & 0 & 0 \\ a_8 & a_9 & a_{10} & (d + a_{11}) \end{bmatrix}.$$

Applying this matrix to a general point gives

6. Matrices: Tools for Manipulating Space

$$\begin{bmatrix} b_0 & b_1 & b_2 & b_3 \\ b_4 & b_5 & b_6 & b_7 \\ * & * & * & * \\ b_8 & b_9 & b_{10} & b_{11} \end{bmatrix} \begin{bmatrix} x_0 \\ x_1 \\ x_2 \\ 1 \end{bmatrix} = \begin{bmatrix} (b_0x_0 + b_1x_1 + b_2x_2 + b_3) \\ (b_4x_0 + b_5x_1 + b_6x_2 + b_7) \\ * \\ (b_8x_0 + b_9x_1 + b_{10}x_2 + b_{11}) \end{bmatrix},$$

enabling the extraction of

$$w = (b_8x_0 + b_9x_1 + b_{10}x_2 + b_{11}),$$
$$x'_0 = (b_0x_0 + b_1x_1 + b_2x_2 + b_3)/w,$$
$$x'_1 = (b_4x_0 + b_5x_1 + b_6x_2 + b_7)/w,$$
and $\quad x'_2 = w - d$

as the coordinates of the resulting point. x'_0 and x'_1 are the (x, y) coordinates of the original point (x_0, x_1, x_2) after transformation into the view window coordinates, x'_2 is the z 'depth' of (x_0, x_1, x_2) from the window after it has undergone modelling and viewing transformations. The unchanged value of w would be just as useful for hidden surface/line and depth cueing purposes, giving the 'depth' of the model vertex from the centre of projection, which represents the location of an observer of the model. This is described in more detail in chapter 10.

If the **origin is at the centre of projection,**

$$\begin{bmatrix} d & 0 & 0 & 0 \\ 0 & d & 0 & 0 \\ 0 & 0 & d & 0 \\ 0 & 0 & 1 & 0 \end{bmatrix} \begin{bmatrix} a_0 & a_1 & a_2 & a_3 \\ a_4 & a_5 & a_6 & a_7 \\ a_8 & a_9 & a_{10} & a_{11} \\ 0 & 0 & 0 & 1 \end{bmatrix} = \begin{bmatrix} da_0 & da_1 & da_2 & da_3 \\ da_4 & da_5 & da_6 & da_7 \\ da_8 & da_9 & da_{10} & da_{11} \\ a_8 & a_9 & a_{10} & a_{11} \end{bmatrix}.$$

Again, if B is the overall result of applying this perspective transformation after a transformation (possibly composite) represented by A, B can be encoded as

$$\begin{bmatrix} b_0 & b_1 & b_2 & b_3 \\ b_4 & b_5 & b_6 & b_7 \\ * & * & * & * \\ b_8 & b_9 & b_{10} & b_{11} \end{bmatrix} = \begin{bmatrix} da_0 & da_1 & da_2 & da_3 \\ da_4 & da_5 & da_6 & da_7 \\ da_8 & da_9 & da_{10} & da_{11} \\ a_8 & a_9 & a_{10} & a_{11} \end{bmatrix},$$

as information in the third row (replaced by asterisks '*') is only a multiple of the fourth by d. Applying this matrix to a point gives

$$\begin{bmatrix} b_0 & b_1 & b_2 & b_3 \\ b_4 & b_5 & b_6 & b_7 \\ * & * & * & * \\ b_8 & b_9 & b_{10} & b_{11} \end{bmatrix} \begin{bmatrix} x_0 \\ x_1 \\ x_2 \\ 1 \end{bmatrix} = \begin{bmatrix} (b_0x_0 + b_1x_1 + b_2x_2 + b_3) \\ (b_4x_0 + b_5x_1 + b_6x_2 + b_7) \\ * \\ (b_8x_0 + b_9x_1 + b_{10}x_2 + b_{11}) \end{bmatrix},$$

enabling the extraction of

$$x'_2 = (b_8x_0 + b_9x_1 + b_{10}x_2 + b_{11}),$$

and
$$x'_0 = (b_0x_0 + b_1x_1 + b_2x_2 + b_3)/x'_2,$$
$$x'_1 = (b_4x_0 + b_5x_1 + b_6x_2 + b_7)/x'_2$$

as the coordinates of the resulting point. x'_0 and x'_1 are the (x, y) coordinates of the original point (x_0, x_1, x_2) after transformation into the view window coordinates, x'_2 is the z 'depth' of (x_0, x_1, x_2) after it has undergone modelling and viewing transformations.

Perspective Transformation Applied Directly to a Vertex (Usually after Clipping)

If the **origin is placed on the view plane**, with centre of projection $(0, 0, 0, -d)$,

$$\begin{bmatrix} d & 0 & 0 & 0 \\ 0 & d & 0 & 0 \\ 0 & 0 & 0 & 0 \\ 0 & 0 & 1 & d \end{bmatrix} \begin{bmatrix} x_0 \\ x_1 \\ x_2 \\ 1 \end{bmatrix} = \begin{bmatrix} dx_0 \\ dx_1 \\ 0 \\ d + x_2 \end{bmatrix};$$

division by the weight $w = (d + x_2)$ gives the transformed point

$$\begin{bmatrix} dx_0/(d + x_2) \\ dx_1/(d + x_2) \\ 0 \\ 1 \end{bmatrix}.$$

The value of $(w - d) = x_2$ should be associated in a data structure with the 2D view window coordinates $(dx_0/(d + x_2), dx_1/(d + x_2))$ as it is used to give depth priorities for hidden surface or line removal.

If the **origin is at the centre of projection**,

$$\begin{bmatrix} d & 0 & 0 & 0 \\ 0 & d & 0 & 0 \\ 0 & 0 & d & 0 \\ 0 & 0 & 1 & 0 \end{bmatrix} \begin{bmatrix} x_0 \\ x_1 \\ x_2 \\ 1 \end{bmatrix} = \begin{bmatrix} dx_0 \\ dx_1 \\ dx_2 \\ x_2 \end{bmatrix};$$

division by the 'weight' $w = x_2$ gives the transformed point $\begin{bmatrix} dx_0/x_2 \\ dx_1/x_2 \\ d \\ 1 \end{bmatrix}.$

The value of $w = x_2$ should be associated in a data structure with the 2D view window coordinates $(dx_0/x_2, dx_1/x_2)$ as it is used to give depth priorities for hidden surface or line removal.

Summary

Users of computer graphics systems can continue blissfully unaware of the use of matrices within the routines that enable their creations. This chapter has developed the concept of matrices, with particular reference to the needs of computer graphics based on the transformations discussed in chapter 4, but with a sideways look at the historically important use of matrices in solving sets of linear equations. This is not wasted; it, too, is used in specialized computer graphics methods, such as rendering using radiosity (chapter 10). We have discussed how purely mathematical matrix methods can be subverted to produce more efficient computer usage.

Does this help the casual user? As a car driver may drive more efficiently with some basic knowledge of the workings of a car, the intention is that computer graphics creators may create better images and do so more efficiently if they have at least a vague knowledge of how their routines work. At least, they should understand why some things take a relatively long time to perform, even with continuing advances in the speed and memory capacity of modern computers.

7. Vectors: Descriptions of Spatial Relationships

Introduction

We have already informally developed a number of algebras: those of the different number systems and the algebra of matrices. Now we introduce another: the algebra of vectors.

Vectors are very important descriptors of spatial relationships: they can be used to compare locations of one object with another, and to identify orientations of objects relative to each other. These have major importance in, for example, identifying the amount of light cast upon an object from a known light source, contributing strongly to the computer graphics activity of 'rendering', or finding the correct locations and colours to give sections of objects when depicted.

As before, the treatment is descriptive rather than rigorous, with methods justified rather than proved. It is worth, however, considering the fundamental definitions of vectors, which is where we begin.

Definition of a Vector

A vector is an object that has a magnitude and a direction and adds according to the parallelogram law. The last phrase is often omitted, but it will be explained and justified later. Vectors are distinguished from scalars, or scalar quantities, which have magnitude alone. A scalar is represented as a standard number and is really a more fancy word for 'number'. A vector may be represented as a directed line segment or arrow, whose length represents its magnitude. For this purpose, length is always taken to be non-negative. Two vectors are equal if and only if they have the same magnitude and the same direction. This means that two vectors satisfying these conditions are equal regardless of their positions in space (fig 7.1). A vector can be considered as a measure of the displacement incurred in moving from one point to another. The same displacement is incurred in moving along any of the arrows of fig 7.1, regardless of the starting point.

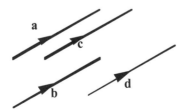

Figure 7.1 Equal vectors

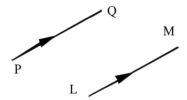

Figure 7.2 Vectors referred to points

Notation

In printed texts, vectors are usually given as bold lower case characters, such as '**a**' in fig 1. For handwritten text, this is difficult to achieve, so the underlined lower case character a̲ is used. The magnitude or length of a vector (a scalar quantity) is denoted by the equivalent plain lower case character 'a' or in 'modulus' form as |**a**| in printed text or |a̲| in handwritten text. This stresses that the magnitude of a vector is always a non-negative quantity. If we are tempted to have a negative magnitude, the modulus of this value should be used and the direction of the vector reversed. Upper case characters are normally reserved for the locations of points. The vector representing displacement from a point P to another point Q can then be written as $\overrightarrow{PQ}$. This does not imply locating the vector in a particular position; for example, fig 7.2 shows that we can have $\overrightarrow{PQ} = \overrightarrow{LM}$. In this case the magnitude of the vector $\overrightarrow{PQ}$ can be written as $|\overrightarrow{PQ}|$ or, more simply, as PQ.

Addition of Vectors: The Parallelogram and Triangle Laws

Now we come to the explanation of the mysterious extra clause in the definition of vectors: 'they add according to the parallelogram law'. To find the sum of two vectors **a** and **b**, represent them with their arrows leaving the same point (fig 7.3). Draw the parallelogram with these representations of **a** and **b** as sides (a parallelogram has opposite sides parallel to each other). The diagonal of this parallelogram leaving from the same point as **a** and **b** represents the sum of **a** and **b**, **a** + **b**. It is important to choose the correct diagonal, so that all arrows in the diagram leave from the same point. This vector sum is also known as the resultant of **a** and **b**.

The triangle law gives another way of defining vector addition. The vectors to be added are drawn with one leaving the termination point of the first, the arrows continuing consecutively (fig 7.4). An arrow from the start point of the first vector to the end point of the second then represents their sum in magnitude and direction.

7. Vectors: Descriptions of Spatial Relationships

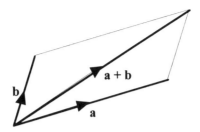

Figure 7.3 Vector addition: the parallelogram law

This ties in with the idea of vector as a measure of displacement. If an object is displaced or translated through **a** and then through **b**, the two displacements can be replaced by the single displacement **a** + **b**. It is important to realize that the parallelogram law and the triangle law are precisely equivalent – they both give identical results. Figure 7.4 also illustrates Minkowski's inequality,

$$|\mathbf{a} + \mathbf{b}| \leq |\mathbf{a}| + |\mathbf{b}|,$$

which should be clear from the diagram. It is a formal way of expressing the common-sense concept that one side of a triangle is shorter than the sum of the other two. Equality is only obtained if **a** and **b** are in the same direction. Otherwise, it is always shorter to travel along one side of a triangle than along the other two.

The commutativity of vector addition,

$$\mathbf{a} + \mathbf{b} = \mathbf{b} + \mathbf{a},$$

should be clear from figs 7.3 and 7.4. The arrow representing **a** + **b** would be the same, regardless of which of the two vectors **a** and **b** were drawn first. Commutativity is one of the consequences of the parallelogram law in the definition of vectors.

Consider rotation as a potential vector: we can rotate by a set amount in a right-hand screw about a directed axis, so it can be given a magnitude (in radians or degrees) and direction. Imagine holding a book horizontally in your hands as though you are about to start reading it (front cover uppermost, spine to the left) – or

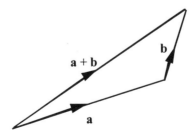

Figure 7.4 Vector addition: the triangle law

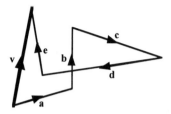

Figure 7.5 Addition of several vectors by an extension of the triangle law

actually do this if you're not in a position where it will embarrass you. Now rotate by 90° so that the part of the book nearest you rises and moves away from you. Follow that by another rotation of 90°, this time so that the highest part of the book moves to your left and down. The front cover should now face away from you with the spine at the bottom. If these two operations are performed in reverse order, the spine of the book ends up facing you, with the front cover to the left. Composition of two rotations is not commutative, so rotations are not vector quantities, although they can be defined to have a direction and a magnitude. This shows why it is necessary to include the clause about the parallelogram law in the definition of vectors.

Figure 7.5 shows how several vectors can be added together by an extension of the triangle law. By considering **a** + **b**, then taking (**a** + **b**) + **c**, adding one more vector at each stage, the triangle law can be extended into a 'chain rule' for the multiple addition of vectors. By drawing the chain of vectors such that the arrows follow in sequence, remembering that it is valid to draw a vector in any position, the vector represented by the arrow from the starting point of the chain to the terminal point of the chain is the sum or resultant of all the vectors. In fig 7.5 **v** is the resultant or sum of vectors **a**, **b**, **c**, **d** and **e**.

v = **a** + **b** + **c** + **d** + **e**.

Multiplication of a Vector by a Scalar

What happens if we add a vector to itself several times? It would be natural to define, for example,

a + **a** + **a** = 3**a**,

the final expression indicating the product of a vector by a scalar. More generally, if λ is a real valued scalar quantity (a number) and **v** is a vector, λ**v** is the vector with magnitude $|\lambda||\mathbf{v}|$ and direction the same as that of **v** if λ is positive or opposite to that of **v** if λ is negative. This rather involved definition of a simple idea ensures that λ**v** has a correctly defined non-negative magnitude. Note that if λ is zero we have a vector of zero magnitude, so the direction ceases to have any meaning. In the expression $|\lambda||\mathbf{v}|$, the same symbols are used to have two different meanings. $|\lambda|$ is the modulus of the scalar quantity or number λ, found by stripping off the negative

7. Vectors: Descriptions of Spatial Relationships

sign if $\lambda < 0$. $|\mathbf{v}|$ is the magnitude of the vector $\mathbf{v}$. The different function domains (real numbers and vectors) indicate which meaning is appropriate, so the symbols can be used without ambiguity.

We can now define the subtraction of vectors as

$$\mathbf{a} - \mathbf{b} = \mathbf{a} + (-1)\mathbf{b}.$$

The more complex issue of multiplication of a vector by a vector will be considered later; it is complicated enough to need two different methods.

Examples of Vector Quantities

Now that the basic concepts of vectors have been discussed, it is possible to identify some examples of vectors that are useful in computer graphics. The simple consideration of vectors as displacements enables their use for locating objects in space through translations (see also transformations in chapters 4 and 6). A 'virtual camera' can be positioned in this way and its orientation also controlled by vectors. This is typically done by defining a 'view direction' vector along which to point the camera lens and a 'view up' vector to give an angle of tilt about this direction. Sophisticated animation methods may be based on the laws of physics, using concepts such as velocity, acceleration and force (all three are vectors) to identify location at any time. Speed is the scalar, pure magnitude part of a velocity and pressure, too, is a scalar.

In generating pseudo-realistic images through computer graphics, the way in which light interacts with the surfaces of objects is mimicked. The direction and intensity of a parallel light beam may be modelled as a vector, and the direction of a plane surface element identified by its 'surface normal', an outward pointing vector at right angles to the surface element. The interaction between these two vectors is the basis of shading methods such as Gouraud or Phong shading. These will be discussed later, but we stress that vector operations on these concepts is central to the methods for evaluating the colour to give surface elements. It is hoped that this brief section will be enough to convince readers that it is worth persevering with this topic as relevant to the core of computer graphics, in its uses to position, control and interpret object models and viewer models.

Vectors in 2D Cartesian Spaces

Vectors can easily be referred to Cartesian coordinate spaces. For example, in a 2D Cartesian space (fig 7.6), suppose we have two points P (p_x, p_y) and Q (q_x, q_y). The vector $\overrightarrow{PQ}$ is represented in magnitude and direction by a displacement of ($q_x - p_x$) in the x direction and a similar displacement of ($q_y - p_y$) in the y direction. Note the order of the coordinates of P and Q and that negative coordinate values are properly taken care of by the formulae. We can represent $\overrightarrow{PQ}$ in Cartesian form as

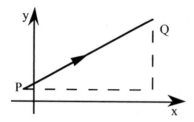

Figure 7.6 A vector in a 2D Cartesian space

$$\vec{PQ} = (q_x - p_x, q_y - p_y).$$

This gives us all the information we need to draw a vector of this magnitude and direction. We must take care to consider this as a displacement rather than identifying a particular point, as a general point (x, y) in 2D Cartesian form is also expressed as an ordered pair of real values in a similar way. Some texts distinguish clearly between representation of a vector and a point through use of unit vectors (having magnitude one) $\hat{\mathbf{i}}$ and $\hat{\mathbf{j}}$ in the directions of the x- and y-axes respectively. The 'hat' symbol over a bold character is used to indicate a unit vector. Then,

$$\vec{PQ} = (q_x - p_x)\hat{\mathbf{i}} + (q_y - p_y)\hat{\mathbf{j}}.$$

We follow the widely used practice of using ordered pairs to denote two-dimensional vectors. Simplicity will compensate for possible ambiguity. In general, we refer to the vector **v** within a 2D Cartesian space as (v_x, v_y), where v_x and v_y are the components of **v** in the x and y directions respectively. Using the triangle law, it should be clear that, with obvious notation,

$$\mathbf{v} + \mathbf{w} = (v_x + w_x, v_y + w_y),$$

and $\quad \lambda\mathbf{v} = (\lambda v_x, \lambda v_y).$

Applying Pythagoras' theorem to fig 7.6 gives us an expression for the magnitude of a vector. We have

$$PQ = \sqrt{\{(q_x - p_x)^2 + (q_y - p_y)^2\}},$$

which is the Cartesian formula for the length of the line segment between the two points P and Q. More generally,

$$v = |\mathbf{v}| = \sqrt{(v_x^2 + v_y^2)}.$$

It is often useful to be able to create a unit vector in the direction of a known vector as a pure or standardized indication of direction. As the vector **v** has magnitude v, multiplying **v** by the scalar 1/v will give a unit vector in the same direction. Thus,

$$\hat{\mathbf{v}} = \frac{1}{v}\mathbf{v} = \left(\frac{v_x}{v}, \frac{v_y}{v}\right),$$

the hat symbol being used to denote a unit vector. By comparison with fig 7.6, the two components of the vector $\hat{\mathbf{v}}$ are the cosines of the angles between **v** and the x- and y-axes respectively. This property is only true for a unit vector.

Vectors in 3D Cartesian Spaces

The extension to three-dimensional spaces is fairly straightforward. In mathematical terms, it is also possible to consider vectors in four and higher-dimensional spaces through extrapolation (meaning extension beyond the normally used range) of the rules used in lower-order systems, although it is difficult to give a proper tangible interpretation of such spaces. Higher-dimensional objects can be visualized through the device of projecting them into a 2D space that can be mapped onto a piece of plotter paper or a VDU screen. This is equivalent to the way that 3D scenes are depicted in 2D environments, so the problem of extension to higher dimensions is one of degree rather than principle. We concentrate here on 3D spaces.

A 3D Cartesian space is referred to three mutually orthogonal axes in the x, y and z directions. These are usually oriented in the form of a 'right-handed set', although a left-handed set is frequently used in one computer graphics application. Figure 7.7 shows a right-handed set of axes (x, y, z) if the z-axis is considered to come out of the page towards the reader (there is always ambiguity in interpreting two-dimensional interpretations of such images). If the thumb of the right hand points in the z direction, the natural direction of finger curl is from x towards y. We define

$\mathbf{r} = \overrightarrow{OP}$

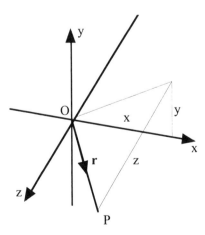

Figure 7.7 Position vector of a point in 3D Cartesian space

as the position vector of P. This represents the displacement undertaken in moving from the origin to P, so

$$\mathbf{r} = (x, y, z).$$

As in the 2D case, there is an alternative 3D notation, using unit vectors $\hat{\mathbf{i}}$, $\hat{\mathbf{j}}$ and $\hat{\mathbf{k}}$ in the directions of the x-, y- and z-axes respectively. This form gives

$$\mathbf{r} = x\hat{\mathbf{i}} + y\hat{\mathbf{j}} + z\hat{\mathbf{k}}.$$

This text uses the less cumbersome method. The magnitude r of $\mathbf{r}$ can be found through two applications of Pythagoras' theorem to fig 7.7. The right-angled triangle with short edges x and y in the x-y plane has hypotenuse $\sqrt{(x^2 + y^2)}$. This is a short side of a right-angled triangle with hypotenuse r = OP and third side z, so

$$r = \sqrt{\{[\sqrt{(x^2 + y^2)}]^2 + z^2\}} = \sqrt{(x^2 + y^2 + z^2)}.$$

A vector in a 3D space can be expressed as

$$\mathbf{v} = (v_x, v_y, v_z).$$

The real (scalar) values v_x, v_y and v_z are the components of $\mathbf{v}$ in the directions of the x-, y- and z-axes respectively. For example, if $\mathbf{v}$ represents a force, v_x is the effective force pulling in the x-axis direction or if $\mathbf{v}$ represents a displacement, v_x is the amount of displacement parallel to the x-axis. Just as vector addition gives us a way of composing vectors, use of components enables us to decompose them into constituent parts. The magnitude v of $\mathbf{v}$ is used to find a unit vector in the direction of $\mathbf{v}$,

$$\hat{\mathbf{v}} = \frac{1}{v}\mathbf{v} = \left(\frac{v_x}{v}, \frac{v_y}{v}, \frac{v_z}{v}\right),$$

where $v = \sqrt{(v_x^2 + v_y^2 + v_z^2)}$.

The components of $\hat{\mathbf{v}}$ (these are v_x/v, v_y/v and v_z/v) are also known as the direction cosines of $\mathbf{v}$; they represent the cosines of angles between $\mathbf{v}$ and the three axis directions x, y and z respectively. For example, v is the hypotenuse of a triangle with v_x as the adjacent side to give the cosine of the angle with the x-axis. These two results are simple extensions of the equivalent formulae for vectors in 2D space. Other easy extensions give us the coordinate form of the vector representing a displacement from point P to point Q as

$$\overrightarrow{PQ} = (q_x - p_x, q_y - p_y, q_z - p_z)$$

and the sum of two vectors in coordinate form as

$$\mathbf{v} + \mathbf{w} = (v_x + w_x, v_y + w_y, v_z + w_z).$$

Multiplication of Vectors: The Scalar or Dot Product

We have seen two ways of describing vectors in 3D: as objects with magnitude and direction or as ordered triplets of real values expressing their components relative to a set of Cartesian axes. The two forms of vector multiplication link both these modes of description in a powerful way through alternative but equivalent definitions. The first, and easier, form of vector multiplication is the scalar product. This is a 'binary operator' (merging *two* vectors) which produces a scalar solution. The 'dot' refers to the symbol used to indicate this form of product. In Cartesian form, we define, with obvious notation,

$$\mathbf{v}.\mathbf{w} = v_x w_x + v_y w_y + v_z w_z.$$

We could equally well have defined the scalar product from the precisely equivalent form

$$\mathbf{v}.\mathbf{w} = vw \cos(\theta),$$

where v and w are the magnitudes of $\mathbf{v}$ and $\mathbf{w}$ respectively and θ is the angle between $\mathbf{v}$ and $\mathbf{w}$. Each of these results can be derived from the other, but, in the general spirit of this book, we accept that someone else has proved this remarkable result. Note that the second form is independent of any coordinate axes, whereas the first contains values v_x, w_x, etc. that depend on the particular axes chosen. This must mean that for any two vectors $\mathbf{v}$ and $\mathbf{w}$ referred to any arbitrary set of Cartesian coordinate axes (provided they use the same measuring scale), the value of

$$\mathbf{v}.\mathbf{w} = v_x w_x + v_y w_y + v_z w_z$$

does not vary for different sets of axes. The value is a property of the vectors alone. Combining the two results, we have

$$vw \cos(\theta) = v_x w_x + v_y w_y + v_z w_z.$$

Given the two vectors $\mathbf{v}$ and $\mathbf{w}$ in coordinate form, we have seen how the magnitudes v and w can be calculated. This leaves θ as the only unknown quantity in the equation, which can be rearranged to give

$$\cos(\theta) = \frac{\mathbf{v}.\mathbf{w}}{vw} = \frac{v_x w_x + v_y w_y + v_z w_z}{vw}$$

so

$$\cos(\theta) = \frac{v_x w_x + v_y w_y + v_z w_z}{\sqrt{(v_x^2 + v_y^2 + v_z^2)(w_x^2 + w_y^2 + w_z^2)}},$$

giving us a method for calculating the angle between any two vectors. This result is a basis of several methods for shading solid objects (chapter 10), as the cosine of the angle of incidence of light on a surface is proportional to the amount of light cast on the surface. From either of the above definitions, it is easy to see that scalar product is a commutative operation, as reordering the elements within the formulae would have no effect on the result. Thus

v.w = w.v.

It is worth noting that if θ is 90° then $\cos(\theta)$ is zero, making **v.w** zero. This is useful as a check for orthogonality of two vectors or lines.

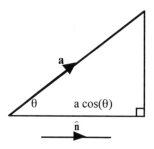

Figure 7.8 Component of **a** in the direction of $\hat{\mathbf{n}}$.

The scalar product also gives us a method for finding components or projections of vectors in arbitrary directions. Given vectors **a** and **n**, we can find the unit vector in the direction of **n** as $\hat{\mathbf{n}}$ from the method described above. Figure 7.8 shows that the 'amount' of **a** 'acting' in the direction of **n**, or the component of **a** in the direction of **n**, is a $\cos(\theta)$, where θ is the angle between the directions of **a** and **n**. From a definition of scalar product, we have

a.$\hat{\mathbf{n}}$ = a $\cos(\theta)$,

as the magnitude of $\hat{\mathbf{n}}$ is one, so **a.$\hat{\mathbf{n}}$** gives the required component directly. This result has use in, for example, the motion control of dynamical systems, when forces, velocities and accelerations can be decomposed into their components in required directions.

Multiplication of Vectors: The Vector or Cross Product

The **vector product**, the second form of vector multiplication, gives a vector quantity as its result. A cross symbol is used to distinguish this form from the scalar or dot product. Again, there are two alternative but equivalent definitions. In Cartesian form, the definition is cumbersome.

v×w = $(v_y w_z - v_z w_y, v_z w_x - v_x w_z, v_x w_y - v_y w_x)$.

As in the discussion of rotation earlier, a cyclic symmetry of x to y to z to x ... may be seen in the way the terms are generated from each other. The individual components of this relate directly to the coordinate axes chosen, but the constancy of the overall form for any coordinate system is shown by its exact equivalence to the alternative definition,

$$\mathbf{v} \times \mathbf{w} = vw \sin(\theta)\, \hat{\mathbf{n}},$$

where θ is the angle between $\mathbf{v}$ and $\mathbf{w}$ and $\hat{\mathbf{n}}$ is a unit vector perpendicular to both $\mathbf{v}$ and $\mathbf{w}$ in the direction of a right-hand screw from $\mathbf{v}$ towards $\mathbf{w}$ (fig 7.9) This, again, is an astounding result. The clumsy, but straightforward, manipulation of coordinate values used in the first formula gives a new vector that is orthogonal to the original vectors. This is particularly useful in finding normals to surfaces, particularly planes, in computer graphics.

Both definitions indicate the non-commutativity of the vector product. If the order of $\mathbf{v}$ and $\mathbf{w}$ is exchanged, then the vector $\hat{\mathbf{n}}$ is reversed in direction. This should also be clear by exchanging the orders of components in the Cartesian form of the definition. In either case, it is clear that

$$\mathbf{v} \times \mathbf{w} = -\mathbf{w} \times \mathbf{v}.$$

As $\hat{\mathbf{n}}$ is a unit vector, it contains all the directional information for $\mathbf{v} \times \mathbf{w}$, which has magnitude $vw \sin(\theta)$. If θ is zero or 180°, the direction of $\mathbf{v} \times \mathbf{w}$ is undefined, but in either case $\sin(\theta)$ is zero, giving a vector of zero magnitude, so the direction in which it points is meaningless. This gives us a test for parallel vectors. If the cross product of two vectors is zero, they are parallel. Given fixed magnitudes for $\mathbf{v}$ and $\mathbf{w}$, the vector product $\mathbf{v} \times \mathbf{w}$ has greatest magnitude when θ is 90° as $\sin(90°) = 1$ is the largest possible value of $\sin(\theta)$ (chapter 3).

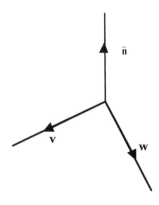

Figure 7.9 Vector product: direction of $\hat{\mathbf{n}}$ for $\mathbf{v} \times \mathbf{w}$ if $\mathbf{v}$ and $\mathbf{w}$ point 'out' of the page

Representation of Lines Using Vectors

Position vectors are used to express construction elements such as lines and planes succinctly. Suppose we have a line through two points A and B with position vectors **a** and **b** respectively (fig 7.10). Suppose point P with position vector **r** is a general point on the line AB. It is clear that AP is always in the same or in the opposite direction as AB. This means that if t is some scalar value, then

$$\overrightarrow{AP} = t\overrightarrow{AB}.$$

When t is zero, P is at A; when t is 1, P is at B. The value of t indicates the location of P on the line with respect to the positions of A and B (fig 7.10). For t between 0 and 1 ($0 < t < 1$), P lies between A and B. If $t < 0$, P is outside AB beyond A, if $t > 1$, P is outside AB beyond B. Using the triangle rule for addition of vectors, we see that

$$\mathbf{r} = \mathbf{a} + \overrightarrow{AP} \quad \text{and} \quad \mathbf{b} = \mathbf{a} + \overrightarrow{AB},$$

so
$$\overrightarrow{AB} = \mathbf{b} - \mathbf{a} \quad \text{and} \quad \overrightarrow{AP} = t(\mathbf{b} - \mathbf{a}).$$

This gives an equation for **r**, the position vector of a point on the line AB,

$$\mathbf{r} = \mathbf{a} + t(\mathbf{b} - \mathbf{a})$$

or
$$\mathbf{r} = (1 - t)\mathbf{a} + t\mathbf{b}.$$

This second version, after a little algebraic rearrangement, is equivalent to that used in interpolating between two numerical values. It is a 'weighted mean' with weight $(1 - t)$ allocated to **a** and t allocated to **b** (the condition for this is that the weights add to one, which they do in this case as $1 - t + t$ must be one). We have effectively set up an axis in t with origin at A and unit length equal to AB. This is a 'parametric form' of the line equation with t as the parameter. If t is taken to represent time, the point P will sweep along the extended line AB, passing through A at $t = 0$ and through B at $t = 1$.

This is a vector equation for a straight line. If **a** and **b** are given in Cartesian coordinates as (a_x, a_y, a_z) and (b_x, b_y, b_z), then

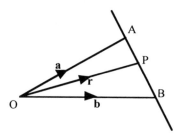

Figure 7.10 P is a point on the line through A and B

$$x = a_x + t(b_x - a_x),$$
$$y = a_y + t(b_y - a_y),$$
$$z = a_z + t(b_z - a_z),$$

gives all three Cartesian components of the vector form

$$\mathbf{r} = \mathbf{a} + t(\mathbf{b} - \mathbf{a}) = (x, y, z).$$

The line equation can be used to set up a line of sight, for example, with A as the position of an observer and B as a point under observation. If the intersection of this line with a set of 'objects' in the scene is found, the value of t immediately indicates if the object is potentially visible. Negative t places it behind the observer, the smallest positive value of t indicates the closest object to the observer along the line of sight, and intersection points with higher values of t must be obscured by this nearest object. This will be useful in considering rendering in computer graphics (chapter 9).

Classification of Points against Planes Using Vectors

A plane can be defined by identifying one point on it and giving a normal vector. Other ways of defining planes can all be reduced to this method. For example, suppose a plane is defined as passing through the three points A, B, C. The vector product

$$\overrightarrow{AC} \times \overrightarrow{AB}$$

is orthogonal to the plane of A, B and C, so the plane could be defined by any one of the three points and this normal vector. We also know that if the plane has an inside and outside, the normal given points outwards if ABC is created to be traversed clockwise from the outside. Such priorities are used to identify the 'sense' of a surface with respect to inside and outside. In the following discussion of planes, we assume 'outward' pointing normals are used to define planes.

Suppose a plane is defined by a point A with position vector $\mathbf{a}$ and an outward unit normal vector $\hat{\mathbf{n}}$ (fig 7.11). P, with position vector $\mathbf{r}$, is a general point. O, as usual, is the origin of the coordinate system; the axes have not been drawn to simplify the figure. Using the triangle law for addition of vectors, we have

$$\mathbf{a} + \overrightarrow{AP} = \mathbf{r},$$

so

$$\overrightarrow{AP} = \mathbf{r} - \mathbf{a}.$$

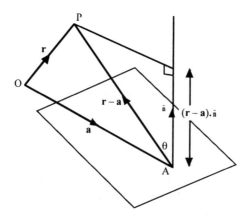

Figure 7.11 A point P tested against a plane through A with normal $\hat{n}$

The distance of P from the plane is the component of $\overrightarrow{AP}$ in the direction of $\hat{n}$, equal to the scalar product $(\mathbf{r} - \mathbf{a}) \cdot \hat{n}$. If P is on the side of the plane pointed at by $\hat{n}$ (considered the 'outside'), then the angle θ between $(\mathbf{r} - \mathbf{a})$ and $\hat{n}$ is less than 90° and $(\mathbf{r} - \mathbf{a}) \cdot \hat{n}$ is positive (as $\cos(\theta)$ is positive). If P is 'inside' the plane, $(\mathbf{r} - \mathbf{a}) \cdot \hat{n}$ is negative as $\theta > 90°$, with $(\mathbf{r} - \mathbf{a}) \cdot \hat{n}$ zero if and only if $\mathbf{r} - \mathbf{a}$ is perpendicular to $\hat{n}$, when P lies on the plane. This gives

$$(\mathbf{r} - \mathbf{a}) \cdot \hat{n} = 0$$

as an equation for the plane. If P is at the origin, $\mathbf{r} = 0$, so the sign of $-\mathbf{a} \cdot \hat{n}$ gives us a test of whether the origin is outside (positive), on (zero) or inside (negative) the plane, and its magnitude indicates the distance of the plane from the origin. This result is used in deriving a test for points against a standardized plane representation in the following section.

Representation of Planes in Standard Form

In Cartesian coordinate geometry, planes can be represented by equations of form

$$ex + fy + gz + d = 0,$$

where e, f, g and d are scalar constants representing the plane (we use these arbitrary symbols rather than the more usual a, b, c, ... to avoid confusion with other uses in this section). Multiplying this equation throughout by a constant does not change its validity, so we could, for example, multiply the equation by -1 to represent the same plane as

$$-ex - fy - gz - d = 0,$$

or by 2 to give

$$2ex + 2fy + 2gz + d = 0.$$

It would be convenient to have one 'standard' method of representing the plane. We saw above that a point P with position vector $\mathbf{r} = (x, y, z)$ lies on the plane through A, position vector $\mathbf{a} = (a_x, a_y, a_z)$, with unit normal vector $\hat{\mathbf{n}} = (n_x, n_y, n_z)$ considered 'outwards' when

$$(\mathbf{r} - \mathbf{a}).\hat{\mathbf{n}} = 0$$

Evaluating this in terms of vector components gives

$$(x - a_x, y - a_y, z - a_z).(n_x, n_y, n_z) = 0,$$

which reduces to an equation in x, y and z as

$$n_x x + n_y y + n_z z - (a_x n_x + a_y n_y + a_z n_z) = 0.$$

This is our standard form with (e, f, g) set to be the components of (n_x, n_y, n_z), the unit outward normal to the plane and

$$d = -(a_x n_x + a_y n_y + a_z n_z) = -\mathbf{a}.\hat{\mathbf{n}}$$

as the distance from the origin to the plane as shown in the last section. This distance is positive if the origin lies 'outside' the plane. Thus, we can describe the defining values of any plane in standard form as ($\hat{\mathbf{n}}$, d). The plane equation then reduces to

$$\mathbf{r}.\hat{\mathbf{n}} + d = 0.$$

Suppose, for example, we are given a plane equation

$$2x - 3y + z - 4 = 0,$$

and are told that the origin is to lie 'outside' this plane. From the analysis above, we know that the vector (2, -3, 4) is a normal to the plane. To reduce this to unit length, we find its magnitude as $\sqrt{(2^2 + (-3)^2 + 1^2)} = \sqrt{14}$. Dividing the equation by $\sqrt{14}$ makes the constants in positions e, f, g in our original form

$$ex + fy + gz + d = 0$$

represent a unit normal vector, reducing the equation to

$$\frac{2}{\sqrt{14}}x - \frac{3}{\sqrt{14}}y + \frac{1}{\sqrt{14}}z - \frac{4}{\sqrt{14}} = 0.$$

The value of 'd' in this form is negative, but we require it to be positive if the origin is to lie 'outside' the plane. Multiplying throughout by -1 gives the required change of sign, giving the standard form of the equation as

$$-\frac{2}{\sqrt{14}}x + \frac{3}{\sqrt{14}}y - \frac{1}{\sqrt{14}}z + \frac{4}{\sqrt{14}} = 0.$$

The information defining the equation is found in the surface normal vector

$$\hat{n} = \left(-\frac{2}{\sqrt{14}}, \frac{3}{\sqrt{14}}, -\frac{1}{\sqrt{14}}\right)$$

and the signed distance from the origin

$$d = \frac{4}{\sqrt{14}}.$$

Any form of plane definition can be reduced to this standard form. If we are given a point in the plane and an outward normal, the unit normal can be found and the distance parameter d calculated by the method used above. If the plane is defined as passing through three points A, B, C, the vector product

$$\overrightarrow{AC} \times \overrightarrow{AB}$$

gives a normal to the plane, so we can take any one of the three points A, B or C with this normal vector to reduce to the required form. This particular normal will point outwards if ABC is traversed clockwise when viewed from the outside. Given that any plane can be represented as ($\hat{n}$, d) and that useful information is readily available from this form, it makes sense to standardize plane definitions to this form.

The classification of a point P, position vector $\mathbf{r} = (x, y, z)$, with respect to the plane can be performed by substituting the point's position vector directly into the expression on the left-hand side of the plane equation,

$$\hat{n}.\mathbf{r} + d = n_x x + n_y y + n_z z + d.$$

If the expression is positive, zero or negative, then P lies outside, on or inside the plane, respectively.

Intersection of a Line with a Plane

This is one of the most frequently required tasks of a computer graphics system. In observing a scene, the vector line of sight is to be intersected with planes representing surfaces of objects in the scene. In creating a 'blend' of two overlapping polyhedral objects (these have plane surfaces and straight edges), the edges of each have to be intersected with the faces of the other. The method given uses the parametric equation of a line to give explicit information on the location of the point of intersection with the line through the line parameter t.

Suppose we are given a plane in standard form ($\hat{\mathbf{n}}$, d) and a line through two points A and B. The line has equation

$$\mathbf{r} = \mathbf{a} + t(\mathbf{b} - \mathbf{a}).$$

If we take $(\mathbf{b} - \mathbf{a}) = \mathbf{v}$, then the line can be defined by the (point, vector) pair $(\mathbf{a}, \mathbf{v})$, giving a point through which the line passes and a vector in the direction of the line. This method is used in the routines given later. The point of intersection (if it exists) of this line with the plane can be found by substituting components of points on the line

$$(x, y, z) = \left(a_x + t(b_x - a_x), a_y + t(b_y - a_y), a_z + t(b_z - a_z)\right)$$

into the plane equation

$$n_x x + n_y y + n_z z + d = 0,$$

giving

$$a_x n_x + a_y n_y + a_z n_z + d + t\left((b_x - a_x)n_x + (b_y - a_y)n_y + (b_z - a_z)n_z\right) = 0.$$

To find t, we divide the equation by the coefficient of t (the value that multiplies it) and do a little rearrangement. This can only be done if the division is not by zero. In other words,

$$(b_x - a_x)n_x + (b_y - a_y)n_y + (b_z - a_z)n_z \neq 0.$$

We should build a check for this into any routines; the solution for this special case is given below. Provided the divisor is not zero, then,

$$t = -\frac{a_x n_x + a_y n_y + a_z n_z + d}{(b_x - a_x)n_x + (b_y - a_y)n_y + (b_z - a_z)n_z}$$

or

$$t = \frac{a_x n_x + a_y n_y + a_z n_z + d}{(a_x - b_x)n_x + (a_y - b_y)n_y + (a_z - b_z)n_z}.$$

The latter form simply absorbs the external minus into the denominator (the lower part of the fraction) by multiplying all its brackets by -1. This can be written more simply in vector form as

$$t = \frac{\mathbf{a}.\hat{\mathbf{n}} + d}{(\mathbf{a} - \mathbf{b}).\hat{\mathbf{n}}}.$$

We should know by now to check the divisor is not zero, so we need $(\mathbf{a} - \mathbf{b}).\hat{\mathbf{n}}$ to be non-zero. This should be included as an error trap in any routine based on this method. If this occurs, $\mathbf{a} - \mathbf{b}$ is at right angles to the normal to the plane, which means the line AB is parallel to the plane's surface. Either AB never meets the plane, so there is no point of intersection, or the line lies in the plane, when there is an infinity of points of intersection as all points of the line lie in the plane. In neither case can we return a *single* value of t to give *the* solution. Either there is none, or there are too many. This problem is revisited in the following section.

Unless deliberately set up, it is unusual for the parallel case to occur, so most cases would give a value of t that can be substituted into the line equation

$$\mathbf{r} = \mathbf{a} + t(\mathbf{b} - \mathbf{a})$$

to give the single point of intersection

$$\mathbf{r} = \mathbf{a} + \frac{\mathbf{a}.\hat{\mathbf{n}} + d}{(\mathbf{a} - \mathbf{b}).\hat{\mathbf{n}}}(\mathbf{b} - \mathbf{a}).$$

It may seem tempting to 'cancel' the $(\mathbf{b} - \mathbf{a})$ in this expression with the $(\mathbf{a} - \mathbf{b})$ giving factor -1, but that is invalid. $(\mathbf{a} - \mathbf{b})$ is an intrinsic part of a scalar product in the denominator and $(\mathbf{b} - \mathbf{a})$ is a free-standing vector at the end of the expression. Remembering that the scalar or dot product yields a scalar quantity (just a number), the rightmost expression in the equation is of form 'scalar times vector divided by scalar', which is essentially a vector quantity to match other values in the equation. If the $(\mathbf{a} - \mathbf{b})$ were erroneously 'cancelled' with the $(\mathbf{b} - \mathbf{a})$, this would leave the rightmost expression as the incorrect form

$$-\frac{\mathbf{a}.\hat{\mathbf{n}} + d}{\hat{\mathbf{n}}},$$

'scalar divided by vector'. We have no way of dealing with division of a scalar by a vector. Indeed, neither does vector algebra – it is not a valid operation. As we have created new rules and concepts apparently at will so far, readers may be surprised by this. The real reason for not having a definition for division by a vector is that there is no real need for it. It has no practical uses. This is not a valid operation.

Now we return to the value of t found in all 'usual' cases, when the line intersects the plane in a single point,

$$t = \frac{\mathbf{a}.\hat{\mathbf{n}} + d}{(\mathbf{a} - \mathbf{b}).\hat{\mathbf{n}}}.$$

The value of t tells us if the point of intersection lies 'behind' A ($t < 0$), between A and B ($0 < t < 1$) or 'in front of' B ($t > 1$). This information is important in several cases. If AB is used as a line of sight for viewing with an observer placed at A looking towards B, $t < 0$ indicates that the plane lies behind the viewer, so it should not be visible. If several planes are intersected by AB, the plane that gives the lowest positive value of t is the plane visible along AB, as all others must lie beyond it from the point of view of the observer. If AB is an edge of one object tested against a plane face of another, the edge intersects the face if $0 \leq t \leq 1$).

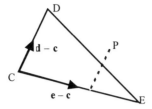

Figure 7.12 Comparison of a point P with a triangle CDE

Inclusion of a Point in a Triangle

Suppose we have a point P in the same plane as three other points C, D and E forming the vertices of a triangle. How can we tell if P lies inside the triangle CDE (fig 7.12)? This would be useful if P is found from a line/plane intersection as above, and we need to find if it lies inside an object with bounding lines given by the edges of CDE. This would enable us to decide if a triangular face of an object is potentially visible along a particular line of sight, for example. If **c**, **d** and **e** are the position vectors of C, D and E respectively, two of the edges of the triangle have vector representations (**d** − **c**) and (**e** − **c**). As shown in fig 7.12, we draw a line through P parallel to CD so that it cuts the edge CE (possibly extended). This shows that we can represent the position vector **r** of P uniquely as

$$\mathbf{r} = \mathbf{c} + \lambda(\mathbf{e} - \mathbf{c}) + \mu(\mathbf{d} - \mathbf{c}),$$

where λ and μ are scalar values. This can be interpreted as a way to reach P. Starting at C (position vector **c**) move a fraction λ along CE, and then move parallel to CD for a fraction μ of the length of CD. Imagine this figure to be distorted by rotation, shearing and scaling so that the points C, D and E are transformed to C' (0, 0), D' (0, 1) and E' (1, 0) (fig 7.13). In this version, the coordinates of the point P' are (λ, μ). This shows more clearly that P' will lie inside the triangle C'D'E' (and correspondingly P will lie inside the triangle CDE) if and only if *all* three conditions

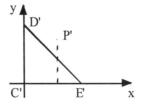

Figure 7.13 Affine transformation of triangle CDE to fit the coordinate axes

$$\lambda \geq 0,$$
$$\mu \geq 0,$$
$$\lambda + \mu \leq 1$$

hold. The last condition is devised from the equation for the line through D' and E' as
$$x + y = 1.$$

Using the methods developed in chapter 3 for inequalities in a 2D space, points lying on or below this line satisfy
$$x + y \leq 1.$$

We see that $(x, y) = (0, 0)$ is below the line, and $(0, 0)$ makes this inequality true, so all points below the line must satisfy the inequality.

To find λ and μ, the position vector **r** of P can be broken into its three components as
$$x = c_x + \lambda(e_x - c_x) + \mu(d_x - c_x),$$
$$y = c_y + \lambda(e_y - c_y) + \mu(d_y - c_y),$$
$$z = c_z + \lambda(e_z - c_z) + \mu(d_z - c_z).$$

Only two of these are needed to solve for λ and μ, the third redundant equation represents the condition that P lies in the plane of C, D and E – the assumption made at the start of this section. Many methods are available for solving such simultaneous equations. We simply state the result of a standard determinant method applied to the first two of these equations, giving

$$\lambda = \frac{(x - c_x)(d_y - c_y) - (y - c_y)(d_x - c_x)}{(e_x - c_x)(d_y - c_y) - (e_y - c_y)(d_x - c_x)},$$

$$\mu = -\frac{(x - c_x)(e_y - c_y) - (y - c_y)(e_x - c_x)}{(e_x - c_x)(d_y - c_y) - (e_y - c_y)(d_x - c_x)},$$

provided the divisor is non-zero. It is zero if C, D and E are in a straight line, so there would be no triangle and no plane defined, making the problem intractable. If needed, substitution of these values into the third equation checks the validity of the original assertion that P lies in the plane of C, D and E.

We can now test if a line through points A and B passes through the interior or boundary of a triangle CDE:

- use $(\mathbf{d} - \mathbf{c}) \times (\mathbf{e} - \mathbf{c}) = \mathbf{n}$ to give a normal to the triangle, hence find standard plane form ($\hat{\mathbf{n}}, d$) if it is not already known;
- find the intersection point of AB with the plane, if it exists, as shown above;
- find values of λ and μ using the formulae given above;
- the point of intersection lies inside or on the boundary of the triangle if *all* the conditions $\lambda \geq 0, \mu \geq 0, \lambda + \mu \leq 1$ are satisfied.

7. Vectors: Descriptions of Spatial Relationships

This sequence also gives us as incidentals the actual point of intersection, the outward unit normal to the triangle and the standard form of the plane of the triangle. These were not explicitly required, but could be useful for other purposes. If the second element above shows that no single intersection point exists, it may be because the line AB lies in the same plane as the triangle CDE. The line may, in this case, pass through the triangle, crossing its boundary at edges and/or vertices. If this test is being done for the purpose of ray tracing and such crossing exists, it may appear that we should return the *nearest* point of intersection with the triangle, which would be at an edge or vertex. In most systems, however, this is unnecessary, as the ray or line AB will also intersect with the edge of a surface facet that lies alongside that currently under investigation. Hence, the return of 'no intersection' should be a suitable result from the test.

Reflected and Refracted Rays

A vector line equation can be used to represent a ray of light. Such rays can be reflected from shiny surfaces or refracted through transparent surfaces. Both these circumstances can be modelled using vector methods. Suppose we wish to reflect a ray represented by the line through A and B from a surface with unit normal vector $\hat{\mathbf{n}}$ (fig 7.14). The 'ray in' or incident ray has equation

$$\mathbf{r} = \mathbf{a} + t(\mathbf{b} - \mathbf{a}),$$

its directional component being given by the vector $(\mathbf{b} - \mathbf{a})$. Suppose $\hat{\mathbf{u}}$ is a unit vector in this direction. The unit versions of the incident ray $\hat{\mathbf{u}}$, reflected ray $\hat{\mathbf{v}}$ and the surface normal $\hat{\mathbf{n}}$ all lie in the same plane, and the angle θ between the incident ray and the normal as shown in fig 7.15 is the same as that between the reflected ray and the normal. The projections of $\hat{\mathbf{u}}$ and $\hat{\mathbf{v}}$ onto $\hat{\mathbf{n}}$ are both equal to $-\cos(\theta)$, so $\cos(\theta) = -\hat{\mathbf{u}}.\hat{\mathbf{n}}$ as the magnitudes of both unit vectors are 1.

Figure 7.16 shows the vectors of fig 7.15 rearranged in space. Again, remember that we are allowed to do this as vectors are independent of position. From fig 7.16,

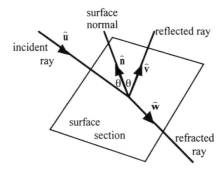

Figure 7.14 Reflected and refracted rays

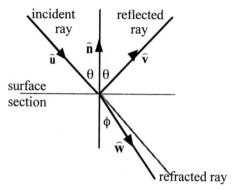

Figure 7.15 Cross-section through fig 7.14 in the plane of the light rays

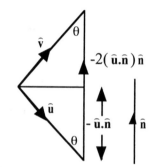

Figure 7.16 Finding the reflected ray

the unit reflected ray direction is given by

$$\hat{v} = \hat{u} - (2\hat{u}.\hat{n})\hat{n}.$$

To establish this, draw $\hat{u}$ and $\hat{v}$ leaving the same point (fig 7.16); the projection of $\hat{u}$ onto $\hat{n}$ is then seen to have length $-\hat{u}.\hat{n}$. The minus term is used as the two vectors have angle greater than 90° between them (in fig 7.16, $\hat{u}$ points broadly downwards, $\hat{n}$ points upwards). The vertical vector joining the tail of $\hat{u}$ to the tail of $\hat{v}$ in this configuration has length $-2\hat{u}.\hat{n}$ in the direction of $\hat{n}$, so it is given as $-2(\hat{u}.\hat{n})\hat{n}$. The result given above for the reflected ray can now be found from fig 7.16 using the triangle law for addition of vectors.

If the surface is partly transparent, the refracted ray passes through the surface, leaving it at an angle ϕ to $-\hat{n}$ (fig 7.15). The physical law of refraction states that

$$\sin(\theta) = k \sin(\phi),$$

where k is the 'index of refraction' between the two light transmitting media to each side of the surface (typically we would be moving from air to glass, or vice versa). If $k > 1$, this means that the ray is refracted towards the surface normal, which typically happens when moving from less to more dense material, such as from air to glass or water. The reverse direction would have refractive index k^{-1}, just as though the ray were retraced. Typical refractive indices are about 1.333 from air to water, 1.5 from air to glass, 1.59 from air to polystyrene, although these vary slightly with the temperature, the wavelength of the light source and the exact specification of the material concerned. At a surface that is reflective and refractive, the incident light is split, part of it being reflected, the remainder being refracted. The percentage of light reflected depends also on characteristics of the light, materials at each side of the surface values, and temperature. For this property, the angle of incidence θ has a major influence; the percentage of light reflected takes values of about 10% for values of θ less than 50°, increasing to about 15% at 60° and 25% at 70° for polished glass. This phenomenon is often ignored in computer graphic simulation, with a blanket value of about 10% to 15% used. Refraction is also ignored in many computer graphics images – rays passing through transparent objects are allowed to do so without deviation.

From fig 7.15, the component of the unit refractive ray direction $\hat{w}$ in the direction of $-\hat{n}$ is $\cos(\phi)$ and the component along the direction of the surface is $\sin(\phi)$. As $\hat{u}$ and $\hat{n}$ are unit vectors, we can use the vector product of $\hat{u}$ and $\hat{n}$ to find $\sin(\theta)$ as its magnitude:

$$\sin(\theta) = |\hat{u} \times \hat{n}|,$$

so $$\sin(\phi) = k|\hat{u} \times \hat{n}|.$$

We previously noted the trigonometric identity valid for any angle,

$$\cos^2(\phi) + \sin^2(\phi) = 1.$$

This enables us to find $\cos(\phi)$ from $\sin(\phi)$ as

$$\cos(\phi) = \sqrt{\{1 - \sin^2(\phi)\}},$$

so both components of $\hat{w}$ can be found. We can find $\hat{w}$ itself by multiplying the components by their directions, and adding these results as vectors. The normal into the surface in the direction of the ray is $-\hat{n}$ so the downward vertical component of $\hat{w}$ in fig 7.15 is

$$\cos(\phi) = \sqrt{\{1 - \sin^2(\phi)\}} = \sqrt{\{1 - k^2|\hat{u} \times \hat{n}|^2\}}$$

Provided θ is not zero (in which case, ϕ would also be zero and the incident, reflected and refracted rays would all be at right angles to the surface), the component direction along the surface can be found from a unit vector $\hat{s}$ in the direction of $(\hat{u} + \hat{v})$, where $\hat{u}$ is the known incident ray and $\hat{v}$ has been found as the reflected ray. So,

$$\hat{s} = (\hat{u} + \hat{v})/|\hat{u} + \hat{v}|$$

and $$\hat{w} = -\hat{n}\cos(\phi) + \hat{s}\sin(\phi) = = -\hat{n}\sqrt{\{1 - k^2|\hat{u} \times \hat{n}|^2\}} + \hat{s}k|\hat{u} \times \hat{n}|.$$

This may seem a relatively complicated process, but each of the component activities in its evaluation is easily achieved from methods indicated in the sections above, and all are achievable using computer methods.

We showed here the evaluation of a unit vector in the direction of the sum of two vectors, $\hat{s} = (\hat{u} + \hat{v})/|\hat{u} + \hat{v}|$. A similar process is widely used in the shading of polyhedral objects, where several faces impinge on a given vertex. To create the effect of a smooth surface, the surface normal at the vertex is calculated as the average of the normals of all faces surrounding it. Suppose these faces have unit normals $\hat{u}_1, \hat{u}_2, \hat{u}_3, \ldots$ and the normal at the vertex is designated as $\hat{u}$. Then,

$$\hat{u} = (\hat{u}_1 + \hat{u}_2 + \hat{u}_3 + \ldots)/|\hat{u}_1 + \hat{u}_2 + \hat{u}_3 + \ldots|.$$

There is no need to divide by the number of faces to produce an average; the process of 'normalization', or reducing to unit length, takes care of that by dividing by the total length of the sum of individual face normals. When dividing by a value, we are usually careful to specify that the value is non-zero. In this case, if the divisor is zero, the object itself would not be a valid 3D object. These methods are put to practical use in Chapter 8 on geometric modelling, the generation of descriptions of 3D objects for computer manipulation, and in discussing smooth shading, an important aspect of computer image generation, in Chapter 10.

Distance between Two Skew Lines

It is sometimes useful to check how close two lines get to each other. Suppose two lines are defined as (point, vector) pairs $(\mathbf{a}, \mathbf{v})$ and $(\mathbf{b}, \mathbf{w})$. This means that the first line passes through point A with position vector $\mathbf{a}$ and vector $\mathbf{v}$ lies in the direction of the line; $\mathbf{v}$ may or may not be of unit length. If the two lines are not parallel, the vector $\mathbf{u} = \mathbf{v} \times \mathbf{w}$ is non-zero and is at right angles or orthogonal to both lines. A unit vector $\hat{u}$ in this direction can be defined as

$$\hat{u} = \frac{\mathbf{v} \times \mathbf{w}}{|\mathbf{v} \times \mathbf{w}|}.$$

The projection of AB onto this direction gives the minimum distance between the two lines (fig 7.17). This is given as

$$d = |(\mathbf{b} - \mathbf{a}) \cdot \hat{u}|,$$

the modulus being taken to ensure the distance is non-negative. If d is zero, it may be useful to find the intersection point of the two lines. We need the point with position vector $\mathbf{r}$ satisfying both line equations,

$$\mathbf{r} = \mathbf{a} + t\mathbf{v}$$
$$\mathbf{r} = \mathbf{b} + s\mathbf{w}$$

7. Vectors: Descriptions of Spatial Relationships

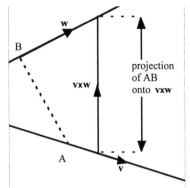

Figure 7.17 The distance between two skew lines as a projection

(the parameters t and s are likely to be different for each line at the point of intersection. Thus, we need to find t (or s) such that

$$\mathbf{a} + t\mathbf{v} = \mathbf{b} + s\mathbf{w},$$

which is simply re-expressed as

$$(\mathbf{a} - \mathbf{b}) + t\mathbf{v} - s\mathbf{w} = 0.$$

This one vector equation is equivalent to three equations, one for each of the coordinates of the vectors,

$$(a_x - b_x) + tv_x - sw_x = 0,$$
$$(a_y - b_y) + tv_y - sw_y = 0,$$
$$(a_z - b_z) + tv_z - sw_z = 0.$$

Using the numerical analysis technique we 'borrowed' previously, we can eliminate s from the first two of these equations to give t as

$$t = \frac{(b_x - a_x)w_y - (b_y - a_y)w_x}{v_x w_y - v_y w_x}$$
$$= \frac{(b_x - a_x)w_y - (b_y - a_y)w_x}{u_z},$$

where u_z is the z coordinate of vector $\mathbf{u} = \mathbf{v} \times \mathbf{w}$ normal to the surface (this is the non-normalized equivalent of the unit surface normal $\hat{\mathbf{u}}$ as defined above). For this formula to work, we must have $u_z \neq 0$. If $u_z = 0$, a search through the coordinates of $\mathbf{u}$ will find at least one non-zero value, as $\mathbf{u}$ itself is only zero if the two lines are coincident, when all points of both lines are points of intersection. If $u_z = 0$, a non-zero alternative formula for t must then be used, either

$$t = \frac{(b_y - a_y)w_z - (b_z - a_z)w_y}{u_x},$$

or $\quad t = \dfrac{(b_z - a_z)w_x - (b_x - a_x)w_z}{u_y}$,

the suffixes being cycled from x to y to z to x, and so on, in changing these formulae. This enables the required point of intersection to be found as

$\mathbf{r} = \mathbf{a} + t\mathbf{v}$.

If s is also needed, it can be found by substituting this point into the second line equation to give

$\mathbf{r} = \mathbf{b} + s\mathbf{w}$.

Separating the components of this equation gives three equations of form

$r_i = b_i + sw_i$,

where i stands for any one of the suffixes x, y or z. Thus,

$s = \dfrac{r_i - b_i}{w_i}$,

where w_i is any non-zero coordinate of $\mathbf{w}$.

Now we return to consider the distance between two parallel lines (fig 7.18); this case is identified when $\mathbf{v} \times \mathbf{w}$ has magnitude zero (remember that the vector product formula involves the sine of the angle between the two vectors; this is zero when the angle between the two vectors is zero). In this case, there is no non-zero $\mathbf{u}$ to use in the distance formula. Now the shortest distance between the two lines is AB $\sin(\theta)$, where θ is the angle between the line AB and the line direction vector $\mathbf{v}$ (fig. 7.18). The definition of vector (cross) product gives us the length of $(\mathbf{b} - \mathbf{a}) \times \mathbf{v}$ as

$|(\mathbf{b} - \mathbf{a}) \times \mathbf{v}| = AB |\mathbf{v}| \sin(\theta)$.

So when the lines are parallel, the required distance is given as

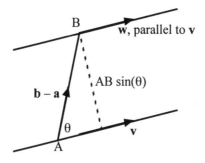

Figure 7.18 The distance between two parallel lines

$$d = \frac{|(\mathbf{b}-\mathbf{a})\mathbf{x}\mathbf{v}|}{|\mathbf{v}|}.$$

This method has a number of potential uses in computer graphics. For example, the two lines may be the axes of two cylinders. If the distance d is less than the sum of the radii of the cylinders, the cylinders intersect. If one line is a view direction and the other represents the axis of a cylinder, the cylinder is potentially visible along this line if d is less than the radius of the cylinder.

Intersection of Two Planes

Two planes generally intersect in a line – this does not happen when the planes are parallel. It is sometimes necessary to identify such an intersection line in computer graphics, for example in finding a blend of two plane faced objects. Suppose we have two planes in standard form (outward unit normal, distance from the origin) as ($\hat{\mathbf{n}}$, d) and ($\hat{\mathbf{m}}$, e). If the scalar product of the two plane normals $\hat{\mathbf{n}}\mathbf{x}\hat{\mathbf{m}}$ is zero, the two planes are parallel. In that case, when d ≠ e, the planes are distinct and never intersect. When d = e, the planes are coincident and they intersect at every possible point. Both cases do not have a *single* solution, both must be identified through error traps.

When ($\hat{\mathbf{n}}$, d) and ($\hat{\mathbf{m}}$, e) represent two different non-parallel planes, they intersect in a straight line. We seek a representation of that line as (**a**, **v**), where **a** is the position vector of a point A on the line and **v** is a vector in the direction of the line. We can simply take

$$\mathbf{v} = \hat{\mathbf{n}}\mathbf{x}\hat{\mathbf{m}},$$

as **v** is perpendicular to both surface normals (fig 7.19). Therefore the vector direction **v** 'lies' in both planes so it must be in the same direction as the common line of both planes. Now we need to find any point A on the intersection line; this means A must lie in both planes. If **r** = (x, y, z) is the position vector of such a point A, it must satisfy both plane equations,

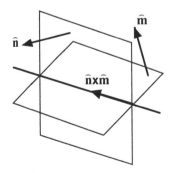

Figure 7.19 Direction of the line at the intersection of two planes

and
$$\mathbf{r}.\hat{\mathbf{n}} + d = 0$$
$$\mathbf{r}.\hat{\mathbf{m}} + e = 0.$$

These expand to

and
$$n_x x + n_y y + n_z z + d = 0$$
$$m_x x + m_y y + m_z z + e = 0.$$

There are as many combinations of x, y and z that satisfy these two equations as there are points on the line. The problem is under-defined: as we have three unknowns, but only two defining equations or constraints, we still have one degree of freedom. We can, almost arbitrarily, choose the value of one of x, y or z. If $v_x = 0$, we can take x = 0, as there is no reason why we shouldn't use an easy option. Then,

and
$$n_y y + n_z z + d = 0$$
$$m_y y + m_z z + e = 0.$$

By the standard determinant method that we have borrowed before, this gives

and
$$x = 0,$$
$$y = -\frac{dm_z - en_z}{n_y m_z - n_z m_y} = -\frac{dm_z - en_z}{v_x},$$
$$z = \frac{dm_y - en_y}{n_y m_z - n_z m_y} = \frac{dm_y - en_y}{v_x}.$$

The condition that the x component of $\mathbf{v} = \hat{\mathbf{n}} \mathbf{x} \hat{\mathbf{m}}$, found as $v_x = (n_y m_z - n_z m_y)$, is not zero was invoked before using this method, so we cannot have a zero divide error. If y can be found in this way, a value for z can generally be found by solving one of the original equations, such as

$$n_y y + n_z z + d = 0$$

to give

$$z = -(n_y y + d)/n_y,$$

saving one multiplication. However, this is only permissible if n_y is not zero, so a further check of this would have to be performed and the time benefit would depend on the particular system's operating characteristics.

If v_x is zero, the line has no points with x = 0, so we must seek another arbitrary defining point. At least one of the components of $\mathbf{v}$ is non-zero, as we have checked that $\mathbf{v} \neq 0$ as a condition for the existence of a well defined intersection of the two planes. In cases when $v_y \neq 0$, we can find an arbitrary point on the line by setting y = 0 to give

7. Vectors: Descriptions of Spatial Relationships

$$x = \frac{dm_z - en_z}{n_z m_x - n_x m_z} = \frac{dm_z - en_z}{v_y},$$

$$y = 0,$$

and $$z = -\frac{dm_x - en_x}{n_z m_x - n_x m_z} = -\frac{dm_x - en_x}{v_y}.$$

If both $v_y = 0$ and $v_y = 0$, we must have $v_z \neq 0$, when we can set $z = 0$ to give

$$x = -\frac{dm_y - en_y}{n_x m_y - n_y m_x} = -\frac{dm_y - en_y}{v_z},$$

$$y = \frac{dm_x - en_x}{n_x m_y - n_y m_x} = \frac{dm_x - en_x}{v_z},$$

and $$z = 0.$$

One of these three cases must be valid, as we have established that $\mathbf{v} \neq 0$ for a valid line to exist. The first case will almost always work, but error checks and compensatory action must be included in a well-defined system This finicky method has given us a useful tool for computer graphics. Checks on intersections of planes are needed in 3D clipping of objects against viewing regions (chapter 10) or in creating the Boolean unions of two objects with plane boundaries (chapter 8).

Summary

This chapter has presented a number of results from vector algebra that are all of direct value in computer graphics systems. Applications include selection from a list of objects of the one visible object that is visible along a line of sight, checking for intersections between objects, and creating 'blends' of objects when such overlaps occur. This is merely a selection of results from vector algebra that are potentially useful, but they should be enough to persuade readers of the power of vectors and their importance in resolving the sorts of spatial queries needed in computer graphics.

8. Geometric Modelling and Fractals: Building Descriptions of Objects

Introduction

From this chapter on, the book concentrates on the methods of computer graphics rather than the mathematical tools previously described, although the precise division is of necessity blurred. A few new mathematical topics will be introduced, but most of the remainder concentrates on applications of what has been introduced previously. There have been occasional hints in previous chapters as to how mathematical concepts can be used. This is where these start coming home to roost.

The subject of geometric modelling is concerned with the creation, storage and manipulation of models of objects for depiction by computers, particularly concentrating on 3D objects. There are several methods for organizing such information, each having particularly useful properties for different forms of specialization. We consider voxel, CSG, B-Rep and isosurface modellers, as well as giving an introduction to fractals as object modelling systems. Most of these use methods from the computer science topic of 'data structures' as well as geometry, for spatial location and orientation, and topology, to define the way spatial structures 'hang together'. We must consider both computer science and mathematical aspects to give a picture of the ways in which computer graphic models are stored and manipulated.

Data Structures

Most computer languages allow users to define data structures in which particular forms of information can be stored. Standard structures for storing particular types of number or for alphabetic characters are provided, but extensions to these can be user defined. For example, we have already seen how an ordered triplet of three numbers (x, y, z) represents the location of a point in a 3D Cartesian system; a user-defined type 'point3D' can be created to hold such information. Functions can then be defined using 'point3D' values as parameters, or inputs, to generate standard results such as vector and scalar products or distances between points.

Arrays are usually supplied as standard data structures. Numbers are held in a fixed-length list. Individual items are identifiable through an array index, a natural number defining the position of a value within the list. If, for example, the character 'a' is defined to hold a list of 10 values in the language C, then a[0], a[1], ..., a[9] represent these individual values. By using two array indices, values can be held in a matrix format. Array element m[1][3] can be considered equivalent to element $m_{1,3}$, appearing in the first row and third column of a matrix. Unlike some other languages, C begins its suffix counts at zero rather than one, so users must decide whether to waste memory by ignoring suffices with value zero, or to use them, in

Figure 8.1 Representation of a linked list 'L'

which case $m_{1,3}$ would be held in array element m[0][2]. Arrays are stored in contiguous memory locations, so tracking along array values is equivalent to moving along successive memory locations within the computer's physical structure.

Various forms of list can be defined using 'pointer' systems. Most procedural computer languages identify memory locations using pointers. This means that ordered lists of values can be held in non-contiguous memory, each value holding a pointer to the memory location of the next object in the list. The linked list data structure has a header that gives the list a name and points at the first value. Subsequent values are pointed at by their predecessors; there is usually a 'null' pointer (that points at nothing) indicating the end of the list. In fig 8.1, the boxes with arrows leaving them represent pointers; the list content is represented by boxes containing 'data'. A doubly linked list also has pointers that reverse this direction, so the list can be easily traversed from head to tail or from tail to head. If required, the tail of a list can also be connected to the header, to create a cyclic list. A stack is a linked list operated in 'last in, first out mode'. New elements are 'pushed' onto the top of the stack, and 'popped' off the top to retrieve them.

If a value is to be inserted into a list held in an array, each element subsequent to the inserted object must be shuffled along the array to maintain the property of holding values in contiguous memory locations. However, insertion into a linked list is easier, as the new object can be held in any memory location if the list pointers are properly adjusted (fig 8.2).

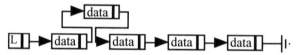
Figure 8.2 Insertion of an element into linked list 'L'

Similarly, deletion of an object from a linked list can be achieved by redirecting pointers around the undesired object (fig 8.3). For these reasons, the linked list is a very important method for holding values describing object data in geometric modelling; there are often relatively long lists of such objects.

Abstract data typing is a popular method in computer science. It is also useful in geometric modelling as a linked list structure can be used to hold a variety of element types, such as vertices or edges of an object. In this method, the list has as content a list of pointers, so the data is held away from the list's own structure, each 'data' section in the examples above being replaced by a pointer to the actual location of the data. Thus, different types of list contents can be readily created and processed, with routines for list manipulation remaining the same (fig 8.4).

Figure 8.3 Deletion of an element from linked list 'L'

8. Geometric Modelling and Fractals: Building Descriptions of Objects

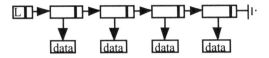

Figure 8.4 Linked list 'L' using abstract data typing

The tree data structure is also important for holding hierarchical data in geometric modelling. As with a real tree, the structure stems from a 'root' element, with branching processes emanating from this base, although for descriptive reasons computer trees are normally depicted to 'hang' below a root element. Figure 8.5 depicts a 'binary tree' data structure. For simplicity, null pointers are not shown and the illustration does not use abstract data typing, although the method is well suited to trees. Unless the tree is degenerate and contains only one data node, the root is a 'parent' node, in that it has at least one successor or 'child node'. A parent node in a *binary* tree has at most *two* child nodes (or children). Nodes that have no children are 'leaf nodes'; they appear at the ends of branches. Metaphors of family and tree are mixed, but the descriptions are readily remembered. The tree of nodes that 'hangs' from the left pointer of an element is its 'left sub-tree', that from its right pointer is its 'right sub-tree'. The tree of fig 8.5 has four levels of data, not all of which are full. As can be seen, leaf nodes can appear at any level of the tree.

Identifying all nodes of a tree is known as 'traversing' the tree. For a binary tree, this can be done by 'pre-order traversal'. Starting at the root node, travel down taking the left sub-branch wherever possible until a leaf node is reached. If a parent node has a null left pointer, then take the right pointer. When a leaf is reached, travel back up the tree. If arriving upwards into a node from a left pointer, and the node has a right pointer, travel down this right pointer and repeat the previous routine down leftmost pointers when possible until a leaf node is reached. Otherwise, continue upwards. Stop when the root is revisited from its right pointer. All nodes in the tree will be visited at least once. Suppose we use the data in nodes only when leaving them in an upward direction. Try tracing this route out with your

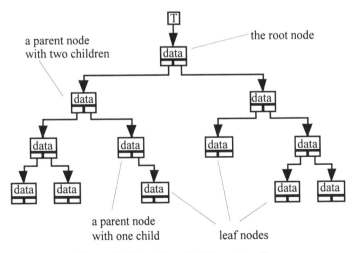

Figure 8.5 Depiction of a binary tree T

finger on the tree of fig 8.5. If the levels are counted as 1, 2, 3, ... from the top (with the root as 1), and nodes from left to right are labelled a, b, c, d, ... within each level, this will visit nodes (1a), (2a), (3a), 4a, (3a), 4b, 3a, (2a), (3b), 4c, 3b, 2a, (1a), (2b), 3c, (2b), (3d), 4d, (3d), 4e, 3d, 2b, 1a in order. Where nodes are bracketed, their data content is not used. 'Post-order traversal' travels down the rightmost branches first, replacing left and right in the above description.

A binary tree is shown in fig 8.5, but trees can be defined to have an arbitrary number of possible children per parent node and parsing rules developed to follow downward paths in preferred node orders. This enables parts of an object to 'inherit' the properties of their parents. For example, in creating a model of a car, structures describing the four wheels can be considered as children of a structure representing the car body. If the car's position is changed, the same movement can be imposed on the wheels, which may then be free to move by rotation relative to this new position. The 'octree', with a maximum of eight children per parent, can be used in describing voxel-based geometric models, as discussed below.

Geometric Modelling Systems

There are often conflicting demands on geometric modelling systems, such as the need to be

- precise,
- compact to store,
- easy to generate,
- quick to depict,
- of guaranteed validity.

Due to these and other competing requirements, many types of data structure are used to store geometric models. The requirement of rapid depiction, for example, means that redundant data, which could be calculated from necessary stored values, needs to be readily available. This means it is stored rather than calculated every time it is needed, so extra memory is used to save on speed. The last bullet point, that of guaranteeing validity, makes sure that no 'devil's tuning forks' (fig 8.6) or similar objects are created. Those familiar with the work of artist Mauritz Cornelis Escher will have seen several such cases where the 2D representation of 3D geometry makes sense in the small scale, but just does not hang together for the overall

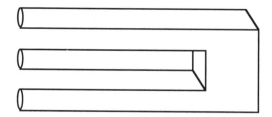

Figure 8.6 A 'devil's tuning fork', an impossible object

8. Geometric Modelling and Fractals: Building Descriptions of Objects

image. Escher's *Belvedere* (1958) is built in an 'impossible style' (the artist's own description[1]): pillars whose bases are at the front have their tops at the back. By creating erroneous links between defining sections of objects, topological errors can create 'non-manifold' objects that could not be manufactured in any real 3D sense.

In using 3D modelling systems, users are often unaware of the underlying method, but it is informative to know how the basic methods work. A description of several methods is given in the remainder of this chapter, taking into account criteria such as those listed above. The multiplicity of methods reflects a multiplicity of needs.

Voxel Modelling Methods

A voxel model characterizes the space in which an object exists, rather than directly addressing the object itself. The method was first thought of in the 1960s, but memory limitations meant that it was not used extensively until the 1990s, when the ready availability of relatively cheap and extensive memory made it possible to implement the memory-hungry, but conceptually simple, method. It has been particularly used for medical modelling.

Images in computer graphics are shown on a 2D display space divided into pixels ('pixel' is a near acronym for 'picture element'): small rectangular regions that are given the necessary colour intensities to define a picture. In voxel modelling, the region in which the scene is composed is divided into small voxels or volume elements, usually cubes, each one tagged with information as to whether it lies inside or outside the object modelled. In more precise systems, data structures that hold information on density, colour and other characteristics of the object that occupies its space may be defined for each voxel. Figure 8.7 shows a simple voxel model, which indicates the simplicity of the method and some of its drawbacks. In order to avoid a boxy representation, a very fine subdivision of space is needed. This kind of level of detail requires large amounts of computer memory. Suppose an object is stored in a three-dimensional array of Boolean variables. In the language C, if array A holds information on spatial occupancy of an object, the element A[i][j][k]

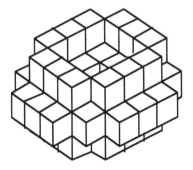

Figure 8.7 A simple voxel model

[1] M.C. Escher (1992) *The Graphic Work*, Taschen, Köln, p. 15.

contains '1' if the cube in position (i, j, k) in the spatial grid lies inside the object, '0' if that cube lies outside the object. For a grid of 100 by 100 by 100 (hardly very precise), 100^3 = 1 000 000 cubes are represented, each by one Boolean value, indicating presence (1) or absence (0) of the object. A Boolean value can be stored in one bit, so the memory needed is a minimum of one million bits. Doubling the linear precision means increasing the memory requirements by a factor of 2^3 = 8. If further information is to be stored, such as colour, density or type of any occupying material, this figure will rise. Even relatively imprecise models require high memory allocations.

One benefit of voxel-based methods is simplicity. If a cube is occupied, it is easy to check if its neighbours are also occupied as they lie adjacent to the cube in question, and their memory locations are easy to find.

'Octree' data structures can improve memory performance at the cost of compromising this spatial simplicity. A node has a maximum of eight children per parent node. Assume the space under consideration is cubic in shape. This is arbitrary, but conceptually simple; other shapes could be dealt with using a little ingenuity. If the space is completely empty or completely full, we know all about it and the root of the octree is encoded as one or zero to indicate these possibilities. In this case, we have a degenerate octree consisting of just one node, the root. This is the only case when the root is itself a leaf node. Otherwise, space is subdivided repeatedly into eight half cubes, slicing at midpoints of cube edges in each of the major axis directions. At any stage, if a cube is identified as being fully occupied or completely empty, it is tagged as such and becomes a leaf node of the octree; it is not itself subdivided. If a cube is only partly occupied, it is further subdivided, its eight component sub-cubes being addressed as child nodes in the octree structure. This is continued until all cubes are either fully occupied or empty, or until a threshold size is achieved. As many of the cubes held are larger than the basic size, it should be clear that there is potential for considerable memory saving in the method. However, operations that are simple in array-based voxel models, such as identifying the status of a neighbour region, can be difficult in an octree, requiring quite complex parsing rules.

Further enhancement involves the storage of limited polygonal surface information at boundary cubes, the leaf nodes of an octree model. In such cases, subdivision continues until the polyhedra contained in all cubes reach a limited level of complexity in terms of the numbers of vertices, edges and faces they contain. This can reduce the boxiness of the interpretation by replacing depicted voxels by small sections of plane surface. With current access to relatively cheap memory, many systems developers today opt for throwing more memory at the conceptually simpler voxel method.

Generation of 3D models from medical scan information is a major application. Scans are typically taken transversely across the body, giving information on the density of structures within the scan plane through the level of grey in the scan image. Lighter parts indicate dense tissue, such as bone, with organs appearing progressively darker as they are less dense. Images within each successive scan are 'pixellated' to indicate densities of tissue in scan sub squares (fig 8.8). Each sub square represents a small cube of tissue; a composite 3D model is created by piling neighbouring scans of cubes one on top of the other. Models can be interrogated by a variety of tools, often implemented on high-powered workstations in real time to assist direct surgical investigation. Depiction at various thresholds of tissue density

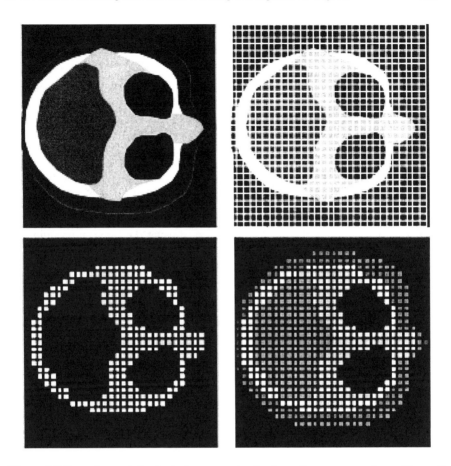

Figure 8.8 From the top left, clockwise: conceptualization of a scan; superimposition of a grid on the scan; pixels of uniform colour in equivalent grid positions; thresholding can produce images of tissue at selected densities, for example bone and cartilage.

can depict layers of surface flesh, muscle, more dense organs and bone structure as 3D models that can be viewed from various angles. Also, the object can be 'sliced' at chosen locations and at chosen angles, which would be impossible from the normal possible range of transverse scan directions. Thresholding can give precise knowledge of the location of tumours in, for example, delicate brain tissue. This is used automatically to control the motion of high-powered X-ray machines to destroy the tumour although causing as little damage as possible to surrounding tissue.

Voxel modelling systems have a limitless range of objects, usually in approximate form. The exceptions are rare objects whose boundaries happen to coincide with voxel boundaries. Voxel systems are relatively easy to program, to interrogate and to depict, but have large memory requirements. They have become increasingly popular when the costs of employing programmers have increased compared to a relative decrease in the cost of memory. Their use is restricted to relatively limited applications areas where high-quality 'glossy' illustration is not an issue.

Constructive Solid Geometry (CSG)

We consider next the method of constructive solid geometry (CSG). Users have access to a range of primitives; these are standard 3D shapes such as cubes, spheres, cones, cylinders and sometimes tori and other geometric shapes. Users can select those needed to make up their required scene, giving each numerical parameters to identify location, size and orientation. The standard affine transformations of translation, scaling and rotation are useful in this context. For example, by scaling a standard cube by different amounts in each axis direction, a rectangular block of any size can be created. By scaling a sphere, ellipsoids can be modelled. A rich set of primitives increases the modelling possibilities; these are further enhanced by allowing various forms of combination of existing objects using methods borrowed from Boolean algebra.

Objects that comprise the scene are constructed using the Boolean operations of union, intersection and difference (fig 8.9). The union of two objects A and B is the set of points that belong to A or B or to both A and B. It may be referred to as 'A ∪ B','A or B' or 'A + B' in different texts. The intersection of A and B is the set of points that is contained in both A and B, known as 'A ∩ B', 'A **and** B' or 'A.B'. Union is known as a 'logical sum', intersection as a 'logical product'. Both union and intersection are commutative, in that

$$A \cup B = B \cup A$$
and $$A \cap B = B \cap A.$$

The difference of A and B is non-commutative; it is an asymmetric operation where the order of A and B matters. Written as 'A − B', it generates the set of points

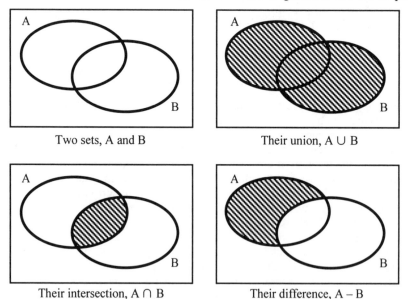

Two sets, A and B   Their union, A ∪ B

Their intersection, A ∩ B   Their difference, A − B

Figure 8.9 The Boolean operators used in constructive solid geometry (CSG)

8. Geometric Modelling and Fractals: Building Descriptions of Objects 223

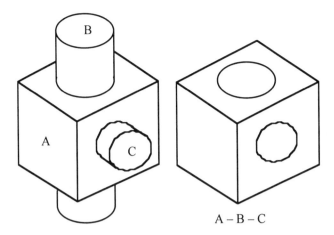

Figure 8.10 A CSG construction by subtracting two cylinders from a block

contained in A that do not belong to B. B − A is the set of points contained in B that do not belong to A. Using combinations of the union, intersection and difference operations on simple primitives, a variety of objects can be constructed in a relatively intuitive way. Figure 8.10 shows three primitives: a block A and cylinders B and C. It is assumed that a user has selected these primitives and has defined parameters to size and locate them as superimposed on each other in the left part of the figure. In the right part, the result of the operation of subtracting the union of B and C from A, A − (B ∪ C), is shown as a block with two through holes. This has alternative Boolean expression A − B − C. This is a very compact way of describing the structure of an object that would be difficult to define in other ways, as we will see later when we introduce B-Rep methods.

CSG methods are relatively easy to use and robust. For a properly set up CSG system, all objects created are valid or manifold. This is an important property not shared by the more popular B-Rep methods, to be discussed later. CSG models are also very compact in comparison with other methods.

The mechanism of performing the Boolean operations is hidden from the user, who has the task of selecting and locating primitives, and defining the types of combination required. In some modelling systems, all stages of this process can be done interactively. Everything in the following description is defined by the system creator; the user should be oblivious to what is going on inside the modelling system.

Each primitive is represented by a set of mathematical inequalities that can be evaluated to identify whether a point lies 'inside' or 'outside' one of the boundaries of the primitive. Thus, for example, a simple unit cube is represented by six inequalities,

$$x \geq 0, \quad x \leq 1,$$
$$y \geq 0, \quad y \leq 1,$$
$$z \geq 0, \quad z \leq 1,$$

one for each plane face of the cube. The equivalent equation for each of these, for example, $z = 0$, describes a plane. The cube is defined in a 3D space with Cartesian points (x, y, z). One constraint or condition on this space reduces the dimension by one, so each of the six equations gives us a 2D space. As the equations are of a simple form, they each represent a plane. For example, the equation

$$z = 0$$

represents all points in the (x, y) plane, for which $z = 0$. Thus, from our discussion of inequalities when we looked at spatial representations, points that satisfy the inequality

$$z \geq 0$$

either lie on this plane or to the side of it that has non-negative z, a 3D region with one closed boundary. Each inequality defines a region known as a half space. Now consider all six inequalities or conditions given above. If a point satisfies all six, then it must lie within or on the surface of the cube. If it fails to satisfy any one, it lies outside the cube. This test defines spatial occupancy of the object directly through identifying interior and exterior points.

Similar sets of inequalities can be used to define other forms of canonical object; spheres, cylinders and cones were discussed in chapter 5. The curved surface of a cylinder of radius 1 with axis of symmetry along the z-axis has equation

$$x^2 + y^2 = 1.$$

To create a 'manifold' cylinder, one that could be manufactured in reality, it is necessary to limit the infinite extent of this against two cutting planes, for example, $z = 0$ and $z = 1$. The three inequalities that determine a cylinder oriented in this way with unit radius and height are (fig 5.25, page 143)

$$z \geq 0,$$
$$z \leq 1;$$
$$x^2 + y^2 \leq 1.$$

If we test the point $(1, 1, 0.5)$ against this, we see that the first two inequalities are satisfied, as $0.5 \geq 0$ and $0.5 \leq 1$, but $x^2 + y^2 = 2$, which is not less than 1, so the third inequality is not valid for this point. Thus, the point $(1, 1, 0.5)$ lies outside the cylinder as it breaches one of the required conditions. This simple example illustrates how easy it is to test points against a CSG model. The test takes just a few function evaluations and numerical comparisons.

Although the cylinder may be considered more complex than a cube, its CSG description requires only three inequalities compared to six for the cube. It should be noted, however, that the inequality for testing against the curved surface uses a more complicated quadratic function rather than a linear one.

The easiest primitive to define is a sphere; a canonical form with centre at the origin and radius 1 has one inequality (fig 5.24, page 141):

$$x^2 + y^2 + z^2 \leq 1.$$

8. Geometric Modelling and Fractals: Building Descriptions of Objects

One further primitive is the cone. A cone shape with axis of symmetry of length one along the z-axis, vertex at the origin and base radius one has defining inequalities (fig 5.26, page 144)

$$z \geq 0,$$
$$z \leq 1,$$
$$x^2 + y^2 \leq z^2.$$

The first inequality prevents the cone extending beyond its apex. It may seem natural for the structure to be bounded in this way, but the equation of the curved surface of a cone extends beyond this to place a nappe, a mirror single cone surface, on the other side of the vertex. The second defines the base plane, where the cone has base radius one. The third defines the curved surface, effectively through a series of circular slices. For a particular value of z, a point inside the cone must lie inside a circle radius z.

It is fairly easy to define simple canonical shapes of these forms. How are objects in more general locations and orientations dealt with? There are two possibilities. When location and orientation parameters are defined, the equations underlying defining inequalities can be manipulated as described earlier in the book. Thus, points can be tested against adapted versions of the inequalities based on these new equations. Alternatively, matrices defining the transformations that change a canonical shape into the required form could be concatenated to one transformation matrix, as described in chapter 6. At the same time as this is done, it is easy to create the inverse of this, the matrix that would return the final shape to its canonical primitive form. So, for example, an arbitrarily oriented ellipsoid shape, like a rugby ball or an American football, can be converted back to a standard sphere, centre (0, 0, 0), radius 1. Testing against this is particularly simple. Thus, to test a point P against the new structure, transform P using the inverse matrix to generate point Q, and then test Q more easily against the canonical form.

To test a point against a complex structure, such as that of fig 8.10, it is necessary to test it against each primitive that comprises the system, creating a value 1 when the point is inside, 0 when the point is outside each primitive. The Boolean expression that defines the composite object is then evaluated with these particular inclusion values for each relevant primitive. If the final evaluation is 1, the point is inside the composite object, otherwise it is outside.

The normal data structure used to hold information on primitives, their modelling transformations and object-defining Boolean operations is a 'CSG tree'. This is a binary tree structure with Boolean operations held at internal nodes and primitives with their parameters held at the leaf nodes, the childless terminator nodes of the structure. The object itself is held at the root of the tree, internal (parent) nodes all hold Boolean operators, and each leaf node contains a primitive with its location and orienting parameters. Figure 8.11 shows a CSG tree for generating the shape α of fig 8.10, using the Boolean expression

$$\alpha = B - (C1 \cup C2),$$

first combining the two cylinders C1 and C2 through a union operation, then subtracting the result of that from the block B. The data content of a leaf node is an indication of the type of primitive being used and the necessary constants to define

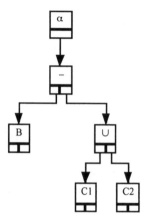

Figure 8.11 A CSG tree defining the object of fig 8.10

its size, location and orientation. Alternatively, the equivalent matrix tranformations could be held, as discussed above. To identify if a point P lies within the shape α, pre-order traversal leads us first to primitive B. Suppose the test of P against block B gives result b, where b is a Boolean value 1 if P lies inside B, 0 otherwise. This value b is returned to the internal node holding a '−' and is held while the right sub-tree is explored. On reaching C1, P is tested against this cylinder with Boolean result c_1. This value c_1 is carried back to the 'union' node, and held as its right sub-tree is investigated. This yields a value c_2 by testing P against the cylinder C2. Now, returning to the union node from its right sub-tree, both elements needed for evaluation of the union function are known, so the Boolean value

$$c_1 \cup c_2$$

is carried out of the union node. This takes value 1 if P is in either of the two cylinders, 0 otherwise. Now the difference node is reached with both its constituent values known, so the final assessment of whether P lies inside or outside α depends on the Boolean value

$$b - (c_1 \cup c_2)$$

which is returned to the CSG tree root to give α a truth value. If this has value 1, P lies inside (or on the surface of) α, else P lies outside α. This is a very simple example, but more complex CSG trees can be dealt with easily by defining the operating characteristics of relatively few forms of tree node. Another data structure, a stack, is used to store intermediate values of the Boolean evaluation so that the correct value is ready to be 'popped' off the stack when an evaluation is reached. The stack is a 'last in, first out' queue that can be operated as a linked list with two major operations. These 'push' a new value onto the head of the queue, and 'pop' existing values from the head. This would be most unfair at post-office counters, but works perfectly to match the right sub-tree values returned up the tree with the relevant left sub-tree values that have been waiting for their partners.

The curved shapes defined in CSG systems are exactly defined through their mathematical definitions. A voxel-based description of a curved surface would be inexact, its precision depending on the spatial subdivision of the voxel grid. In the next method to be discussed, B-Rep, most curved surfaces are described only approximately, although special measures can be taken to deal with some curves exactly. Approximations are normally adequate for the generation of computer graphics images, but if there are other underlying purposes in the geometric model, the exact depiction of the CSG method can be an advantage. For example, if the model is of an engine part that has to roll correctly against another part, this is allowed by the exact CSG description. Also, CSG models mimic some mechanical engineering manufacturing processes. The subtraction of a cylinder is equivalent to drilling a hole through an object (fig 8.10). The method is much used in computer aided design (CAD) systems for engineering, as such precision is important to these applications. This is particularly so for CAD/CAM systems, where the augmented acronym includes 'computer aided manufacturing', so the model information generated in the CAD phase is used directly to drive computer numerically controlled manufacturing machinery.

Although curves are modelled exactly, the depiction of models is difficult and slow in CSG, as the data structure holds no direct surface information. It is easy to test single points against an object, but the particular set of points that form the object's surface, the potentially visible points, is not explicit. The method of 'ray casting' may be used (we discuss the more extensive 'ray tracing' in chapter 10), but this is notoriously slow. Having used the benefits of CSG to create models, some systems convert models to the B-rep method to be discussed in the next section; this is more amenable to rapid rendering.

Hence, in spite of its desirable properties of compactness and guaranteed validity, CSG is not widely used in circumstances where high-quality rendering is needed in fairly rapid time. Considering also that its potential range of models is limited by the choice of primitives, the reader may wonder why it is of any importance. Its importance is in the ease of use; the ability to generate models using the Boolean operators is one that trained users find particularly attractive. Many modellers now allow this mode of interaction, although the data structure within the system is not CSG. Although algorithmically more difficult to achieve in other systems such as B-Rep (to follow), the ease of use of Boolean operators leads systems developers to supply the method to users. Many systems have hybrid operating methods built on experience of several underlying data structures. Offering users a variety of methods for model generation eases what is the most humanly time-consuming part of 3D computer graphics: the generation of models.

Boundary Representation (B-Rep)

This is the most widely used method for the generation of 'quality' images. Unlike voxel and CSG methods, the surface characteristics of an object are explicitly described. This enables surfaces to be rendered rapidly and as accurately as the description allows. This cryptic description suggests a potential defect in the method; none of the methods described here is perfect, or it would have taken over completely from the others. B-rep methods describe objects as polyhedra (plane faced objects), so exact curves are not possible, except by using special methods described in chapter 9. B-Rep models also need large data sets, are complex and do not have guaranteed validity. Implications of this are considered later.

Starting with the points that give the spatial location of an object, the edges that join these points and the faces that lie within these edges are defined. Figure 8.12 shows these elements for the minimal basic 3D structure, a four-faced tetrahedron. No polyhedron contains fewer faces. Vertices (corner points) are labelled with capital letters P, Q, R, S, edges with lower case characters a, b, ... f and faces in bold capitals W, X, Y and Z (arrows show that Y and Z are faces at the 'back' of the object). Face W, for example, is defined by edges a, b and c, and edge a joins the points R and Q. This kind of 'topological' information, indicating how the elements that make up a structure are linked, must be stored in a B-Rep data structure. A typical method is to hold information on vertices, edges and faces in linked lists, with pointers joining elements of different lists to indicate topological relationships. Another popular method is 'Baumgart's winged edge' data structure, in which the edge is the central element, and simple 'loop' lists of vertices define faces. Other methods are based on definitions of faces, with the same vertices stored several times within the structure. This is problematic for structural amendment, when several versions of the same vertex must be changed in order to maintain the integrity of the structure. The method chosen here to illustrate the concepts is a vertex–edge–face linked list structure, shown in fig 8.12 as if implemented for a tetrahedron.

It is worth commenting on the complexity of fig 8.12, which represents the least complicated solid object, with four vertices, six edges and four faces (many may consider the cube 'simpler', but that has eight vertices, six faces and twelve edges). The tetrahedron also has simplicity in that all vertices are joined to each other, and they relate to each other in the same way. With a cube, each vertex is joined to three

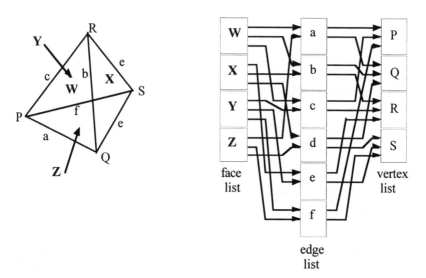

Figure 8.12 A tetrahedron and its representation in a B-Rep system

8. Geometric Modelling and Fractals: Building Descriptions of Objects

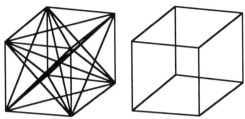

Figure 8.13 The join of all vertices of a cube, and the selection of only those edges that define the cube

others, leaving four to which each vertex does not have a direct join. To define a cube, we have to identify which vertices are to be joined – merely to join up all possible pairings (fig 8.13) gives a figure which loses much of the familiar sense of 'cubeness'. The network of relationships between vertices, edges and faces represents the topology of the object, the way in which its structure hangs together.

Looking at fig 8.12 in more detail, we see that each face in the face list has three pointers out of it and hence three edges. Each edge has two vertices (two pointers out) and lies alongside two faces (both necessary conditions). In particular, **W** is surrounded by edges (a, b, c), **X** by (e, b, d), **Y** by (c, e, f) and **Z** by (a, f, d). Referring this to fig 8.12, we note that each face is scanned anti-clockwise when viewed from outside the tetrahedron. This ordering can be used to define the 'sidedness' of the face, defining not just the boundary of the object, but inside and outside, and hence the spatial occupancy of the object. The choice of 'anticlockwise' here is arbitrary; many applications use a clockwise scan. The importance is consistency within the application.

We could use the pointer network of fig 8.12 to define any tetrahedron; specific instances of tetrahedra would be distinguished by position information in the form of Cartesian coordinates (x, y, z) held as data at the point nodes, P, Q, R and S. The topology of a general tetrahedron can be copied, once we are sure we have a successful 'version'. The same is true of other structures. Once a version has been tested, it can be essentially repeated in terms of topology or pointer structure, with the geometry added by inserting specific values for vertex locations.

Apart from the pointer structure shown, data can be held at each of the nodes. For example, faces can hold information on colour, degree of shininess, whether the surface is transparent and, to speed the rendering process, redundant information such as a surface normal. The use of these is described in chapter 10. By redundant, we mean that the information could be calculated from other features of the data, such as the position of vertices, but its explicit storage avoids it having to be recalculated each time it is needed, enabling faster rendering speeds at the cost of greater memory requirements. A surface normal is a vector giving the direction at right angles to the surface. It is important in rendering as the amount of light cast on a surface, and hence its perceived colour, depends on the angle between the surface normal and the direction of light shining on the surface. Note that a face will only have a single surface normal if it is plane; twisted faces (you can create one by lifting a single corner of this page) may pose problems. Some B-Rep systems only allow triangular faces, as triangles are by definition planes. Quadrilaterals are allowed as two triangles, 'safety' achieved at the cost of greater memory requirements. Edges and vertices may also contain relevant information used in rendering as well as being part of the pointer structure.

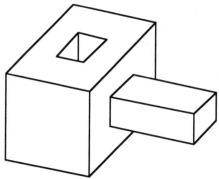

Figure 8.14 A manifold object that satisfies the Euler–Poincaré formula

B-Rep models are widely used due to the ready availability of information that enables fast rendering (where fast is a relative term), but they use considerable memory; models are not that easy to generate and are very difficult to validate. Misplacement of one or two pointers in more complicated structures can lead to invalid or non-manifold objects, which could not be 'built' as real solids. There are a number of error traps that can be built in as checks; for example edges can only have two pointers in (from adjacent faces) and only two pointers out (to their defining vertices), but these are *necessary* rather than *sufficient* conditions. This means that if the conditions are violated we know that the system is in error. However, if they are checked to be valid, the system could still be in error in other ways; this is no more than an error trap that is not exhaustive. Sophisticated systems check the Euler *topological* formula (Euler was responsible for many other formulae as well, so it is necessary to categorize this),

$$v - e + f = 2,$$

where v, e and f are the numbers of vertices, edges and faces respectively. Structures with inner boundaries within faces (known as rings, r), holes directly through them (h) and multiple separable parts (or shells, s) satisfy the extended Euler–Poincaré formula

$$v - e + f = 2(s - h) + r.$$

These, too, are of the 'necessary but not sufficient' kind.

Figure 8.14 shows a single object (s = 1) with one through hole (it is assumed the hole in the top face passes straight through the object, so h = 1). This creates a 'ring' in the top and bottom faces. There is also a ring caused by the 'handle' creating an inner boundary within the right face of the larger block (r = 3). Adding up the numbers of vertices, edges and faces gives v = 24, e = 36, f = 15, so

$$v - e + f = 3,$$
and $$2(s - h) + r = 3,$$

illustrating that the Euler–Poincaré formula holds for this structure.

8. Geometric Modelling and Fractals: Building Descriptions of Objects

Systems have been developed to ensure that these formulae are not violated. Mäntylä's 'Geometric Work Bench'[2] allows only a limited set of 10 essential operations to be used in generating geometric models, each of which ensures the model's conformity with the Euler–Poincaré formula as the object develops. It is intended that these mini-operations should be used within geometric modelling systems by systems developers, with users being given 'macro-tools', unaware of the detail that is involved within the system structure. As a general rule, vigilance is the only way to ensure validity of B-Rep models. However, once a model's validity has been established, its topology can be endlessly repeated by copying its pointer structure.

The kinds of B-Rep method described above allow a limitless range of models, provided some approximation is allowed. Curved surfaces must be approximated by plane surfaced polyhedra (a 'polyhedron' is an object bounded by plane surfaces). More precise aproximations to curved surfaces may be created by increasing the numbers of faces, another example of the kind of 'trade-off' often found in geometric modelling. In chapter 10 on rendering, we will discuss ways of 'smooth shading'. These disguise internal face boundaries when needed to depict adjacent faces as though they are continuous curved sections.

Many B-Rep modellers allow the generation of curved surface patches through the use of splines. These are properly part of the description of geometric modelling systems, but they are described separately in the following chapter.

Isosurface Modelling

This method is also known as 'soft object' modelling, as it is particularly suited for creating 'blobby' or soft-edged objects. The method is based on the concept of a mathematical scalar field. A field function gives a specified value at every point inside some region of space; it is a function of points in space, typically of x, y and z coordinates. If the function takes a scalar value, we have a scalar field. Temperature comprises a physical example: a scalar temperature can be found for every point within a room. Wind velocity and magnetic field are vector field examples, where the field function defines direction as well as magnitude.

An isosurface object is created by defining a scalar field function, $\phi(x, y, z)$ say, in a controlled way (ϕ is the Greek character 'phi'). The object's surface is the set of all points at which the field takes a specified constant value. By imposing one constraint,

$$\phi(x, y, z) = \text{constant}$$

on a 3D set of points, we have produced a 2D surface. On a 2D weather map, the isobars represent points at which the pressure is a constant (the prefix 'iso' means 'the same'). What you see is a slice at ground level through the 3D isosurface that extends up into the atmosphere. The trick in creating an isosurface is in controlling

---

[2] Details can be found in M. Mäntylä (1988) *Introduction to Solid Modeling*, Computer Science Press, Rockville, MD.

the creation of the field function to create the desired structure. We describe one simple method here.

The field creator controls a number of 'key points'. These can be placed in space at positions $P_i = (x_i, y_i, z_i)$, and their intensities and breadth of influence controlled. Each key point has an associated function that can be evaluated as a function of any space position P, for example one of form

$$\phi_i(P) = A_i \exp(-k_i |PP_i|^2),$$

where 'exp' is the decaying exponential function (chapter 5, pages 135–136), $|PP_i|^2$ is the squared distance from P to the key point P_i,

$$|PP_i|^2 = (x - x_i)^2 + (y - y_i)^2 + (z - z_i)^2,$$

A_i and k_i are constants associated with key point P_i (chapter 3 has a description of this kind of function). ϕ_i takes a maximum value of A_i when $|PP_i|$ is zero (that is, when P is coincident with P_i) and P represents the point (x, y, z). As P moves away from P_i, the value of ϕ_i 'decays' towards zero, getting very close to zero for large distances. If k_i is increased, this decay is more severe, so the value of k_i controls the range of influence of ϕ_i, whereas A_i controls its central intensity. If k_i is small, ϕ_i has a wider range of influence.

Suppose n key points, $P_1, P_2, \ldots P_n$, are defined. The overall field function ϕ is the sum of the n separate functions ϕ_1 to ϕ_n,

$$\phi(P) = \sum_{i=1}^{n} \phi_i(P).$$

All points satisfying the equation

$$\phi(P) = \phi_0,$$

where ϕ_0 is a given constant, lie on the isosurface defining the object. The interior of the object is given by the inequality

$$\phi(P) > \phi_0.$$

For a single key point, the isosurface is a sphere. For several key points at long distances from each other compared to the 'range of influence' of each one, their combined isosurface still appears as a collection of spheres with centres at the key points. However, when two key points approach each other to within their ranges of influence, each separate sphere is seen to stretch towards the other until the surface bridges the two sections, the isosurface taking an intermediate 'dumb bell' shape (fig 8.15). When the two key points are coincident, the surface is again a single

Figure 8.15 Two key points approach each other until their surfaces coalesce

8. Geometric Modelling and Fractals: Building Descriptions of Objects   233

sphere, now with larger radius. The combined effect of adding local functions increases the field value between the two key points, so 'stretching' the surface between them until they eventually coalesce smoothly. In the sequence from left to right of fig 8.15, the surfaces are seen as separate balls stretching towards each other as they approach until they bridge in a smooth way.

In practice, structures other than key points may be used. For example, an isosurface based on distance from a line segment (a line between two given points) is like a cylinder with hemispherical caps on the ends. We have seen how to evaluate distance from a line in the chapter on vectors, so it is relatively easy to identify if a point lies within this given range of a line segment. Isosurface functions can also be based on distances from triangles, rectangles, or other plane and solid shapes.

For efficiency, most versions of isosurface modelling do not use the exponential function as described above. Usually, a polynomial in the square of distance $|PP_i|$ is devised to have the exponential properties of a maximum value when $|PP_i|$ is zero, tailing off towards zero as $|PP_i|$ increases. Using the square of $|PP_i|$ avoids the time-consuming evaluation of square roots and ensure the function is symmetric about the key point. Such functions are usually given a maximum 'radius of influence'. If we are evaluating the isosurface function at point P, and P lies further away from P_i than this radius of influence, the value of $\phi_i(P)$ is set to zero. This avoids having to calculate the function $\phi_i(P)$ when its contribution to $\phi(P)$ is negligibly small.

Other forms of evaluation of 'distance' give a wider range of modelling possibilities. The outline form of the die of fig 8.16 is based on a distance calculation using eighth powers rather than squares. This isosurface is then a 'super-sphere' with equation of form

$$x^8 + y^8 + z^8 = \text{constant}.$$

This approaches the shape of a cube compared to the more usual sphere,

$$x^2 + y^2 + z^2 = \text{constant}.$$

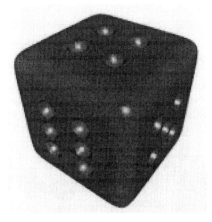
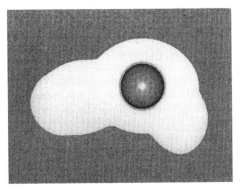

Figure 8.16 Soft object die and fried egg. The die uses negative key points, both use non-spherical isosurface functions

The figure marking holes are created using negative spherical key points to subtract substance, this time with the usual 'power of two' distance measure. The fried egg, also figure 8.16,[3] uses two overlaid isosurfaces. The white part is a smooth blend of three key points; the yolk is a differently coloured separate isosurface that is overdrawn onto the other parts to avoid blending with them. In both these forms, the vertical y coordinate is scaled within the function to give a flatter effect.

The usual method for depiction of isosurface models is the 'marching cubes algorithm'. This is related to voxel modelling, using a similar regular grid imposed on the space within which the model lies. Starting at a key point, known to be in the object, the isosurface function is evaluated at a succession of grid vertices in an axis direction moving away from the key point. As soon as a vertex function value is found to take a value below ϕ_0, the isosurface constant, we know that the isosurface must pass through the edge just traversed, and hence through all four cubes that share that edge. Each cube so identified is pushed onto a stack data structure. When a cube is popped off the stack, all of its eight vertex isosurface values are calculated or identified as having been previously found. Using the temperature analogy, from the field function values at 'hot' (inside) and 'cold' outside vertices, an approximation to the isosurface shape as it passes through the cube is found by interpolating values along cube edges. These 'hot–cold' edges also indicate other neighbouring cubes through which the isosurface passes; these are pushed on to the stack if they have not already been dealt with. In this way, the algorithm follows an irregular march across the isosurface, identifying all cubes through which the isosurface passes and approximating the isosurface within each cube by plane surface sections. In this way, a representation of the isosurface is created as a series of small planar sections that can be passed to a standard rendering system. Searches should start from each key point, to ensure all disjoint surface sections have been identified.

The method is successful in creating images of rounded edged objects, has been used for human and animal skin modelling, and is the technique underlying a delightfully quirky animated video, 'The Great Train Rubbery'.[4]

Fractals

Fractals are not usually considered part of geometric modelling. It is appropriate to deal with them here as systems for producing object models in 2D and 3D.

The word 'fractal' was coined by Benoit Mandelbrot during the 1970s:

> A fractal object is by definition a set for which the Hausdorff Besicovitch dimension strictly exceeds the topological dimension.[5]

---

[3] Images produced by former students of the author include fig 8.15 and the die of fig 8.16 by Dave Baldwin, and the egg of fig 8.16 by Renay Cooper.

[4] G. Wyvill, C. McPheeters and B. Wyvill describe their animation method in 'Animating Soft Objects, *The Visual Computer*, August 1986, **2**(2), 235–242. These three have been responsible for much development of isosurface methods.

[5] You will find this in B.B. Mandelbrot (1983) *The Fractal Geometry of Nature*, W.H. Freeman, New York, p. 15.

8. Geometric Modelling and Fractals: Building Descriptions of Objects

This is unhelpful to the lay person who does not have expertise in point set topology, but in the sections below, the concept of 'fractalness' is explained in more qualitative terms before revisiting this definition. It may be more useful initially to consider the defining properties of fractal objects through exposure to several examples. These properties are

- self-similarity: any subset of the object is, in some sense, a copy of the whole;
- infinitesimal subdivisibility: there is no apparent change in the amount of detail observed at different levels of magnification.

The latter property is achieved in theoretical examples, but is seldom if ever achieved in real-world cases for extremely large levels of enlargement, as objects invariable cannot be subdivided beyond atomic levels. A number of naturally occurring phenomena do, however, have fractal properties to a 'good' level of enlargement.

Many mathematical constructions now identified as truly 'fractal' were created in the period from 1880 to 1930, when point set topologists created the 'mathematical monsters' that disturbed many traditionalists. The apparently paradoxical properties led to a major split and much personal animosity in mathematical circles. Some examples of exactly self-similar fractals are given below. The cagey words, 'in some sense', in the first property indicate that other forms of self-similarity are also permissible.

The Koch Curve and Cantor Set

Published by Helge von Koch in 1904, this fractal 'curve' is generated by recursively replacing line segments by 'polylines' consisting of four line segments each one third of the original length. The first five stages of this process are shown in fig 8.17. Limitless repetition of the process generates a true fractal form. The curve contains four exact copies of itself, each at one third original scale, so at each stage of generation every line segment is replaced by four thirds of its original length. If the original line has length one, the total length after n stages is $(4/3)^n$. A few key presses on a calculator should convince sceptical readers that this becomes indefinitely large, it is said to approach infinity as n approaches infinity. Thus, the pure fractal object has infinite length – there is not enough ink in the universe to draw it properly.

This means that each sub-copy is also of infinite length – one third of infinity must still be infinity! That goes for sub-sub-copies and so on. But the curve itself is completely made up of copies of itself, so any two distinct points on the curve are separated by at least one sub-copy at a given level of subdivision, so any two points on the Koch curve must be separated by an infinite distance. There is not enough ink in the universe to join any two points on a Koch curve; the paint appears to pour into a bottomless hole – no progress can be made.

Now consider the area between the curve and its original defining line. At the first stage, a single triangle is added, and we suppose its area is A. (This is an isosceles triangle with sides equal to 1/3, and A can be found as $\sqrt{3}/36$, but the actual amount is irrelevant to the argument that follows.) At each successive stage, four times as many new triangles are added (fig 8.18), each one ninth the area of the

Figure 8.17 Five stages in generation of a Koch curve

previous stage (when shape lengths are multiplied by 1/3, areas are multiplied by $(1/3)^2$ or 1/9, a consequence of area being two-dimensional). So the added area at any stage is 4/9 that added at the previous stage. We can express the total added area as the sum S of a series,

$$S = A + (4/9)A + (4/9)(4/9)A + (4/9)(4/9)(4/9)A + \ldots,$$

or $\quad S = A + (4/9)^1 A + (4/9)^2 A + (4/9)^3 A + \ldots .$

Each term in the series becomes smaller by a fraction of 4/9; remember that $(4/9)^1$ is the same as 4/9. Series like this with constant multiples are called geometric series. As 4/9 is numerically less than one, the series converges (we do not digress to establish this). The mathematical trick of multiplying the whole equation by 4/9, 'shuffling' the tail along one position, helps to find S. This gives

$$(4/9)S = (4/9)^1 A + (4/9)^2 A + (4/9)^3 A + \ldots .$$

Now subtracting this new equation from the old eliminates the tail of the series

$$S - (4/9)S = A,$$

so $\quad (5/9)S = A$

and $\quad S = (9/5)A = (9/5)(\sqrt{3}/36) = \sqrt{3}/20.$

8. Geometric Modelling and Fractals: Building Descriptions of Objects   237

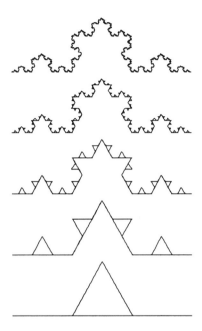

Figure 8.18 Triangular areas added at each stage of Koch curve generation

The important point is that there is a finite area. We have an infinite-length curve that encloses a finite area. A painter could easily find enough paint to colour the interior of a Koch curve, but he or she would not have enough time to complete the precisely detailed form.

The points remaining on the original line when the Koch curve is completed form another famous fractal, the Cantor set. This is created by repeatedly taking out the middle third of all line segments involved (fig 8.19). All points are distinct, forming a 'dust'; this is impossible to show in an image. In fig 8.19, points are 'fused' after the fifth subdivision. If the original line has length one, one third of its length is taken out at the first stage, then two lengths of one ninth, four lengths of one twenty-seventh, and so on, are taken out. The total length eliminated is

$$L = (1/3) + 2(1/3)^2 + 2^2(1/3)^3 + 2^3(1/3)^4 + \ldots$$
$$= (1/3)\{1 + (2/3) + (2/3)^2 + (2/3)^3 + \ldots \}.$$

Noting that we have another geometric series with successive terms multiplied by 2/3, and 2/3 is numerically less than one, we multiply the whole equation by 2/3 (the same trick as used in finding the area below the Koch curve):

Figure 8.19 A Cantor set created by elimination of the middle third of line segments

$$(2/3)L = (1/3)\{(2/3) + (2/3)^2 + (2/3)^3 + \ldots\}.$$

Subtracting this from the previous equation to eliminate the tail,

$$(1/3)L = 1/3,$$

so $L = 1$.

The process extracts the whole of the length of the original line, but still leaves points in the line. Thus, none of these points can be joined by any length of line, they must all be distinct, forming a mathematical 'dust'. The process creates a pattern with two self-similar copies at one third scale. This concept of numbers of repetitions at a particular scale factor is central to the definition of a fractal object.

The Sierpinski Triangle or Gasket

A Sierpinski triangle or gasket can be constructed by recursively extracting triangular forms from within an original triangle. The construction should be evident from fig 8.20. At each stage of recursion, three times as many triangles are extracted, each being a quarter of the area of those used in the previous stage. This leads to another series, which can be summed to show that the total area extracted is equal to the area of the original triangle – no area remains, yet a set of points remains. The resulting fractal object contains three copies of itself at half linear scale. Sierpinski also showed how to generate the same spatial occupancy through recursive redrawing of a curve by a technique similar to that used for the Koch curve.[6] This recursively approaches infinite length and visits all points in the Sierpinski gasket. Yet another method for generating the same object is described by Barnsley,[7] using a specialized form of the iterated function systems (IFS) described below. Given the vertices V_1, V_2 and V_3 of a triangle and an arbitrary starting point P_0, select one of the vertices V_i at random and plot the mid-point P_1 of V_i and P_0. Continue with this process several thousand times (this is where a computer

Figure 8.20 A Sierpinski triangle or gasket

---

[6] First published by Waclaw Sierpinski in 1916, found by the author in W. Sierpinski, (1916, republished 1975) Sur Une Courbe dont tout point est un point de ramification, in *W. Sierpinski: Oeuvres Choisis, Tome II*, PWN – Polish Scientific Publishers, Warsaw, Poland.
[7] In M.F. Barnsley (2000) *Fractals Everywhere (3rd edn)*, Morgan Kauffman.

8. Geometric Modelling and Fractals: Building Descriptions of Objects

Figure 8.21 A stage in generation of a Sierpinski tetrahedron from solid tetrahedra

becomes useful), plotting a sequence of points P_1, P_2, P_3, ... After plotting a few stray initial points, this sequence is 'attracted to' and fills the Sierpinski gasket.

The Sierpinski gasket can be generated by drawing a collection of two-dimensional triangles, by manipulation of an apparently one-dimensional line or by a collection of zero-dimensional points. A similar set of methods can be used to generate its 3D equivalent, a Sierpinski tetrahedron. Figure 8.21 shows a stage in its generation as a set of sub-tetrahedra. At each generation stage, a tetrahedron (triangular-based pyramid) is replaced by four half-sized tetrahedra having a vertex coincident with the original, leaving an octahedral hole in the centre. The 3D object can also be filled using a scatter of random points in the IFS method, described below (plate 8.3). You've probably guessed by now that the whole volume of the original tetrahedron is eliminated, but the shape still remains. The Sierpinski tetrahedron contains four copies of itself at half linear scale. The true dimension of this volume-less object that could be generated from a set of subdivided 3D tetrahedra or a set of zero-dimensional points will be revealed when we return to the formal definition of 'fractal dimension'.

A Space-Filling Curve

In 1890, Giuseppe Peano showed how mathematics could 'outrage common sense' through the construction of a space-filling curve.[8] Sierpinski, David Hilbert and others created similar monsters, generated as curves that visit every point within a two-dimensional region. The first three stages in the generation of such a curve is shown in fig 8.22, the arrows showing how the second stage, and one sub-part of the third stage can be created by tracing a continuous curve (it can be drawn without raising the pen from the paper). When repeated several times, this process 'fills' the diamond-oriented square that bounds the outline of the figure. As in the Koch curve construction, a replacement method is used, recursively replacing each line segment by nine other line segments at one third scale. This generates an object that contains nine copies of itself at one third scale.

We again have a problem in understanding the dimensionality of the object created. We start with a one-dimensional curve, whose length approaches infinity as the process continues, and end with an object that fills a region of two-dimensional space. We have also seen an object in 3D space that can be created by a set of points or by a set of solid tetrahedra, and 2D objects that may be created by triangles, lines or points. What does dimension mean in these contexts?

---

[8] Comment in C.B. Boyer and U.A. Merzbach (1991) *The History of Mathematics (2nd edn)*, Wiley, New York, p. 598.

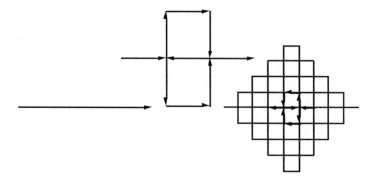

Figure 8.22 Three stages in creation of a space-filling curve

Fractal Dimension

We see from the objects discussed in the last few pages (all of which are 'exactly self-similar' in that they can be subdivided into a number of exact scale copies of the whole object itself) that some apparent paradoxes make it difficult to define their dimensions. We consider first the dimensions of some non-fractal objects, then extend this idea to exactly self-similar fractals and statistically self-similar fractals.[9]

In the examples looked at above, the number of copies and the scale of such copies have been identified. Some non-fractal objects can be similarly subdivided, for example, it is clear that lines, squares or triangles and cubes contain two, four and eight copies respectively at half scale, and three, nine and twenty-seven copies respectively at one third scale (fig 8.23). The number of copies N, the scale factor f

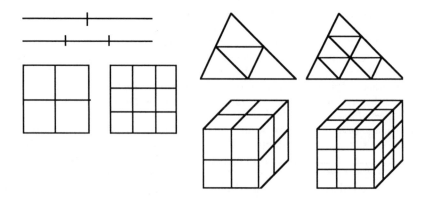

Figure 8.23 Some subdivisions of a line, square, triangle and cube

[9] These concepts are thoroughly discussed in B. Mandelbrot (1983) *The Fractal Geometry of Nature*, W.H. Freeman, New York. Only a brief overview is given here.

8. Geometric Modelling and Fractals: Building Descriptions of Objects

and the dimension of the object D, can be connected for these examples by a power rule,

$$N = (1/f)^D.$$

This equation can be rearranged to give the dimension D as

$$D = \log(N)/\log(1/f).$$

(Those unfamiliar with logarithms should not fret over this, it simply gives a way of rearranging the equation to put D as its subject; a logarithm reverses the effect of raising to a power.) If this is accepted as a formula for the dimension of non-fractal objects, we can extend the concept to our problem objects that are built up completely from scaled copies of themselves. The values of (N, f) for the Cantor set, Koch curve, Sierpinski gasket, Peano curve and Sierpinski tetrahedron are (2, 1/3), (4, 1/3), (3, 1/2), (9, 1/3) and (4, 1/2). The formula for D is applied to these.

Cantor set	$D = \log(2)/\log(3) = 0.6309$ (to 4 decimal places)
Koch curve	$D = \log(4)/\log(3) = 1.2619$ (to 4 decimal places)
Sierpinski gasket	$D = \log(3)/\log(2) = 1.5850$ (to 4 decimal places)
Peano curve	$D = \log(9)/\log(3) = 2$
Sierpinski tetrahedron	$D = \log(4)/\log(2) = 2$

The first two have non-integer solutions – the concept of a non-integer dimension was one of the causes of dispute amongst mathematicians early in the 20th century. The Cantor set is, in some sense, 'more' than a mere point, the Koch curve 'more' than a curve, whilst the Sierpinski gasket is 'less' than the triangle from which it originated. The Peano space-filling curve is much more than the curve it originated from, its dimension of two showing that it fills a region of two-dimensional space. There is no obvious explanation for the exact integer dimension of the Sierpinski tetrahedron. It was mentioned above (and will be explained below) that a Sierpinski tetrahedron can be generated as a three-dimensional version of an IFS. If the four vertices used in the algorithm are arranged at the corners of a square (rather than the four vertices of a regular tetrahedron), the algorithm generates a scatter of points that evenly fills the square. This gives a one-to-one correspondence between the points of a Sierpinski tetrahedron and the points of an obviously two-dimensional square.

Fractal Dimension for Statistically Self-Similar Objects

Many naturally occurring objects, such as clouds, terrain and coastlines, exhibit statistical self-similarity. Sections of these objects are not exact copies of the whole, but their general features are, on the whole, indistinguishable from the overall form. There is an absence of scale about such fractals. If a coastal outline is displayed, stripped of any indications of scale such as towns, roads, vehicles or creatures, it is impossible to identify whether the image was taken from a distant satellite or from a low-flying aircraft. The 'wiggliness' of the outline is independent of the magnification of the image. The fractal property of subdivisibility does not hold up to infinite regress in such real examples – we eventually end up with a single grain

 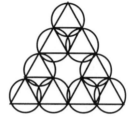

Figure 8.24 Circles that 'cover' the Sierpinski triangle

of sand – but holds up for reasonable lengths. The fractal properties of self-similarity and subdivisibility are maintained.

A property of fractal coastlines was identified in a study by Lewis Fry Richardson.[10] The recorded length of a coastline increases as the length of the measuring instrument decreases. A striding giant cuts off more corners than does a walking child, so the child travels a greater distance. A similar property is observed in boundaries between countries – it is said that the length of the Portuguese/Spanish border was once stated as longer in Portuguese statistics than in the Spanish equivalent. It is reasonable to assume that the smaller country would have used a map at larger scale, equivalent to using a smaller measuring instrument. This effect is more severe for coastlines or borders with larger fractal dimensions.

Evaluation of the fractal dimension depends on the 'neighbourhood' concept defined by Hausdorff, the 'high priest'[11] of point set topology. In very simple terms, a neighbourhood is a small region completely surrounding a particular point. For our purposes, we can take a neighbourhood to be a small circle in 2D or sphere in 3D. Suppose the object under consideration can be 'covered' by a minimum of N_1 neighbourhoods of a particular size. Now reduce the scale of the neighbourhoods by a factor f, say, and once more count the number, N_2, of neighbourhoods that 'cover' the object. The fractal dimension is then the limiting value of the ratio

$\log(N_2/N_1)/\log(1/f)$

as the size of the neighbourhoods approaches zero. Consider the situation for a Sierpinski gasket. It can be covered by three circles of a particular size or nine half-sized circles (fig 8.24). Thus, its fractal dimension is

$\log(9/3)/\log(2)) = \log(3)/\log(2)$,

as found above. For this regular form, ratios remain the same as the sizes of neighbourhoods decrease, so an accurate result is found from large neighbourhoods.

For less regular forms, the limiting value is approximated using relatively small neighbourhoods. Consider the simulated coastline of fig 8.25. 58 circles of one

---

[10] Richardson's results are described in detail in in B.B. Mandelbrot (1983) *The Fractal Geometry of Nature*, W.H. Freeman, New York.
[11] Again, a comment from C.B. Boyer and U.A. Merzbach (1991) *The History of Mathematics (2nd edn)*, Wiley, New York, p. 621. Neighbourhoods were explained by Hausdorff in his publication of 1914, *Basic Features of Set Theory*.

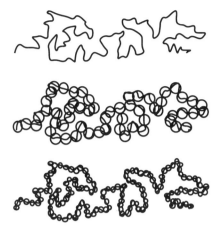

Figure 8.25 Counting neighbourhoods to estimate fractal dimension

scale cover the curve, whereas it takes 134 half-sized circles. For a straight line, we would need about twice as many half-sized circles. The circles are relatively small compared to the length of the curve, so we can estimate the fractal dimension as

$\log(134/58)/\log(2) = 1.21$ (to 2 decimal places).

This empirical method is generally used for estimating fractal dimensions of irregular objects. For convenience, counts of the numbers of squares crossed by the curve on different scales of squared paper can be used. Any loss of accuracy is due to the pre-set positions of squares; they are not moved to match the curve. The relative size matters rather than the shape of the neighbourhoods.

Fractals Based in the Complex Plane: Julia and Mandelbrot Sets

These are widely documented elsewhere. It is necessary to mention them briefly here because of their inspirational importance to the recent development of fractal theory. Only a few examples of the full glory of the pictorial forms developed from these fractal sets are shown here,[12] but the method of depiction will be explained.

The real and imaginary parts of complex numbers (chapter 2) represent x and y coordinates in the Argand plane (chapter 5). A complex number z can be written as

---

[12] Two books played a huge part in explaining and popularizing images of these fractal forms: H.-O. Peitgen and P.H. Richter (1986) *The Beauty of Fractals*, Springer-Verlag, Berlin, and H.-O. Peitgen and D. Saupe (1988) *The Science of Fractal Images*, Springer-Verlag, New York. The first of these has an excellent 'how to do it' section, giving ranges of values within which to seek interesting Mandelbrot set outlines.

$x + iy$, where x and y are real numbers and i is an imaginary number such that $i^2 = -1$.

Consider the complex valued function of $z = x + iy$,

$$f(z) = z^2 + c = (x^2 - y^2 + a) + i(2xy + b)$$

where $c = a + ib$ is some complex constant. This function is iterated to create a sequence $z_1, z_2, z_3, \ldots$ from some starting value z_0,

$$z_k = f(z_{k-1}).$$

This sequence may diverge, meaning that the sequence of points z_i moves further and further from the origin – the point (0, 0) representing the complex number 0 – as k increases. Alternatively, it may converge on a single fixed location or may remain in a cycle fairly close to the origin. To illustrate a Julia set, define a particular value of $c = a + ib$. Suppose a display surface on a computer monitor represents a region of the complex plane. For each display element (or pixel), generate a sequence z_k with z_0 set to the pixel's equivalent complex value. If the sequence moves far away from the origin (beyond a distance d) before the cycle is of length n, it is considered to diverge. Otherwise, z_0 lies inside the Julia set for the given value of c. The pixel can be coloured according to whether it is inside or outside the set (black for in, white for out is typical). This method shows the 'filled-in' Julia set; technically, the boundaries of the shapes formed comprise the

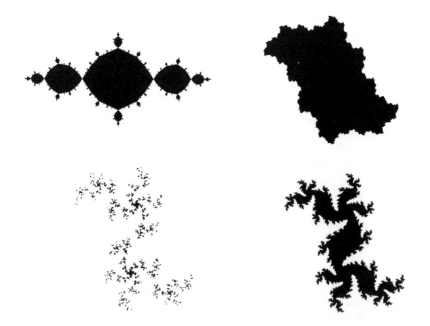

Figure 8.26 Filled-in Julia sets for values of c (clockwise, from top left) -1, 0.5i, 0.3 + 0.5i, 0.36 + 0.5i

8. Geometric Modelling and Fractals: Building Descriptions of Objects

true Julia set. Figure 8.26 shows results from this algorithm for different values of c, showing how slight differences in values of c have considerable influence on the structure formed. In these illustrations, loop cycle n is 30, divergence distance d is 10. To save evaluating square roots, the distance squared is compared with 100.

There are as many Julia sets as there are values of c. Some generate symmetric but irregularly bounded island-like structures. These forms are 'connected', in that a point in the Julia set can be joined to any other without leaving the set. Others are mathematical dusts, with scattered points on the plane like firework displays in the sky. Some images of these forms may show islands, but this is due to the imprecise nature of depiction. All points should be individually isolated, but the gaps between them are too small to be depicted.

An intermediate stage between island and firework forms generates 'dendrites': filaments joined in intricate patterns. The direct method above is inadequate to deal with these as filaments seldom pass through pixel centres. They usually miss and the structure appears blank. This is the opposite of the problem of depicting separate points: if the filament structures miss pixel centres, the objects themselves rather than the gaps between them are too narrow to be seen.

An alternative 'inverse' method is more successful. Suppose you start with some arbitrary point z_k and imagine this is part of a sequence $z_0, z_1, \ldots, z_n$ used to determine if a point is inside a Julia set. Either this sequence moves outwards away from the set's boundary, or it may move inwards towards an attracting point or set of points that it cycles around. In either case, the movement is away from the boundary. If we can reverse this process, we will generate a sequence that moves towards the boundary. Values in the sequence are created using the equation

$$z_k = f(z_{k-1}) = z_{k-1}^2 + c.$$

To reverse the sequence, we need to express z_{i-1} as a function of z_i. It is easy to show that

$$z_{k-1}^2 = z_k - c.$$

This has two possible solutions; formally f(z) is a function that cannot be inverted. However, the diversity of solutions is the basis of the algorithm. We must choose at each stage whether to take the positive or negative possibility from

$$z_{k-1} = \pm\sqrt{\{z_k - c\}}.$$

The algorithm takes this form:

```
Start with arbitrary value z_n
Loop for k = n down to 1
    z_{k-1} = √{z_k - c}
    Select random value from [0, 1]
    If this value is 0, multiply z_{k-1} by -1
End loop
z_0 is a value near the boundary of the Julia set
```

As a reminder of work from chapter 5, to find the square root of a complex number, it must first be put into its polar form,

$$z = r(\cos\theta + i\sin\theta)$$

where $r = \sqrt{(x^2 + y^2)}$ and $\theta = \arctan(y/x)$. The two possible square roots can then be found from

$$\sqrt{r}\{\cos(\theta/2) + i\sin(\theta/2)\} \quad \text{or} \quad \sqrt{r}\{\cos(\pi + \theta/2) + i\sin(\pi + \theta/2)\},$$

giving the two roots in $x + iy$ form

$$x_1 + iy_1 = \sqrt{r}\cos(\theta/2) + i\sqrt{r}\sin(\theta/2))$$

and $$x_2 + iy_2 = -\sqrt{r}\cos(\theta/2) - i\sqrt{r}\sin(\theta/2)),$$

using the results $\cos(\pi + \theta/2) = -\cos(\theta/2)$, $\sin(\pi + \theta/2) = -\sin(\theta/2)$.

Results of this process are seen in fig 8.27, showing (clockwise from the top left) a dendrite, a dendrite with rings like beads on a string, and two of the Julia sets of fig 8.26 reworked using this inverse method. These are in equivalent positions in the bottom row of fig 8.27 for comparison with fig 8.26. It should be clear that the method generates the true Julia set in the form of the boundary points of the filled-in Julia set, but its random nature means that gaps may appear in boundaries where they should not. Increasing the number of iterations n in the algorithm can help to fill these gaps, at the cost of increased execution time.

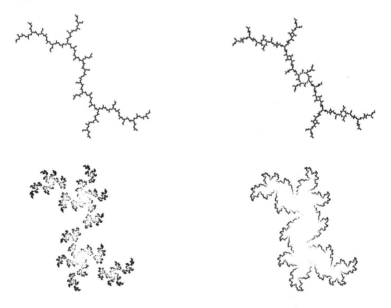

Figure 8.27 Julia sets by the inverse method, clockwise from top left, for constant c set to i, -0.15652 + 1.03225i, 0.3 + 0.5i and 0.36 + 0.5i

8. Geometric Modelling and Fractals: Building Descriptions of Objects

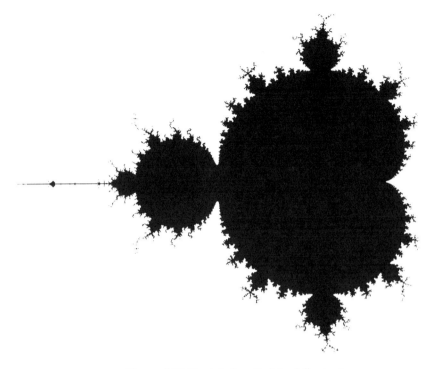

Figure 8.28 The interior of a Mandelbrot set

Figure 8.29 A banded version of the Mandelbrot set

The Mandelbrot set is created using a similar looping method and an almost identical algorithm to the direct method for creation of Julia sets described on page 244. Within a loop that visits every pixel of the display, initialize z_0 arbitrarily to zero (both x_0 and y_0 are set to zero), but this time set c to be the complex number equivalent of the current pixel. Now follow the same sequence to test whether or not it converges or diverges. Practically, for both Julia and Mandelbrot sets, the decision as to convergence is made by testing z_k to see if it lies within a large circle (radius 10 is usually adequate) from the origin $z = 0$. Typically, z_k^2 is tested against the square of the circle radius to avoid calculating a square root. As soon as z_k lies outside this circle, the sequence is considered to have diverged and the originating pixel lies outside the Mandelbrot set. If this does not happen after a large number of loop iterations (a value set at 30 or more), the point is considered to lie inside the Mandelbrot set.

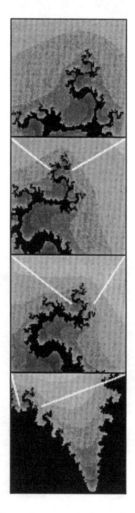

Figure 8.30 Increasing enlargement of a region in 'seahorse valley'

8. Geometric Modelling and Fractals: Building Descriptions of Objects 249

Familiar coloured images of the Mandelbrot and Julia sets are formed by shading interior points as a given colour (usually black) and colour coding exterior points by the value of k for which the sequence was found to diverge. Figure 8.28 shows a monochrome version, fig 8.29 a banded version, alternate black and white bands representing odd and even values of the number of iterations before external values are considered to have 'escaped'. The boundary of the set in fig 8.29 has a more fine-grained appearance than that of fig 8.28. It was run for n = 100 loop iterations to test divergence, so more points are determined as outside the set in this more rigorous test. Figure 8.28 used only 30 iterations; the greater coarseness allows the line to the left to be seen. This is a useful reminder that any image of a fractal object is only an approximation.

The set is truly fractal, with more and more detail being exposed as the display surface is mapped onto decreasing sections of the complex plane. Figure 8.30 shows increasing enlargement of a region of 'seahorse valley', the portion of the Mandelbrot set in the 'groove' between the two main structural forms. Many of its structures reveal shapes rather like seahorse tails. The forms revealed by repeated enlargement show no loss of detail, as well as revealing self-similar forms.

It is chastening to remember that most of the properties of Julia sets were published by French mathematicians Gaston Julia and Pierre Fatou in 1918, long before they could develop any images of the sets' fractal boundaries. Research into the properties of these structures still continues. It was discovered in the 1990s that the boundary of the Mandelbrot set has fractal dimension 2 and that its apparently disconnected islands are connected by dendrites: filaments too fine to show in normal modes of depiction. There is also one major connection between the Julia and Mandelbrot sets. Some Julia sets are connected, others are not. The Mandelbrot set consists of all values c for which the Julia set is connected.

The function $f(z) = z^2 + c$ is not the only form to generate fascinating shapes. Functions other than the f(z) defined above can be used to generate equally interesting objects with chaotic features, as shown in plate 8.1.[13]

Fractals in Simulation of Natural Phenomena

Many naturally occurring objects have fractal properties, even if they are not truly fractal. The example of coastlines has been cited above. Other objects, such as trees, shrubs and coral, exhibit repetitive branching structures that have some sense of self-similarity. Clouds and terrain have irregular structures that are susceptible to fractal modelling. A variety of specialized techniques have evolved to generate simulations of such objects. Many of these methods also involve stochastic or random methods to model irregularities. In the following sections, a variety of fractal modelling

---

[13] 'Biomorphs', forms that look like zoological cellular structures are shown in C.A. Pickover (1986) Biomorphs: Computer displays of biological forms generated from mathematical feedback loops, *Computer Graphics Forum*, 5, 313–316. Reeve finds forms by varying n in $f(z) = z^n + c$ in D.E. Reeve (1991) Mandelbrot, Julia sets and non-linear mappings, in R.A. Earnshaw, A.R. Crilly and H. Jones (eds), *Fractals and Chaos*, Springer, New York, pp. 35-42.

techniques are touched on to indicate the types of modelling applications for which they are suited. Treatment is necessarily sketchy and any imprecision in definition of the methods is entirely the fault of this author. Readers should refer to footnoted texts for more exact details of methods.

L-Systems

L-systems (or Lindenmayer systems) were originally developed in the 1960s by the biologist Aristid Lindenmayer to describe the branching structures of plants and similar objects. Stunning images of a wide variety of such structures have been produced by Lindenmayer's collaboration with Przemyslaw Prusinkiewicz.[14] The similar technique of 'graftals' was developed by Alvey Ray Smith after he heard a lecture by Lindenmayer. Structures are generated by repeatedly applying replacement rules or 'productions' to the elements of a defined alphabet, starting with an initial word or 'axiom'. These rules, applied simultaneously to all characters, generate relatively complicated words after a few iterations of the process. The characters of the alphabet are generally interpreted as geometric features, usually related to turtle graphic commands used in the language LOGO. In this, the 'turtle' is controlled by simple commands to move around the screen leaving a trail to draw (with 'pen down) or not ('pen up'). The 'turtle state' is defined by its (x, y) location and its angle of orientation (given here as an anti-clockwise angle with the x-axis at zero. Typical alphabet characters and their turtle interpretations are

    F   a forward drawing step,
    f   a step forward without drawing,
    +   a left turn of given angle δ,
    −   a right turn of the same angle,
    [initiation of a branch,
    ]   termination of a branch.

Symbols '[' and ']' denoting the start and end of a branching process are implemented as pushing and popping the current turtle state (its location and direction) onto or from a stack structure. Figure 8.31 shows the effect of the production

Figure 8.31 The result of applying F → F [− F + F] F F four times to axiom F

---

[14] P. Prusinkiewicz and A. Lindenmayer (1990) *The Algorithmic Beauty of Plants,* Springer-Verlag, New York. This is a beautifully illustrated classic, reissued in paperback form in 1996.

8. Geometric Modelling and Fractals: Building Descriptions of Objects

F → F [– F + F] F F

applied four times to the axiom F with angle δ set at 45°. This means that F is replaced by the string to the right of the arrow at each drawing stage. By convention, if no production is defined to change a character, it remains unchanged. Thus, after the first 'clock tick', the structure is encoded as

F [– F + F] F F

Using the production again to replace each character F, the second stage is

F [– F + F] F F [– F [– F + F] F F + F [– F + F] F F] F [– F + F] F F F [– F + F] F F

and the third stage

F [– F + F] F F [– F [– F + F] F F + F [– F + F] F F] F [– F + F] F F F [– F + F] F F [– F [– F + F] F F [– F [– F + F] F F + F [– F + F] F F] F [– F + F] F F F [– F + F] F F + F [– F + F] F F [– F [– F + F] F F + F [– F + F] F F] F [– F + F] F F F [– F + F] F F] F [– F + F] F F [– F [– F + F] F F + F [– F + F] F F] F [– F + F] F F F [– F + F] F F F [– F + F] F F [– F [– F + F] F F + F [– F + F] F F] F [– F + F] F F F [– F + F] F F

This shows how a complex structure can be generated from relatively simple rules. At each stage, the length of drawing of F is reduced by one third to keep subsequent drawings at the same scale. A plant-like shape appears from this relatively simple rule, but it is too regular in its structure or topology and its shape to be realistic. Use of stochastic productions, where a choice of results is available under certain probabilities, and stochastic variation of drawing rules produce greater irregularity. Most computer languages have built-in random number generators to assist in stochastic choice or variation. Figure 8.32 shows structures subject to both forms of change, giving a more naturally irregular appearance. The production for F is

Figure 8.32 L-systems 'plants' with irregular structures and drawing characteristics

$$F \begin{cases} \xrightarrow{0.95} F[+F][<F]F[-F]F \\ \xrightarrow{0.05} \% \end{cases},$$

new characters '<' indicating a left turn of 30° and '%' indicating that the whole branch structure depending on this character is pruned.[15] This is invoked randomly on 5% of occasions, otherwise the upper production is used to develop the structure. When the L-system is drawn, the turtle angle is given a random deviation biased towards the upward direction, hence the slightly curved appearance of some branches simulating the orthotropic growth of some plants.

Use of three-dimensional drawing rules and a wider variety of drawing primitives to interpret more extensive alphabets and productions enable realistic depictions of plant-like structures. Particular extensions include the use of context-sensitive productions, where neighbouring characters in the string affect the application of productions, environmentally sensitive productions, where the production depends on the physical location of the character being processed, and parametric L-systems, in which characters' data structures carry parameters that can themselves be used to control the productions. Typical parameters include the number of 'clock ticks' since the character was formed, or some indication of size, colour or content of the character.

This has been used to simulate the growth of a mould fungus, *Aspergillus nidulans*. Small sections of the fungus are represented as string characters, each having a nutrient content parameter, which is distributed along the string from the fungus spore at each growing stage. When the content of particular elements reaches a threshold, this triggers growth as the creation of a new branch or the extension of a growing tip (fig 8.33). The method was extended to model plant growth by adding a reverse distribution of energy (biologically this takes the form of starches and sugars created by photosynthesis) backwards down the L-system string from leaves to a forward distribution of nutrient from the root. Combinations of energy and nutrient were used to control thickening, extension of branches and creation of

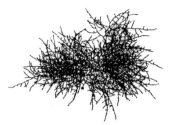

Figure 8.33 Simulation of the fungus *Aspergillus nidulans* using parametric L-systems by Andrew Tunbridge

---

[15] This is after a method of Prusinkiewic and his collaborators in one of their many excellent follow-up papers to Prusinkiewic and Lindenmayer (1990), the book previously cited. The full reference is P. Prusinkiewicz, M. James and R. Měch (1994) Synthetic topiary. *Proceedings of ACM SIGGRAPH '94*, 351–358.

8. Geometric Modelling and Fractals: Building Descriptions of Objects   253

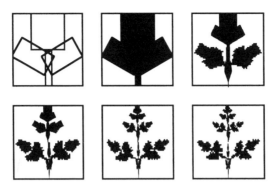

Figure 8.34 Five stages in generation of an IFS from the transformations of the top left image

leaves. Plate 8.2 shows a tree with fallen leaves and growing shoots from reseeding. The leaves to the left of the image get less sunlight than those to the right, having an effect on the overall growth. Most productions simply change the nutrient and energy parameters; structural change or growth is achieved when these parameters reach thresholds, triggering the application of topology-changing productions, or thickening trunk and branch sections.[16]

Iterated Function Systems (IFS)

This method has been widely developed by Michael Barnsley[17] and his collaborators. There is considerable interest in using this method for image data compression, as complex images can be encapsulated by a much reduced data set and a simple image regeneration rule. An iterated function system (IFS) consists of a set of contractive transformations, usually taken as affine transformations, generally combinations of scalings, rotations and translations that reduce the distances between points. These transformations are simultaneously applied to an original image, and then repeatedly applied to the results of the previous application. After a few stages, the resulting image stabilizes to the 'attractor' of the IFS, the set of points that will be unchanged by simultaneous application of the affine transformations. Figure 8.34 shows the first five stages of applying an IFS with four contractive affine transformations applied to an outer square. The first image shows this square and the outlines of the transformed regions, representing the trunk, lower left branch, lower right branch and upper tree of a tree-like attractor. After relatively few stages, the IFS has approached very close to its attractor; the changes seen in the final step are quite minor. The functions are built using affine

---

[16] A summary of these methods is found in H. Jones, A. Tunbridge and P. Briggs (1997) Modelling of growing biological processes using parametric L-systems, in *Visualization and Modeling*, R.A. Earnshaw, H. Jones and J. Vince (eds), pp 303–317, Academic Press, San Diego. Images are by the author's former students, Andrew Tunbridge (fig 8.33) and Paul Briggs (plate 8.2). Jaap Kaandorp has developed a geometric growth rule method for modelling sponges and corals: J.A. Kaandorp (1994) *Fractal Modelling: Growth and Form in Biology*. Springer.
[17] M.F. Barnsley (2000) *Fractals Everywhere (3rd edn)*, Morgan Kauffman.

construction rules discussed in chapters 4 and 6. If the y-axis passes down the centre of the square, with the x-axis along its base (the lower left corner is (-0.5, 0), the upper right corner is (0.5, 1), they are:

- Lower trunk   Scale (0.1, 0.5).
- Lower left branch   Scale (0.3, 0.5), Rotate (60°), Translate (0, 0.3).
- Lower right branch Scale (0.3, 0.5), Rotate (-60°), Translate (0, 0.3).
- Upper tree   Scale (0.5, 0.5), Translate (0, 0.5).

An alternative rule, often called the 'chaos game', generates the deterministic attractor using a stochastic method. Suppose each of the affine transformations can be chosen according to a given probability. Best results are obtained by making the probability of a transformation proportional to its contraction ratio. Given an arbitrary initial point, apply a selected transformation to that point and plot the resulting position. By repeatedly selecting transformations, applying them to the point generated in the previous stage and plotting the new point, the attractor is gradually filled. Figure 8.35 shows the result of using this technique on the four transformation IFS of fig 8.34, running the process for 10 000, 100 000 and 1 000 000 iterations, this time with equal probabilities of choice of transformations. This tends to give greater 'fill' of smaller transformations, such as the lower trunk, and less dense 'fill' of the larger, such as the upper tree. However, if enough points are generated, this random process will eventually completely fill the deterministic object; the same result will be seen each time. These may seem like long loop runs, but images appear in the blink of an eye. This shows how, as the number of iterations is increased, filling of the attractor becomes more complete. The method of fig 8.34 closes in on the attractor from outside; that of fig 8.35 fills it in from the inside. The true attractor is somewhere between the two versions.

The Sierpinski triangle shown earlier (fig 8.20) can be generated by a particularly easy version of the chaos game. Given the three triangle vertices, V_0, V_1, V_2 and an arbitrary starting point P_0, the sequence of points P_1, P_2, P_3, ... to be plotted is generated by the following simple algorithm:

```
Loop for i from 1 to a large number
   Select V_i at random from V_0, V_1, V_2
   Set P_i = (P_{i-1} + V_i)/2
   Plot P_i
End I loop
```

The points P_i plotted are midway between the previous point P_{i-1} and a randomly selected vertex V_i. An arbitrary starting point generates a few stray points before the attractor is reached; the process cannot escape from it. To avoid such stray points,

Figure 8.35 The IFS of fig 8.34 using the stochastic point method

the process can be run several stages before plotting begins, or a starting point known to be inside the attractor can be chosen. Any of the vertices V_0, V_1 or V_2 could have been chosen in this case.

It is clear that the transformations used affect the final image. There is an 'art' to defining transformations to produce specific effects, and experience is useful in recognizing how sub-parts of an object can be modelled from the whole. The 'collage theorem' is useful: an IFS can be created as a collage of the original image, overlaying possibly overlapping copies of the original as an artist would overlay pieces of paper. Interactive methods for defining IFS transformations have been developed, and automatic methods are available for IFS image compression, but these are still relatively slow.[18]

The examples shown above lie in a 2D space, but the method can be extended to depict objects in a 3D universe. A 3D version of the stochastic point generation method can be implemented relatively easily; transformations in 3D can be composed using affine matrix methods. 3D points depicted using z-buffer and shadow buffer algorithms (chapter 10) and given colour codes according to the selected transformation give particularly attractive effects. A Sierpinski tetrahedron (fig 8.19) can be generated using the same algorithm as for the Sierpinski triangle, but applied to four vertices V_0, V_1, V_2 and V_3 in a 3D environment (plate 8.3). Similar methods can be used to generate abstract fractals based on other geometric shapes such as cubes or octahedra. By using non-affine transformations, the range of options is extended. Plate 8.4 shows a copse of 10 trees using non-affine transformations in the branch generation, as does the fern of plate 10.4. Other plant-like features and clouds have been modelled by the method.

Figure 8.23 showed how a square contains four half-scale copies of itself. By extending the random point generation algorithm of page 254 to the vertices of a square, the square is evenly filled, as the algorithm places four half-scale copies of the square into the appropriate places. This is how the floor plane of plate 8.4 was generated. The author has used this to create models of 3D objects in a computer graphics environment. We saw in chapter 5 that surfaces of objects such as cones, cylinders and spheres can be represented by equations using two parameters. By mapping the coordinates of points created in an IFS square to these two parameters, the surfaces of the objects are filled by an IFS algorithm. This is how the ball and sections of the plant pot were created in plate 10.4. A simple adaptation of this method can also be used to fill triangles (fig 8.23 shows how the triangle also contains four half-scale copies of itself) and any quadrilateral, so the surface of any polyhedral object can be created using this method, through subdividing all surface polygons into triangles and quadrilaterals. The IFS method is applied separately to each of these regions. This technique is used by the author for creating smoothness from a fractal modelling technique, when the normal outcome of a fractal technique is to create objects that are the antithesis of smooth. The method is slower than established techniques described in chapter 10, but only one generated point at a time is rendered, so many of the problems of rendering surfaces in classical computer graphics, such as clipping and depth comparisons, are simplified. Plate 8.3 was created by the author's former student Aurelio Campa.

---

[18] See, for example, Y. Fisher (1994) *Fractal Encoding: Theory and Applications to Digital Images*, Springer, or N. Lu (1997) *Fractal Imaging*, Academic Press.

Terrain Modelling

The modelling of terrain has become a specialized area. Vast data sets are needed to hold information on landscapes. A popular technique is to store coarse-grained information on regions, simulating more detailed aspects algorithmically. A variety of advanced methods is used to generate the detail outline curves. A simple and popular technique uses 'mid-point displacement', where random displacements are added to the mid points of line segments, thus creating two new line segments which are similarly (and recursively) subdivided. Given two initial end points, a terrain outline curve can be created. At each stage of the subdivision, the range of the random dispersion is halved, so the deviation is in scale with the length of line being perturbed. By controlling the random displacement, the effects of smooth (smaller displacements) or jagged terrain (having larger displacements or larger fractal dimension) can be simulated. The method is illustrated in fig 8.36. Deviation ranges are shown as vertical bars; the outline of the previous stage is also shown for comparison. This method can also be used to generate simulated coastlines.

Surfaces can be synthesized applying this technique to grid heights (fig 8.37). Given four known heights at the corners of a grid square, mid-point displacement heights for each side of the square are found, and four new grid squares created using these heights, the original grid vertices and a new vertex at the grid centre. This is allocated a height equal to the mean of the four original grid points (as in fig 8.37), which may itself be randomly deviated. The five points so created, plus the original four, define the vertices of the four new half-scale grid squares, which can be similarly subdivided to create as much detail as is required in the terrain model, as in plates 8.5 and 8.6. For rendering, the mostly non-planar sections created are usually triangulated by drawing a diagonal for each surface section (the sections are hyperbolic paraboloids, but this is too much of a digression to explore). The plane triangles so created form a B-Rep surface that can easily be displayed. This method can give rise to a phenomenon of 'creasing', by 'clinging' to old extreme boundaries that remain visible as subdivision continues. An improvement is achieved by alternating grid replacement points between a horizontal and diagonal grid layout to avoid such artefacts. Figure 8.38 shows a plan view of the alternate horizontal and diagonal arrangements.

Figure 8.36 Generation of a curve using mid-point displacement

8. Geometric Modelling and Fractals: Building Descriptions of Objects 257

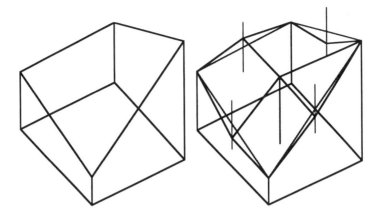

Figure 8.37 Creating surfaces by mid-point subdivision using an averaged central height and randomized dispersions of grid edges

The basic mid-point displacement method is simple and can give visually pleasing effects, although geologists may note that it does not give convincing geological models. For example, turning a model upside down generates a similar appearance as there is no bias between up and down. This does not occur with natural terrain. Some landscapes have jagged peaked mountains and smooth valleys (formed through deposition of soil from the eroded higher land). Such features can be simulated by amending the method either by lowering the range of the random deviation for lower heights, or by post-processing the heights found according to a power rule. A simple method replaces the terrain height by its own square or cube, and then scales the height down to achieve the required height ranges. This gives very pointed peaks, so other functional adjustments may be implemented in an ad hoc way. Plate 8.5 shows an example with broad valleys and not too pointed peaks, coloured to show tree and snow lines. Another method of simulating coastlines is to slice a 3D terrain model with a horizontal plane at the required sea level as in plate 8.6. The author's former students Denis Crampton and Semannia Luk Cheung created plates 8.5 and 8.6 respectively.

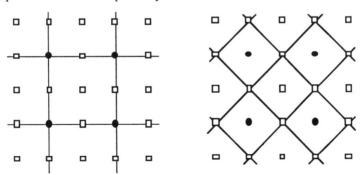

Figure 8.38 Arrangement of grid points to avoid creasing. Black circles are original reference points, white squares are new points inserted at the first stage using the grid division of the first image. At the second stage, new points will be inserted by the same method based on the diagonal grid division.

Summary

This chapter discusses a number of ways of creating models for display by computers. First, standard methods for geometric modelling are covered, such as voxel, CSG and B-Rep methods. These are all discussed in terms of their benefits and disadvantages. A selection of the most popular fractal modelling methods is also presented, in the context of the idea of fractal dimension. Other methods, such as diffusion limited aggregation, are not discussed here.

Just as an inspiring scene is necessary to create a landscape painting or photograph, interesting geometric models are needed to be interpreted as images in computer graphics. In current methods, the creation of the geometric model is often the most human time-consuming part of image generation, renderers, as discussed in chapter 10, being available as standard packages.

9. Splines: Generation of Curves and Surfaces

Introduction

Many B-Rep modellers allow the generation of 'spline surface patches': curved surface sections defined by formulae that are controlled relatively simply through user manipulation of sets of control points. The shapes of surfaces generated are relatively intuitively related to the locations of control points. Spline 'patches' can be treated topologically as distorted faces, and can be linked to other patches with a controlled level of smoothness at joins. In this way, complex smooth surfaces can be created from collections of spline patches, or patches can be joined with deliberate irregularities to give a wide variety of possibilities for object modelling. As there are many such possibilities, there is also a range of methods for defining and controlling spline surfaces. NURBS (an acronym for non-uniform rational B-splines) can be patched together to give exact models of spheres, cylinders, cones and tori as well as more 'free-form' surfaces. Simpler methods such as Bézier splines are popular, more easily understandable and very effective. These were separately and simultaneously developed for car body design at French manufacturers by Pierre Bézier at Renault and Paul de Casteljau at Citroën.

Reasons for Splines

Spline curves are 'piecewise continuous' curves. It is difficult to devise a single formula to describe a complicated modelling curve; even if this is possible, the formula will also be complicated. Spline methods subdivide such curves into more manageable sections. By modelling in a relatively intuitive way, formulae are devised for curve sections that have readily understood properties. These can be fitted together continuously to reproduce the effect of a single curve. 'You can't see the join' is a good motto for spline curve users. These may be designers of motor cars, shoes, ships, aeroplanes or fonts. Pioneer developers of spline curve methods from the late 1950s onwards include James Ferguson (Boeing), Pierre Bézier (Renault) and Paul de Faget de Casteljau (Citroën). Their early work was often not published and was preceded by theoretical studies by Lobaschevsky in the 19th century based on differential geometry developed by Euler, Monge, Gauss and others in the 18th and 19th centuries. Systems that control complex 'free form' curves and surfaces using relatively few control points and parameters are powerful design tools. NURBS are used in many computer graphics and CAD systems. The following sections describe spline curve formulations through the fundamental concept of interpolation, leading through Bézier curves and interpolating splines to the description of NURBS and their properties. The uses of splines in controlling animation and for 'free form' surface construction are also described.

There are many different ways of defining splines, but most have similar properties or are no more than reformulations of more standard types. The selection of techniques introduced here should give readers enough understanding of splines to enable them to follow other methods. Most of the following is in the context of 3D space. This is not a constraint: splines can be drawn in 2D space or conceived in four or more dimensions. The conceptual device of 4D space will be used as a temporary stepping stone in devising NURBS. There is inevitably some algebraic manipulation involved in this work. Those who find this exceptionally difficult, in spite of their best efforts, should try to understand the concepts involved at first reading. The development below is intended to be mathematically sound, but also readable at a descriptive level. Some of the theory underlying splines, such as considerations of higher-level continuity, curvature and torsion, is not covered. Those who wish to specialize in spline curve and surface development should see this as no more than an introductory text. Others, who want merely to use or implement simple forms of splines, may find the following descriptions adequate for their needs.

Interpolation

Interpolation is the art of approximating intermediate values from a set of incomplete data. It is typically required if a sequence of positional readings is taken at regular time steps. To create an animated sequence at a better time precision, some of the gaps need to be filled in. Extending beyond the domain of a known set of values is a related problem known as extrapolation. As a simple example, suppose we wish to find the mid-point of two Cartesian points A (4, -2, 10) and B (12, 14, 6). It seems fairly obvious to average the coordinates to give (8, 6, 8) as the mid-point. This can be represented in shorthand form as

$$\tfrac{1}{2}(A + B) = \tfrac{1}{2}A + \tfrac{1}{2}B = (8, 6, 8),$$

where the factor of $\tfrac{1}{2}$ is applied to all components of A, B, and so on. How would we find the point $\tfrac{1}{4}$ of the way from A to B? This contains 'more' A than B. The correct combination is

$$\tfrac{3}{4}A + \tfrac{1}{4}B = (6, 2, 9).$$

This is quite easy to justify. The difference between A and B is the vector from A to B,

$$B - A = (8, 16, -4).$$

A quarter of this difference is (2, 4, -1), so the required point is

$$A + \tfrac{1}{4}(B - A) = \tfrac{3}{4}A + \tfrac{1}{4}B.$$

9. Splines: Generation of Curves and Surfaces

Here, we have used A and B for the position vectors (chapter 7) of points A and B as well as the points themselves. No ambiguity is caused by this lax use of notation, and we continue with this method for the rest of the chapter.

Now consider a more general case: suppose we wish to find a general point P somewhere between A and B.

$$P(u) = A + u(B - A) = (1 - u)A + uB$$

is a point a fraction u of the way from A to B when u is a number between 0 and 1 inclusive (formally stated as $0 \leq u \leq 1$). If u changes evenly from 0 to 1, the point P(u) will move at a steady pace from A to B. This indicates how u could be used to indicate time in plotting steady motion from one point to the next. This form of interpolation is called linear interpolation. This is not because the point P(u) moves in a straight line, but because its motion is uniform as u changes – the upper part of fig 9.1 shows how constant changes in u give equal changes in position of P(u). The formula for P(u) can be written

$$P(u) = f_a(u)A + f_b(u)B,$$

where $f_a(u) = (1 - u)$ and $f_b(u) = u$. These are linear functions; they do not contain squares or higher powers of u. This is what gives P(u) its even steps. The functions $f_a(u)$ and $f_b(u)$ are called blending functions. They give the relative 'amounts' of A and B that are blended to give the resulting point P(u), as if they are ingredients in a recipe. Note that, regardless of the value of u,

$$f_a(u) + f_b(u) = 1.$$

Also, $f_a(u) \geq 0$ and $f_b(u) \geq 0$ when $0 \leq u \leq 1$. These properties make P(u) a weighted average of the points A and B. If masses $f_a(u)$ and $f_b(u)$ are placed at A and B respectively, the position of P(u) is their centre of mass – P(u) is the balance point for these masses.

If blending functions are non-negative and add to one, the point

$$P(u) = f_a(u)A + f_b(u)B$$

lies in the segment of straight line between A and B. Consider the point

$$Q(u) = g_a(u)A + g_b(u)B,$$

where $g_a(u) = (1 - u^2)$ and $g_b(u) = u^2$. For $0 \leq u \leq 1$, $g_a(u) \geq 0$, $g_b(u) \geq 0$ and $g_a(u) + g_b(u) = 1$, so Q(u) also lies on the line between A and B. However, equal

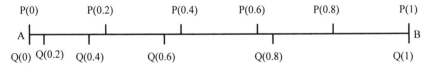

Figure 9.1 Linear (P) and non-linear (Q) interpolation

changes in u will not step Q(u) evenly along the line, as illustrated in fig 9.1. We see that even changes in u give relatively short steps of Q(u) when u is small. These steps increase as u approaches 1. Although Q(u) stays in the straight line AB, this is not 'linear interpolation', as the blending functions used involve squared terms. In the following sections, we begin with linear interpolation, but it is important to remember that this is not the only way to define intermediate values or points. As an exercise, it would be useful for the reader to calculate the positions of P(u) and Q(u) for u = 0, 0.2, 0.4, and so on, when A and B are defined as above.

Bézier Splines for Curve Drawing

We have seen how linear interpolation can be used to 'sweep' a point evenly along a line segment. The process can be extended to create smooth curved forms, called Bézier spline curves. Pierre Bézier's method was developed at Renault for use in the design of car bodies. The curves are controlled by a sequence of points called 'control points' and can be described in terms of blending functions or through a recursive geometric algorithm that carries the name of de Casteljau. Paul de Faget de Casteljau's method was developed (possibly before Bézier's) at Citroën. This technique has an intuitively convincing pictorial explanation (fig 9.2), which will be developed before showing its interpretation using blending functions. We concentrate on the cubic Bézier spline curve, which is the most generally used form. This has blending functions that involve cubic terms and is defined in terms of four control points. The second-order form is simpler to deal with, but is too restricted to be of general use; for example, second-order curved sections lie in a plane, so it is not possible to create convincing free form 3D curves. Higher-order curves can be created, but collections of cubic curves are generally powerful enough to approximate any required form without needing the increased complexity of higher powers. Manipulation of control points enables manipulation of the curves to match most circumstances.

A cubic Bézier spline curve is based on a set of four control points: P_0, P_1, P_2 and P_3 say. The open 'polyline' (this describes a linked set of lines) joining these points is called the control polygon. In the illustrations of fig 9.2, these lie in the 2D space of the page, but in general the points can be considered to lie in 3D space.

The first stage involves linear interpolation along the three lines of this 'polygon', to define three points, one for each of these lines:

- $P_0'(u) = (1 - u)P_0 + uP_1$,
- $P_1'(u) = (1 - u)P_1 + uP_2$,
- $P_2'(u) = (1 - u)P_2 + uP_3$.

Figure 9.2 illustrates these first-stage points for $u = \frac{1}{4}$. As u changes from 0 to 1, the points move simultaneously, $P_i'(u)$ moving from P_i to P_{i+1}. These points are joined to form the open polyline $P_0'(u)P_1'(u)P_2'(u)$. For the second stage, further interpolation on the two lines involved in this construction gives two new points:

9. Splines: Generation of Curves and Surfaces 263

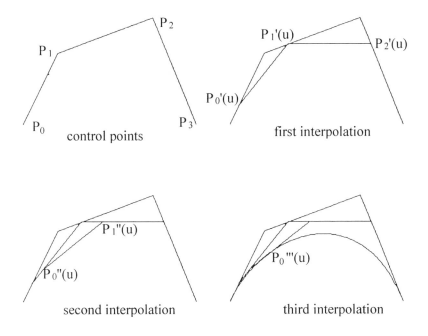

Figure 9.2 Generation of a cubic Bézier curve with u = 0.25

- - $P_0''(u) = (1 - u)P_0'(u) + uP_1'(u)$,
- - $P_1''(u) = (1 - u)P_1'(u) + uP_2'(u)$.

It is important that the same value of u is used in all interpolations. P_0, P_1, P_2 and P_3 remain fixed during this process, but the trellis hanging from the structure in fig 9.2 moves generally from left to right of the picture as u increases. The two points $P_0''(u)$ and $P_1''(u)$ have now been derived from the four original points. The third stage of interpolation is performed on these points to give the final Bézier point:

- - - $P_0'''(u) = (1 - u)P_0''(u) + uP_1''(u)$.

It is more convenient to use C(u) to define a general point on the curve,

$C(u) = P_0'''(u)$.

As u changes from 0 to 1, C(u) defines the Bézier cubic curve for control points P_0, P_1, P_2 and P_3. Figure 9.2 shows all stages in the generation of C(0.25), which is the same as $P_0'''(0.25)$. A typical algorithm for drawing a Bézier curve calculates a sequence of points for small steps in u and joins these up by very short line segments, to give the appearance of a smooth curve as in the final image of fig 9.2.

There are obvious similarities in the formulae for each of these interpolation stages. They can be recast in general form as

$P_i^{(k)}(u) = (1 - u)P_i^{(k-1)}(u) + uP_{i+1}^{(k-1)}(u)$, where $P_i^{(0)}(u) = P_i$. The point $P_i^{(k)}(u)$ results from the kth stage of interpolation (the superfix k in brackets indicates k dashes) between the points with suffices i and i + 1 in the $(k - 1)^{th}$ stage. This formula, together with the definition of the control points P_i as terminating values, is the basis of de Casteljau's algorithm for generation of Bézier splines of any order. A Bézier spline of order k is based on k + 1 control points and can be found by extension of the cubic algorithm.

The same curve can be generated using blending functions, which can be found by working through the recursive algorithm stage by stage. The point on the curve for parameter u is given as

$$C(u) = P_0'''(u) = (1 - u)P_0''(u) + uP_1''(u).$$

Substituting for $P_0''(u)$ and $P_1''(u)$ into this formula gives

$$\begin{aligned}C(u) &= (1 - u)\{(1 - u)P_0'(u) + uP_1'(u)\} + u\{(1 - u)P_1'(u) + uP_2'(u)\} \\ &= (1 - u)^2 P_0'(u) + 2(1 - u)u P_1'(u) + u^2 P_2'(u).\end{aligned}$$

The next stage substitutes for $P_0'(u)$, $P_1'(u)$ and $P_2'(u)$ to give C(u) in terms of the control points.

$$\begin{aligned}C(u) &= (1 - u)^2 P_0'(u) + 2(1 - u)u P_1'(u) + u^2 P_2'(u) \\ &= (1 - u)^2 \{(1 - u)P_0 + uP_1\} + 2(1 - u)u\{(1 - u)P_1 + uP_2\} \\ &\quad + u^2\{(1 - u)P_2 + uP_3\} \\ &= (1 - u)^3 P_0 + 3(1 - u)^2 u P_1 + 3(1 - u)u^2 P_2 + u^3 P_3.\end{aligned}$$

This relates C(u) directly to its control points P_0, P_1, P_2 and P_3. Implicitly, this is a parametric curve. C(u) has three coordinate components: x(u), y(u) and z(u). In discussing functions earlier, we stressed that functions had to be 'single valued', so here one value of u produces one and only one possible value of x (and of y and z). However, x and y and z are linked as all three use the parameter u. Given a value u, we can find one and only one point $C(u) = (x(u), y(u), z(u))$. Strictures on curves not 'doubling back' on themselves are now freed. Although values of x cannot 'double back' on u (there is only one value of x for any value of u), the combined effect is that parametrically defined curves, through astute combinations of their separate functions linked by the parameter u, may 'double back' x upon y or upon z; the curves may even cross through themselves. Parametric curves are free from some of the restrictions imposed by definition on simple functional curves.

The method can be extended to give similar definitions for Bézier curves of other orders, but the structure of this solution hints at a general formula for any order of curve. In C(u), control points are multiplied by blending functions; P_0 by $(1 - u)^3$, P_1 by $3(1 - u)^2 u$, P_2 by $3(1 - u)u^2$ and P_3 by u^3. A standard result from algebra gives

$$(a + b)^3 = a^3 + 3a^2 b + 3ab^2 + b^3.$$

Replacing a by (1 − u) and b by u, we have

$$[(1 - u) + u]^3 = (1 - u)^3 + 3(1 - u)^2 u + 3(1 - u)u^2 + u^3.$$

9. Splines: Generation of Curves and Surfaces

The blending functions for a cubic Bézier curve (commonly known as Bernstein polynomials, although some mathematicians insist on a slight distinction between the Bézier blending functions and true Bernstein polynomials) are terms from the expansion of $[(1 - u) + u]^3$, hinting at an extension to other powers. Bézier splines of power n have blending functions taken from the expansion of $[(1 - u) + u]^n$.

As $[(1 - u) + u]^n = [1 - u + u]^n = 1^n = 1$,

the blending functions always add to 1 regardless of the values of u and n. All terms in the blending functions are multiples of $(1 - u)$ and u, whose values remain between 0 and 1 inclusive when $0 \leq u \leq 1$. As these are non-negative, the blending functions are always non-negative.

These properties imply that C(u) is a 'weighted mean' of the control points P_0, ... P_n. In physical terms, this can be interpreted as locating C(u) at the centre of mass of a collection of (non-negative) masses of values equal to the blending functions at each respective control point. For the cubic case, there are four such masses,

$(1 - u)^3$ at P_0,
$3(1 - u)^2 u$ at P_1,
$3(1 - u)u^2$ at P_2,
and u^3 at P_3.

This interpretation of the location of C(u) implies that it lies within the 'convex hull' of the control points. The convex hull of a set of objects is the smallest convex shape that contains them. For four control points in 3D space, their convex hull is generally a tetrahedron with the points as vertices – imagine 'shrink wrapping' the points in space. Default cases occur when the points are in the same plane or line. The centre of mass of a collection of masses must lie 'inside' them, showing the validity of the 'convex hull property'.

Setting u = 0 and u = 1 into the curve equation

$$C(u) = (1 - u)^3 P_0 + 3(1 - u)^2 u P_1 + 3(1 - u)u^2 P_2 + u^3 P_3$$

gives $C(0) = P_0$ and $C(1) = P_3$.

The curve starts and ends at the start and end points respectively of the control polygon.

If the curve equation is differentiated with respect to u, we get

$$\frac{dC(u)}{du} = -3(1 - u)^2 P_0 + \{3(1 - u)^2 - 6(1 - u)u\} P_1 + \{6(1 - u)u - 3u^2\} P_2 + 3u^2 P_3.$$

Some readers will find this difficult – the concept is explained in chapter 5. This example uses the 'function of a function' rule. The equation could be 'tidied up'; we do not bother. We concentrate on the meaning of this term, which is not too difficult to understand. $\frac{dC(u)}{du}$ is the velocity of $C(u)$ if u is interpreted as time. The properties of $C(u)$ at $u = 0$ and $u = 1$ are important. Even if the above formula for $\frac{dC(u)}{du}$ is taken on faith, it is not difficult to substitute $u = 0$ and $u = 1$ to give

$$\frac{dC(0)}{du} = -3P_0 + 3P_1 = 3(P_1 - P_0)$$

and $$\frac{dC(1)}{du} = -3P_2 + 3P_3 = 3(P_3 - P_2).$$

Remembering that $C(0) = P_0$, and $C(1) = P_3$, this shows that the initial velocity of $C(u)$ as it leaves P_0 is in the direction of $(P_1 - P_0)$, which is the line P_0P_1 joining P_0 to P_1. The final velocity of $C(u)$ as it approaches P_3 is in the direction of $(P_3 - P_2)$, the line P_2P_3 from P_2 to P_3 (fig 9.2). The implication is that the curve $C(u)$ is tangential to P_0P_1 at P_0 and to P_2P_3 at P_3. This gives designers a concept of how the curve relates to the control points. We also see that the values of the velocities at these end points are three times the lengths of the first and last line segments of the control polygon respectively. This is useful in planning animations, as discussed below. The cubic curve is used here, but analogous properties are available for other powers. For example, the 'velocity' at P_0 of an nth order Bézier spline is $n(P_1 - P_0)$.

More manipulation of the formula for $\frac{dC(u)}{du}$ in conjunction with other formulae from above gives

$$\frac{dC(u)}{du} = 3\{P_1"(u) - P_0"(u)\},$$

showing that the curve at $C(u)$ is tangential to the 'trellis' line from which it is found in the de Casteljau algorithm, $P_0"(u)P_1"(u)$ in fig 9.2. Expert mathematicians may rise to the challenge of developing this result; others are asked to take it on trust. In any case, the statement is again valuable for designers in generating a mental image of the shape of the curve. Some typical cubic Bézier curve shapes are illustrated, with control polygons, in fig 9.3. Cubic Bézier curves can trace simple bends, can 'cross over' themselves or can generate 'S bends' (remember the discussion a few pages back on this form of parametric curve). The limitation of the cubic form means that 'triple bends' cannot be generated by a single curve.

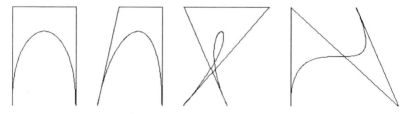

Figure 9.3 Some typical cubic Bézier curve shapes with their control polygons

9. Splines: Generation of Curves and Surfaces   267

This is not really a limitation as, knowing the end point properties of a single cubic Bézier curve section, several such curves can be linked smoothly to produce composite curves with any number of bends. We consider first how to join two Bézier cubic curve sections: one defined with control points P_0, P_1, P_2 and P_3, the other with respect to points Q_0, Q_1, Q_2 and Q_3 (fig 9.4). If the junction is 'smooth', there should be no way of detecting where it occurred by viewing the composite curve alone. If the end of the first curve at P_3 is to coincide with the start of the second at Q_0, we must have $P_3 = Q_0$. For continuous tangents (first-order smoothness with no change of direction in passing from one cubic section to the next), we must also arrange for P_2P_3 to be in the same direction as Q_0Q_1, so $P_2P_3Q_1$ (or, equivalently, $P_2Q_0Q_1$) are collinear (fig 9.4). This shows why there is no great need to extend beyond the cubic form. With control of smoothness at the junctions, long sequences of cubic Bézier curves can be joined to produce composite curves with any required degree of complexity. If the links are properly controlled, there should be no way of detecting the junction points by viewing the curve alone. This is a powerful technique for designers to approximate any required form of space curve.

One property of Bézier curves in general is worth mentioning. Bézier curves are 'affine invariant'. The affine transformations, formed by combinations of translations, scalings and rotations, were discussed earlier. If we are given a set of control points P_0, P_1, P_2 ..., calculate a point on a Bézier curve $C_P(u)$ from these points and then subject $C_P(u)$ to an affine transformation T, we arrive at a point $T\{C_P(u)\}$, say. Now suppose we apply the affine transformation first to the control points, giving a new set $T(P_0)$, $T(P_1)$, $T(P_2)$, ..., and then generate a new Bézier curve point from these as control points, $C_{TP}(u)$, say, then the result will be the same for both processes. If we wish to move the curve according to an affine transformation, it is easier to apply the transformation to the relatively few control points and then reconstitute the curve according to these new points.

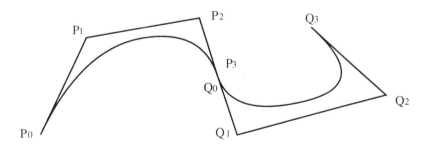

Figure 9.4 Smooth join of two cubic Bézier curves

Animation Control Using Cubic Bézier Curves

The topic of animation was discussed briefly in chapter 5. Spline curves, with their known properties, are useful for creation of animation along smooth paths. Some of the mathematical details here may be difficult on first reading, so readers should concentrate initially on understanding concepts rather than concerning themselves with detailed mathematical expansion. This can come later.

We have seen how to produce complex curves with spatial continuity. To create convincing animations, we need to create continuity in time. If an object is to be animated to simulate natural motion, its path and, just as importantly, its velocity must not contain jumps. This involves linking curves with positional and tangent continuity, but also controlling the spline generation point to approach the end of one curve section at the same velocity as it starts the next curve section (rather like a smooth transfer of the baton in a relay race). Such a spatially and temporally seamless link can be formed for a sequence of Bézier spline sections.

Consider two curves as in fig 9.4 with the curve parameter u directly representing time. The velocity of the curve point on leaving the first curve is $3(P_3 - P_2)$ and the velocity on entering the second curve is $3(Q_1 - Q_0)$, where $(P_3 - P_2)$ is the line segment from P_2 to P_3 and $(Q_1 - Q_0)$ is the line segment from Q_0 to Q_1. For spatial continuity and equal velocities at the point of transfer from one curve to the next, P_3 and Q_0 are coincidental, line segments P_2P_3 and Q_0Q_1 are equal in length as well as in direction, and the second curve starts with local parameter $u_q = 0$ when $u_p = 1$ for the first curve. This method also ensures continuity of curvature at the join. The severity of the bend is smoothly changed, an important property for road designers.

However, the situation may not always be as simple. Suppose an animator wants an object to pass through P_0 at time t_0, through P_3 (and Q_0) at t_1 and finally through Q_3 at time t_2. u_p changes from 0 to 1 for the 'P curve' as time t passes from t_0 to t_1 and u_q changes from 0 to 1 for the 'Q curve' as time t passes from t_1 to t_1. This is achieved by defining

$$u_p = \frac{t - t_0}{t_1 - t_0} \quad \text{for } t_0 \leq t \leq t_1, \text{ used for the 'P curve'},$$

$$u_q = \frac{t - t_1}{t_2 - t_1} \quad \text{for } t_1 \leq t \leq t_2, \text{ used for the 'Q curve'}.$$

The time taken to pass through the 'P curve' is $t_1 - t_0$, and that taken to pass through the 'Q curve' is $t_2 - t_1$. If $t_1 - t_0$ is large, the approach to P_3 will be slow; if $t_1 - t_0$ is small, this approach will be fast. The velocity on arriving at P_3 is $\frac{3(P_3 - P_2)}{t_1 - t_0}$ and that on leaving P_3 (or Q_0) is $\frac{3(Q_1 - Q_0)}{t_2 - t_1}$. These are equal if

$$\frac{(P_3 - P_2)}{t_1 - t_0} = \frac{(Q_1 - Q_0)}{t_2 - t_1}.$$

9. Splines: Generation of Curves and Surfaces

This is a vector equation. We have already insisted on the directions of these vectors being the same in order to ensure tangent continuity. Now we need to impose equality of length, so considering lengths alone,

$$\frac{P_2P_3}{t_1 - t_0} = \frac{Q_0Q_1}{t_2 - t_1},$$

or

$$\frac{P_2P_3}{Q_0Q_1} = \frac{t_1 - t_0}{t_2 - t_1}.$$

The ratio of the length P_2P_3 to the length Q_0Q_1 is equal to the ratio of the times taken to cover the 'P curve' and the 'Q curve'. This condition, with $P_3 = Q_0$ and the collinearity of $P_2P_3Q_1$ (or $P_2Q_0Q_1$), ensures continuity of velocity as we pass across the curves. To create an animation sequence of an object moving along a spline curve, or to control a pseudo-camera moving along such a curve, control points must be set to satisfy these conditions. The time variable is reparameterized at each stage to the variable for the relevant curve section to find individual curve points, enabling the object to be redrawn in the correct position for each animation frame.

Drawing Bézier Curves

The normal method of drawing a Bézier curve is to calculate repeated locations of $C(u)$ for short intervals of u, joining the points so produced by a series of straight lines. This calculation can be done directly from the recursive formula, giving a compact coding task. In some circumstances, it may be better in time performance to use the blending function approach. Both methods involve repeated evaluation of points $C(u)$ for short steps in the parameter u across the domain of u from 0 to 1.

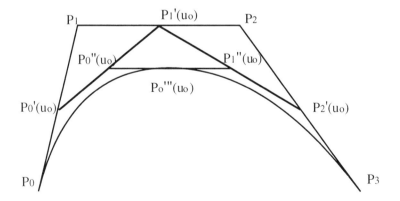

Figure 9.5 The de Casteljau trellis for a cubic Bézier curve at $u = u_0$

The method of curve subdivision is an alternative. The control polygon for a Bézier curve in some sense approximates the curve itself. Curve subdivision splits the curve into two parts, creating two new control polygons lying closer to the curve. This is followed recursively until a required degree of approximation to the curve is achieved. The de Casteljau 'trellis' suggests a possible control polygon (fig 9.5).

Suppose the curve is to be split at the point with parameter $u = u_0$ ($0 < u_0 < 1$). The two lines P_0P_1 and $P_0''(u_0)P_1''(u_0)$ in contact with the ends of the first sub-curve are tangential to it and have lengths three times the 'speed' of the curve point with respect to u (the speed property for $P_0''(u_0)P_1''(u_0)$ was not stated above but can be derived in the same way as the speeds at the end points were found). Subsections of these lines, $P_0P_0'(u_0)$ and $P_0''(u_0)P_0'''(u_0)$ are scaled down tangents by a factor u_0. To define a cubic Bézier curve to cover this subsection alone, we must introduce a new parameter $v = u/u_0$ that changes from 0 to 1 as the curve point travels from P_0 to $P_0'''(u_0)$. v responds to a 'faster clock' than u. This recalibration of time affects the speed of the curve sweeping point which can be compensated for by recalibrating distances accordingly. The control polygon for the sub-curve should have lengths scaled by u_0 compared to those of the original curve. This is exactly the scaling down achieved by the lines $P_0P_0'(u_0)$ and $P_0''(u_0)P_0'''(u_0)$, which are also end point tangents for the sub-curve. Thus, P_0, $P_0'(u_0)$, $P_0''(u_0)$ and $P_0'''(u_0)$ have all the properties required to make them Bézier control points for the sub-curve.

Using symmetry arguments it should be clear that $P_0'''(u_0)$, $P_1''(u_0)$, $P_2'(u_0)$ and P_3 are equivalent control points for the second sub-curve. This could also be established formally by reparameterizing with $w = (u - u_0)/(1 - u_0)$, so this time the scaling factor used is $(1 - u_0)$.

Calculation can be made very simple by choosing $u_0 = 0.5$. Given the set P_0, P_1, P_2 and P_3, the new control points for both sub-curves are found from the following sequence:

- P_0
- P_3

- $P_0'(0.5) = (P_0 + P_1)/2$
- $P_2'(0.5) = (P_3 + P_2)/2$

$P_1'(0.5) = (P_1 + P_2)/2$

- $P_0''(0.5) = [P_0'(0.5) + P_1'(0.5)]/2$
- $P_1''(0.5) = [P_2'(0.5) + P_1'(0.5)]/2$

- $P_0'''(0.5) = [P_0''(0.5) + P_1''(0.5)]/2$

Figure 9.6 A cubic Bézier curve and its polygons after two splitting stages

9. Splines: Generation of Curves and Surfaces   271

Items identified by • are those which constitute the two control polygons. Calculations involve only six divisions by 2, which can be implemented extremely efficiently by memory shift operations in some systems. This is particularly so if the spline is generated in integer arithmetic, which is the case if the curve is based on control points located in a raster display, such as a computer display screen.

The subdivision shown above is continued recursively for a set number of stages or until a given level of precision is achieved. Figure 9.6 shows how closely the curve can be approximated by the split control polygons after only two subdivisions. Note that splitting at the point with parameter 0.5 does not necessarily give even length subdivision – this is particularly obvious if the first line segment of the control polygon has vastly different length than the last.

This latter method is very good for drawing complete curves or sections, but is not recommended if only one parametrically defined point is required, when direct evaluation from a formula (either recursive or blending function form) would be more efficient.

Interpolating Splines for Curve Generation

The Bézier spline is an excellent tool for designers who wish to generate and control curves in space, especially if they have access to an interactive system that allows them to change positions of control points and see the effects these have on the curve. An alternative technique creates a 'smooth' curve that passes through all of a given set of control points. Clearly, there are many curves that can pass through one such set of points, so other information is needed to specify the precise curve to be generated. Tangents to the curve as it passes through control points can be used for this purpose. Tangents define curve directions and their lengths can be used, in some sense, to control how close the curve stays to the tangent as it moves away from or approaches the relevant control point. This is equivalent to the way in which the length of a Bézier control polygon section has an effect on the curvature of the curve produced. The end point velocity directions define the initial and final tangents and their magnitudes indicate, in some sense, how close the curve stays to this tangent as it moves away from or approaches the relevant control point. The longer the tangent, the lower the curvature near the control point, and the closer the curve 'hugs' the control polygon edge.

A cubic Bézier curve with control points P_0, P_1, P_2 and P_3 has end 'velocity'

$$T_0 = 3(P_1 - P_0) \quad \text{and} \quad T_3 = 3(P_3 - P_2).$$

These end velocities can be used as tangents, the same curve can be redefined in terms of its end points and these tangents. P_1 and P_2 are eliminated from the cubic Bézier curve equation from a rearrangement of the last two equations,

$$P_1 = P_0 + T_0/3 \quad \text{and} \quad P_2 = P_3 - T_3/3.$$

Thus, the Bézier curve with control points P_0, $P_0 + T_0/3$, $P_3 - T_3/3$, P_3 satisfies these conditions. This Bézier form is

$$C(u) = (1 - u)^3 P_0 + 3(1 - u)^2 u(P_0 + T_0/3) + 3(1 - u)u^2(P_3 - T_3/3) + u^3 P_3.$$

This is a perfectly reasonable formulation of the solution, but it is developed further by separating P_0, P_3, T_0 and T_3 into the cubic Hermite interpolation form.

$$C(u) = \{(1 - u)^3 + 3(1 - u)^2 u\} P_0 + (1 - u)^2 u T_0 \\ - (1 - u)u^2 T_3 + \{3(1 - u)u^2 + u^3\} P_3,$$

or, $$C(u) = (1 - u)^2(1 + 2u)P_0 + (1 - u)^2 u T_0 - (1 - u)u^2 T_3 + (3 - 2u)u^2 P_3.$$

The blending functions that multiply P_0, T_0, T_3 and P_3 in this expression are the Hermite polynomials and they enable us to move directly to the curve equation from the end points P_0 and P_3, and the corresponding end derivatives, T_0 and T_3.

This gives a method for deriving a spline curve equivalent to the cubic Bézier form from the start and end points and start and end tangents. This formulation is related to the Ferguson curve, named after the pioneering work of James Ferguson at Boeing. This is a convenient alternative way to define a composite spline curve. It is relatively easy to check that $C(0)$ is still P_0 and $C(1)$ is still P_3. Again, we ask those who find the algebraic manipulation difficult to take the above results on faith. Those with greater mathematical ability should check through these results and may wish to confirm through differentiation that the end point tangents are given as T_0 and T_3. In implementing this form of curve generation, efficient methods for evaluating the relevant functions can be devised.

Although the curve so created is equivalent to the Bézier form, it is no longer affine invariant, as transformations of the end tangents give different results after some forms of affine transformation. This may lead to difficult artefacts if used to animate a curve shape. In order to maintain affine invariance, the lengths of the end point tangents must be adjusted if the domain of the curve is transformed, rather in the way that the end line sections of Bézier control polygons are rescaled when different time intervals are used for transfer across a Bézier curve. If the system is reformulated initially in terms of Bézier control points (as indicated above) and then subjected to affine transformations (combinations of translation, scaling and rotation), this problem does not occur. The following section continues to develop the interpolating spline from the Hermite method, but this property of affine invariance should be held in mind if animation is to be performed.

The points P_1 and P_2 from the Bézier definition are no longer used, so the problem is reformulated more conveniently. Given two points P_i, P_{i+1} and curve tangents T_i and T_{i+1} respectively at those points, the curve fitting these values is

$$C_i(u_i) = (1 - u_i)^2(1 + 2u_i)P_i + (1 - u_i)^2 u_i T_i \\ - (1 - u_i)u_i^2 T_{i+1} + (3 - 2u_i)u_i^2 P_{i+1},$$

where the domain of u_i is from 0 to 1 inclusive ($0 \leq u_i \leq 1$). We are now in a position to derive a composite curve that passes through a set of points P_0, P_1, ... P_n with associated tangents T_0, T_1, ... T_n through piecewise definition of curve sections $C_0(u_0)$, $C_1(u_1)$, ... $C_{n-1}(u_{n-1})$ as given above. It may be tedious to have to define all tangents for interior points – a designer may wish to specify only the control points through which the curve passes, having some unspecified concept of 'smoothness'

9. Splines: Generation of Curves and Surfaces

for the curve. In this case, an automatic formula for the tangents T_i can be given. A simple but effective method defines

$$T_i = k(P_{i+1} - P_{i-1})$$

for some positive constant k and for i = 1 to n − 1. The tangent is in the direction of the join of the two neighbouring control points. Taking k = 1 is usually satisfactory, but introducing it into the formula gives more control of curve 'flatness' near the control points – the higher k, the 'flatter' the curve at the control point. A simple implementation of this is shown in fig 9.7. It is assumed that the end point tangents have been defined. If not, formula definitions can be adopted. For example, sensible results are obtained from

$$T_0 = (P_1 - P_0) + (P_1 - P_2)/2 = -P_0 + (3P_1 - P_2)/2$$

and $\quad T_n = (P_{n-2} - P_{n-1})/2 + (P_n - P_{n-1}) = P_n + (-3P_{n-1} + P_{n-2})/2.$

Vector interpretations of these show that they combine the end line segment with half of the neighbouring line segment in reverse. More controllable, but more complicated, end tangents can be created through fine tuning the constants a and b in

$$T_0 = a(P_1 - P_0) + b(P_1 - P_2)$$

and $\quad T_n = a(P_{n-2} - P_{n-1})/2 + b(P_n - P_{n-1}),$

or similarly devised formulae. The Catmull-Rom spline is an interpolating spline identical to that defined above, except that it is only drawn between P_1 and P_{n-1}, thus avoiding the problem of defining derivatives at P_0 and P_n. Adopting a formula-based approach to defining the tangents at all control points leaves the curve designer with a simpler task, but gives less control over the final curve shape.

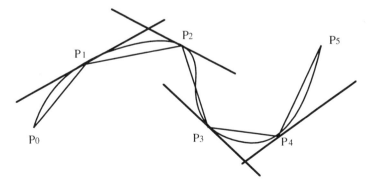

Figure 9.7 An interpolating spline with its interior tangents $T_i = (P_{i+1} - P_{i-1})$

Animation Control Using Interpolating Splines

Methods described in the sections above give smooth interpolating curves, but the form of smoothness is extended into the time domain for animation. Let times for passing through each control point be set to $t = t_i$, for $i = 0$ to n, where $t_0, t_1, \ldots t_n$ is strictly monotonically increasing ($t_{i+1} > t_i$). We now need to match velocities rather than tangents at control points. Consider the position at an internal control point P_i (where $i = 1, 2, \ldots n-1$) and for an interpolating curve $C(t)$. When $C(t)$ is between P_{i-1} and P_i inclusive, $t_{i-1} \leq t \leq t_i$. Local curve parameters u_{i-1} and u_i for curve sections immediately before and after P are

$$u_{i-1} = \frac{t - t_{i-1}}{t_i - t_{i-1}}, \text{ for } t_{i-1} \leq t \leq t_i, \quad \text{and} \quad u_i = \frac{t - t_i}{t_{i+1} - t_i}, \text{ for } t_i \leq t \leq t_{i+1}.$$

This formulation ensures that u_{i-1} changes from 0 to 1 as t changes from t_{i-1} to t_i and u_i changes from 0 to 1 as t changes from t_i to t_{i+1}. The composite spline curve will have tangent T_i as it arrives at P_i and as it leaves P_i. However, consider the effect of differing time intervals on the velocity of approach and departure from P_i. Using the argument developed in the section on Bézier splines, the velocities are scaled inversely with the time intervals for each section – the longer it takes, the slower the velocity. The velocities on approach to P_i and on leaving P_i are

$$V_{i-} = \frac{T_i}{t_i - t_{i-1}} \quad \text{and} \quad V_{i+} = \frac{T_i}{t_{i+1} - t_i} \quad \text{respectively.}$$

Take care to distinguish lower case 't' for time from upper case 'T' for a tangent. For these to be the same and both equal to the given velocity V_i at P_i, separate approach and departure tangents for P_i must be used. We set

$$T_{i-} = (t_i - t_{i-1})V_i \quad \text{and} \quad T_{i+} = (t_{i+1} - t_i)V_i$$

on approach to and departure from P_i respectively. Note that this does not change the directions of the tangents, so directional continuity is ensured. Their lengths are adjusted to give the required time domain continuity. T_{i+} and T_{i-} can be inserted into the spline formulae for the spline sections linking P_{i-1} to P_0. We now have a procedure for moving a point $C(t)$ along a smooth curve through points $P_0, P_1, \ldots P_n$ at times $t_0, t_1, \ldots t_n$ with velocities $V_0, V_1, \ldots V_n$.

- Define local time variables $u_i = \dfrac{t - t_i}{t_{i+1} - t_i}$, for $t_i \leq t \leq t_{i+1}$ and $i = 0$ to $n - 1$.

- Set $T_{i-} = (t_i - t_{i-1})V_i$ for $i = 1$ to n and $T_{i+} = (t_{i+1} - t_i)V_i$ for $i = 0$ to $n - 1$.

- $C(t) = C_i(u_i)$
  $= (1 - u_i)^2(1 + 2u_i)P_i + (1 - u_i)^2 u_i T_{i+} - (1 - u_i)u_i^2 T_{(i+1)-} + (3 - 2u_i)u_i^2 P_{i+1}$,
  for $t_i \leq t \leq t_{i+1}$.

Once more, stepping through values of t from t_0 to t_1 at the required time intervals for animation generates a point that behaves in the required way.

If the interior velocities are not specified, but the curve is to move smoothly in time and space through points P_0, P_1, ... P_n at times t_0, t_1, ... t_n, the method used earlier can be adapted to define approach and departure tangents at the point P_i as

$$T_{i-} = k(t_i - t_{i-1})(P_{i+1} - P_{i-1}) \quad \text{and} \quad T_{i+} = k(t_{i+1} - t_i)(P_{i+1} - P_{i-1})$$

for i = 1 to n – 1. These can be entered into the formula for C(t) given above to give time and space continuity. Adjustment of k affects not only the shape of the curve, but also the way that the tracking point moves along the curve. Larger k gives larger velocities and less curvature near the control points. Tangents at the end points P_0 and P_n can either be found from specified velocities or defined from the control points, for example as

$$T_{0+} = (t_1 - t_0)\{(P_1 - P_0) + (P_1 - P_2)/2\}$$

and $\quad T_{n-} = (t_n - t_{n-1})\{(P_{n-2} - P_{n-1})/2 + (P_n - P_{n-1})\}$.

Other constants could be introduced for 'fine tuning', as mentioned above. In general, we could have a problem where some, but not all velocities are specified at control points. In this case, T_{i-} and T_{i+} are defined in terms of velocities when V_i is known, and in terms of neighbouring control points when V_i is not known.

B-Splines

The general B-spline method enables curves of greater complexity than the Bézier curve and the interpolating spline form to be created in a single generation phase, contrasted with the piecewise linking previously used. Curves formed are implicitly piece-wise linked, but this is hidden due to the cunning construction of the blending functions. Bézier curves are limited cases of the more general B-spline form. Each B-spline curve has a control polygon, P_0, P_1, ... P_n, which has the same purpose as in the Bézier curve formulation. It gives the designer control over the broad shape of curve, which may be allowed to have tangential relationship with the control polygon at its start and end points. A B-spline depends on a knot sequence, a set of monotonic increasing values u_0, u_1, ... u_m between 0 and 1 ($u_i \geq u_{i-1}$). A knot may be equal to but cannot fall below the previous value in the sequence. A B-spline can be considered as a sequence of pieces of curves, each defined from a subset of the control points. The first p + 1 control points alone, P_0, P_1, ... P_p, have influence at the 'start' of the curve, where p is the power of the curve (lower case p is the curve's power, capital P is the general symbol for a point). As a knot value is passed, the first such point becomes inactive and another 'joins', so P_1, ... P_{p+1} becomes the active set of control points. This process continues as the curve parameter u crosses further knot values. This gives strong local control. If a control point is changed, it only affects the curve for the knot spans in which it has influence. The 'i^{th} knot span' is formally defined as $[u_i, u_{i+1})$, the semi-open interval from u_i to u_{i+1}, the specific form of brackets indicating the inclusion of u_i and exclusion of u_{i+1} to avoid overlap between successive intervals.

Given a knot sequence and a set of control points, the B-spline of power p is defined through its blending functions $f_i(p)$ as

$$C(u) = \sum_{i=0}^{n} f_i^{(p)}(u) P_i ,$$

which means

$$C(u) = f_0^{(p)}(u)P_0 + f_1^{(p)}(u)P_1 + f_2^{(p)}(u)P_2 + \ldots f_n^{(p)}(u)P_n.$$

(The Greek character sigma, Σ, denotes summation of all terms between the specified limits $i = 0$ and $i = n$ inclusive.) The blending functions are defined recursively as

$$f_i^{(p)}(u) = \frac{u - u_i}{u_{i+p} - u_i} f_i^{(p-1)}(u) + \frac{u_{i+p+1} - u}{u_{i+p+1} - u_{i+1}} f_{i+1}^{(p-1)}(u),$$

where

$$f_i^{(0)}(u) = \begin{cases} 1 & \text{for } u_i \leq u < u_{i+1} \text{ and } u_i < u_{i+1} \\ 0 & \text{otherwise.} \end{cases}$$

In evaluating these, the form 0/0 may occur. This is normally indeterminate, but these formulae are made to 'work' by the local convention that 0/0 is defined to be 0. This has to be checked for in any implementation of B-splines. As u changes from 0 to 1, the point defined by $C(u)$ sweeps across the B-spline curve.

Blending functions were explained when Bézier curves were discussed. Knowledge of the simpler forms introduced in that section should help in understanding the more difficult formulae given here. The basic zero power functions $f_i^{(0)}(u)$ act as switches, turning on (value 1) when u lies in the ith knot span and off or 'inactive' (value 0) when u lies outside this span. The span must have more than zero length for this to make sense, so the strict inequality $u_i < u_{i+1}$ is included as a condition. The formula for $f_i^{(p)}(u)$ involves both $f_i^{(p-1)}(u)$ and $f_{i+1}^{(p-1)}(u)$. Clearly, the knot sequence has a major effect on the blending functions for B-splines, even affecting their domains.

Consider what happens as the power p increases. The definition of $f_i^{(1)}(u)$ involves $f_i^{(0)}(u)$ and $f_{i+1}^{(0)}(u)$, so $f_i^{(1)}(u)$ is 'active' across the ith and $(i + 1)^{th}$ knot spans, from u_i to u_{i+2} overall. As p increases, more spans are 'grabbed' by this process, so $f_i^{(p)}(u)$ is active (or non-zero) from the ith to the $(i + p)^{th}$ knot spans, across the domain from u_i to u_{i+p+1} involving $p + 1$ spans overall. In the formula for $C(u)$, $f_i^{(p)}(u)$ multiplies the control point P_i, so this particular control point influences the shape of the curve only when u lies between u_i and u_{i+p+1} (fig 9.8 illustrates this for $p = 3$). This indicates a relationship between the number of control points, the number of knots and the power of the curve. For a curve of power p, the first blending function and first control point P_0 are active across the

9. Splines: Generation of Curves and Surfaces   277

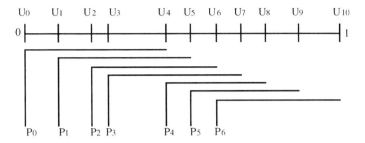

Figure 9.8 Influence of control points in knot spans for a cubic B-spline

domain of $p + 1$ knot spans starting at u_0 and ending at u_{p+1}. Similarly, the last control point P_n has influence from knots u_n to u_{n+p+1}. A knot sequence u_0 to u_m can be used to define a B-spline curve of degree p using control points P_0 to P_n when

$$m = n + p + 1.$$

Some properties found for the Bernstein polynomials used as Bézier curve blending functions also hold for the B-spline blending functions (also often referred to as basis functions). These, and other general properties of B-spline curves, are stated but not proved. B-spline blending functions are non-negative and add to unity for all u between 0 and 1 inclusive. More formally,

$$\left. \begin{array}{l} f_i^{(p)}(u) \geq 0, \\ \sum_{i=0}^{n} f_i^{(p)}(u) = 1 \end{array} \right\} \text{ for } (0 \leq u \leq 1).$$

The implication is that the B-spline curve point C(u) can once more be interpreted as the centre of mass of a collection of (non-negative) masses equal to the blending functions $f_i^{(p)}(u)$ each located at the corresponding control point P_i. This confirms the convex hull property – the curve lies totally within the convex hull of the control points. Strong local control of the curve enables us to be more precise. When u is in the jth knot span $[u_j, u_{j+1})$ for a power p curve, only the blending functions from $f_{j-p}^{(p)}(u)$ to $f_j^{(p)}(u)$ inclusive are non-zero, meaning that zero 'weight' is placed on all control points other than P_{j-p} to P_j (fig 9.8 illustrates this diagrammatically for $p = 3$, $n = 6$ and $m = 10$). Thus, the curve point C(u) lies in the convex hull of P_{j-p} to P_j when u is in the jth knot span. Figure 9.8 also illustrates how altering the position of a single control point will change the shape of the curve only across $p + 1$ knot spans – changing the location of P_i affects the curve between u_i and u_{i+p+1}. This gives designers the powerful ability to adapt the shape of part of the curve without affecting its appearance in other sections.

If arbitrary control points, knots and blending functions are set up as described above, the formula C(u) evaluated for short steps from $u = 0$ to $u = 1$, and short line segments drawn to join these points, a good approximation of the B-spline curve will be seen. In general, this will *not* have the desirable properties of passing through the end control points P_0 and P_n and being tangential to the last polyline segments P_0P_1 and $P_{n-1}P_n$. The knot sequence has to be chosen carefully to give this sort of control. A power p B-spline curve *will* have these properties if the first p + 1 knots are set to zero and the last p + 1 knots are set to one (no proof offered),

$$u_0 = u_1 = ... = u_p = 0 \quad \text{and} \quad u_{m-p} = ... = u_{m-1} = u_m = 1.$$

The knot sequence now has the form

$$\{0, 0, ..., 0, u_{p+1}, ... u_{m-p-1}, 1, 1, ... 1\}.$$

The first control point P_0 is 'active' from u_0 up to u_{p+1}. As u passes each knot, a new point becomes influential in controlling the B-spline, so as u passes through zero, the points $P_0, P_1, ... P_p$ immediately gain influence (fig 9.9 illustrates the domains of influence for a cubic B-spline). P_0 affects the shape of the curve only from u = 0 to u = u_{p+1}, along with $P_1, P_2, ... P_p$. When u reaches u_{p+1}, P_0 becomes inactive and P_{p+1} is switched on. If the middle knots 0, $u_{p+1}, ... u_{m-p-1}$, 1 are evenly spaced, the knot sequence is said to be uniform. If they are unevenly spaced, control points associated with the longer spans exert greater influence on the shape of the curve.

To take a specific example, if we wish to generate a B-spline curve of power 3 controlled by seven control points as illustrated in fig 9.9, we have p = 3, n = 6 (as n + 1 = 7). This means we must define knots u_0 to u_m, where

$$m = n + p + 1 = 10.$$

The first p + 1 = 4 of these are set to 0 and the last 4 to 1, so we have a knot sequence

$$\{0, 0, 0, 0, u_{4e}, u_5, u_6, 1, 1, 1, 1\}.$$

The start and end knots here are said to have 'multiplicity four'. A uniform sequence is created by spacing the central cluster of knots, 0, u_4, u_5, u_6, 1, evenly. This is achieved by giving the 4 (found as m − 2p) central knot spans equal lengths of $\frac{1}{4}$, to give the knot sequence

$$\{0, 0, 0, 0, \tfrac{1}{4}, \tfrac{1}{2}, \tfrac{3}{4}, 1, 1, 1, 1\}.$$

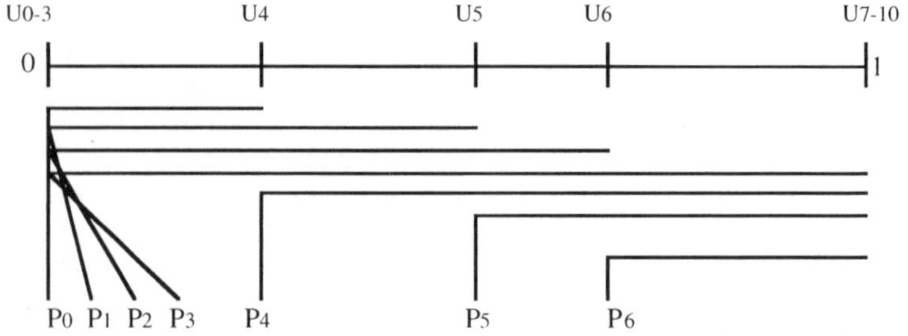

Figure 9.9 Control point influence in knot spans with multiple end knots in a cubic B-spline

9. Splines: Generation of Curves and Surfaces 279

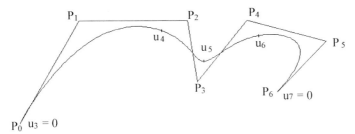

Figure 9.10 A typical cubic B-spline curve, the positions of internal knots shown

Figure 9.10 illustrates a typical result of this formulation, giving end point control, with locations of knots u_3 to u_7 marked on the curve. The effect of changing one control point is shown in fig 9.11. Here P_1 is relocated to Q_1. As illustrated in fig 9.9, P_1 only influences the B-spline up to knot u_5, so the shape of the curve is unchanged beyond this point.

Non-uniform knot sequences affect the shape of the curve by changing the domain of influence of control points. If the ith knot span from u_i to u_{i+1} is lengthened, the corresponding set of active control points P_i to P_{i+p+1} have greater influence on the shape of the curve. The ability to change knot spans gives a curve designer extra control over and above the ability to locate control points. In some circumstances, it may even be beneficial to give interior knot spans zero length, making $u_i = u_{i+1}$.

One special form of knot sequence will be mentioned. Suppose a curve of power 3 is created using four control points with knot sequence (0, 0, 0, 0, 1, 1, 1, 1), containing no 'interior' knots. This is a valid formulation as $p = 3$, $n = 3$ and $m = 7$, giving $m = n + p + 1$. All four control points are involved across the only non-degenerate knot span from 0 to 1. Derivation of the blending functions for this formulation involves some contorted algebra – some readers may prefer to jump to the final paragraph of this section at first reading. The result, however, is intriguing, so this jump is a little like skipping to the last page of a detective novel.

The derivation is based on the formulae for B-spline blending functions. The lowest level of functions is given by the rule

$$f_i^{(0)}(u) = \begin{cases} 1 & \text{for } u_i \leq u < u_{i+1} \text{ and } u_i < u_{i+1} \\ 0 & \text{otherwise.} \end{cases}$$

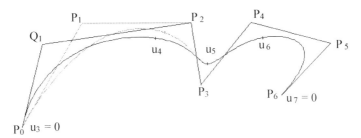

Figure 9.11 Change of location of P_1 to Q_1 affects the B-spline only up to knot u_5

Our example has only one non-zero blending function at this level, as the first three knot spans and the last three have zero length, making

$$f_0^{(0)}(u) = 0, \quad f_1^{(0)}(u) = 0, \quad f_2^{(0)}(u) = 0$$
and
$$f_4^{(0)}(u) = 0, \quad f_5^{(0)}(u) = 0, \quad f_6^{(0)}(u) = 0, \quad \text{all for } 0 \leq u < 1.$$

These knot spans violate the rule $u_i < u_{i+1}$, so their functions are set to zero. The only non-zero function is for i = 3, which is active for $u_3 \leq u \leq u_4$, or $0 \leq u < 1$, so

- $f_3^{(0)}(u) = 1 \quad \text{for } 0 \leq u < 1.$

Higher power blending functions are derived from the recursion rule

$$f_i^{(p)}(u) = \frac{u - u_i}{u_{i+p} - u_i} f_i^{(p-1)}(u) + \frac{u_{i+p+1} - u}{u_{i+p+1} - u_{i+1}} f_{i+1}^{(p-1)}(u).$$

At power 1, there are two non-zero functions. Definitions for $f_i^{(1)}(u)$ involve $f_i^{(0)}(u)$ and $f_{i+1}^{(0)}(u)$, so $f_0^{(1)}(u)$, $f_1^{(1)}(u)$, $f_4^{(1)}(u)$ and $f_5^{(1)}(u)$ are all zero as they involve only the zero-valued lower-level functions $f_0^{(0)}(u)$, $f_1^{(0)}(u)$, $f_2^{(0)}(u)$, $f_4^{(0)}(u)$, $f_5^{(0)}(u)$ and $f_6^{(0)}(u)$. We must pay more attention to $f_2^{(1)}(u)$ and $f_3^{(1)}(u)$. By direct substitution into the formula with i = 2 and p = 1, and using knot values $u_0 = u_1 = u_2 = u_3 = 0$ and $u_4 = u_5 = u_6 = u_7 = 1$,

$$f_2^{(1)}(u) = \frac{u - u_2}{u_{2+1} - u_2} f_2^{(1-1)}(u) + \frac{u_{2+1+1} - u}{u_{2+1+1} - u_{2+1}} f_{2+1}^{(1-1)}(u)$$

$$= \frac{u - u_2}{u_3 - u_2} f_2^{(0)}(u) + \frac{u_4 - u}{u_4 - u_3} f_3^{(0)}(u)$$

$$= \frac{u - 0}{0 - 0} f_2^{(0)}(u) + \frac{1 - u}{1 - 0} f_3^{(0)}(u).$$

There appears to be a problem with the first term through division by zero, but, the $f_2^{(0)}(u)$ is identically zero, so the term reduces to 0/0, and we have a special rule in this context alone that 0/0 = 0. With this cleared up, using $f_3^{(0)}(u) = 1$ for $0 \leq u < 1$,

- - $f_2^{(1)}(u) = (1 - u) \quad \text{for } 0 \leq u < 1.$

Similarly, from the recursive formula,

$$f_3^{(1)}(u) = \frac{u - u_3}{u_4 - u_3} f_3^{(0)}(u) + \frac{u_5 - u}{u_5 - u_4} f_4^{(0)}(u)$$

$$= \frac{u - 0}{1 - 0} f_3^{(0)}(u) + \frac{1 - u}{1 - 1} f_4^{(0)}(u).$$

9. Splines: Generation of Curves and Surfaces

This time, the second term contains a 0/0 as $f_4^{(0)}(u)$ is identically zero. Invoking the special rule $0/0 = 0$,

- - $f_3^{(1)}(u) = u$ for $0 \leq u < 1$.

The method should now be clear; use of the '0/0 rule' will no longer be mentioned. At the second power, we have three non-zero blending functions with $p = 2$ and $i = 1, 2$ and 3. The first of these is

$$f_1^{(2)}(u) = \frac{u - u_1}{u_3 - u_1} f_1^{(1)}(u) + \frac{u_4 - u}{u_4 - u_2} f_2^{(1)}(u),$$

- - - $f_1^{(2)}(u) = (1 - u)^2$ for $0 \leq u < 1$.

$$f_2^{(2)}(u) = \frac{u - u_2}{u_4 - u_2} f_2^{(1)}(u) + \frac{u_5 - u}{u_5 - u_3} f_3^{(1)}(u),$$

- - - $f_2^{(2)}(u) = 2(1 - u)u$ for $0 \leq u < 1$.

$$f_3^{(2)}(u) = \frac{u - u_3}{u_5 - u_2} f_3^{(1)}(u) + \frac{u_6 - u}{u_6 - u_4} f_4^{(1)}(u),$$

- - - $f_3^{(2)}(u) = u^2$ for $0 \leq u < 1$.

Some readers may begin to observe a pattern. Continuing to the final, third power, we set $p = 3$ and find non-zero blending functions for $i = 0, 1, 2$ and 3. They are

$$f_0^{(3)}(u) = \frac{u - u_0}{u_3 - u_0} f_0^{(2)}(u) + \frac{u_4 - u}{u_4 - u_1} f_1^{(2)}(u)$$

- - - - $f_0^{(3)}(u) = (1 - u)^3$ for $0 \leq u < 1$.

$$f_1^{(3)}(u) = \frac{u - u_1}{u_4 - u_1} f_1^{(2)}(u) + \frac{u_5 - u}{u_5 - u_2} f_2^{(2)}(u)$$

- - - - $f_1^{(3)}(u) = 3(1 - u)^2 u$ for $0 \leq u < 1$.

The other two results will be quoted directly, as the process should by now be clear.

- - - - $f_2^{(3)}(u) = 3(1 - u)u^2$ for $0 \leq u < 1$.

- - - - $f_3^{(3)}(u) = u^3$ for $0 \leq u < 1$.

Thus, the B-spline curve function for the knot sequence $\{0, 0, 0, 0, 1, 1, 1, 1\}$ and control points P_0, P_1, P_2 and P_3 is

$$C(u) = (1-u)^3 P_0 + 3(1-u)^2 u P_1 + 3(1-u)u^2 P_2 + u^3 P_3, \quad \text{for } 0 \le u < 1.$$

This is the familiar cubic Bézier curve. We have shown that the cubic B-spline with no interior knots and four control points is the cubic Bézier curve. The principle extends to other powers. B-splines are general curve forms which include Bézier curves as special cases. The extra control given the B-spline form by the use of knot sequences has to be balanced against the extra complexity in its use and the extra understanding needed by designers who use this method.

Non-Uniform Rational B-Splines: NURBS

NURBS have become very popular in many computer graphics and CAD systems, so it is worth explaining what they are and why they have become so widely used. They are more complicated to evaluate and more difficult to understand than Bézier and B-splines, but offer the compensatory property of being able to give exact representation of circles and other classical 'conic sections' such as ellipses, parabolae and hyperbolae. This may be important if, for example, the exact properties of a rolling part is to be investigated in a mechanical engineering CAD problem. In most computer graphics activity, depiction of these forms is only at a low level of integrity in any case, so there is less justification for their use for image generation alone, as the results achieved are often visually indistinguishable from non-rational (that means ordinary) B-splines. B-splines can be manoeuvred to give very good approximations to circles (and hence cylinders, spheres and cones in their surface manifestations); NURBS can give these forms exactly. Perhaps a more important property for practitioners of computer graphics is that NURBS have projective invariance. If a projection transformation is applied to the control points and a new curve reconstructed from these points, the effect is the same as applying the projection to all points on the curve. If curves (and later surfaces) are to be animated in a computer graphics system, and viewed by perspective projection, this ensures integrity of the surface as it is viewed from different angles. B-splines have only affine invariance, which does not include perspective projection. Perspective invariance is a stronger condition that implicitly includes affine invariance.

Control points and knots are used in NURBS exactly as in irrational B-splines – they have all the standard B-spline properties. The 'NU' part of the name refers to the possibility of defining knot sequences in a non-uniform way, as in standard B-splines. Extra control is achieved by attaching a 'weight' to each control point (or equivalently to each blending function). This is achieved through defining curve points in homogeneous coordinate space, which was introduced in chapter 6. As a reminder, the point (x, y, z) can be represented as (wx, wy, wz, w), where w can take any non-zero value. For affine transformations, we normally standardize $w = 1$, but in this case we can give each control point a positive valued weight. Suppose the Cartesian representation of a control point P_i is (x_i, y_i, z_i) and it is allocated weight $w_i > 0$. Then $P^w_i = (w_i x_i, w_i y_i, w_i z_i, w_i)$ is its homogeneous representation. As in the normal formulation, we take $n + 1$ control points, so i can take values from 0 to n. The recursive formulation for a B-spline curve point was given above:

9. Splines: Generation of Curves and Surfaces

$$C(u) = \sum_{i=0}^{n} f_i^{(p)}(u) P_i,$$

or $\quad C(u) = f_0^{(p)}(u)P_0 + f_1^{(p)}(u)P_1 + f_2^{(p)}(u)P_2 + \ldots f_n^{(p)}(u)P_n,$

where the $f_i^{(p)}(u)$ are the power p B-spline blending functions. Replacing the control points by their weighted homogeneous forms gives us the rational B-spline formulation,

$$C^w(u) = f_0^{(p)}(u)P^w{}_0 + f_1^{(p)}(u)P^w{}_1 + f_2^{(p)}(u)P^w{}_2 + \ldots f_n^{(p)}(u)P^w{}_n,$$

where $C^w(u)$ is a homogeneous representation of the rational B-spline curve point. Suppose

$$C^w(u) = (w^*x, w^*y, w^*z, w^*),$$

where w^*, a function of u, is the weight of $C^w(u)$ in this formulation (we are not using the symbol * here to indicate multiplication, as is common in computer programs). From the rational B-spline equation, we can extract the last (fourth) coordinate of each term to give

$$w^* = f_0^{(p)}(u)w_0 + f_1^{(p)}(u)w_1 + f_2^{(p)}(u)w_2 + \ldots f_n^{(p)}(u)w_n$$

or $\quad w^* = \sum_{i=0}^{n} f_i^{(p)}(u) w_i,$

exactly the same formula as used for B-spline points, except that it is now applied to the set of scalar values (in other words, plain numbers rather than control points) $w_0, w_1, \ldots w_n$. To reduce $C^w(u)$ to its Cartesian form, its first three coordinates are to be divided by its fourth, w^*. The first three coordinates for control point $P^w{}_i$ are $(w_i x_i, w_i y_i, w_i z_i)$, which can be written as $w_i(x_i, y_i, z_i) = w_i P_i$. Substituting these directly into the equation for C(u), the first three coordinates for $C^w(u)$ are

$$f_0^{(p)}(u)w_0 P_0 + f_1^{(p)}(u)w_1 P_1 + f_2^{(p)}(u)w_2 P_2 + \ldots f_n^{(p)}(u)w_n P_n.$$

The rational B-spline point $C_r(u)$ with parameter u is found on dividing by w^*,

$$C_r(u) = \frac{f_0^{(p)}(u)w_0}{w^*}P_0 + \frac{f_1^{(p)}(u)w_1}{w^*}P_1 + \frac{f_2^{(p)}(u)w_2}{w^*}P_2 + \ldots \frac{f_n^{(p)}(u)w_n}{w^*}P_n,$$

or $\quad C_r(u) = \sum_{i=0}^{n} \frac{f_i^{(p)}(u)w_i}{w^*} P_i.$

This gives blending functions for NURBS as

$$g_i^{(p)}(u) = \frac{f_i^{(p)}(u)w_i}{w^*} = \frac{f_i^{(p)}(u)w_i}{\sum_{j=0}^{n} f_j^{(p)}(u)w_j} \quad \text{for } i = 0 \text{ to } n.$$

The dummy suffix j is used in the denominator to avoid confusion with the suffix i identifying the particular blending function. This structure is informative. It is the ratio of two polynomials of power p in u (hence the term 'rational'). Once all knots and weights w_i are defined, the denominator w^* is a power p polynomial of u (having form $a_0 + a_1u + a_2u^2 + ... + a_pu^p$). The $f_i^{(p)}(u)$ are all positive for $0 \leq u \leq 1$, as established above, so all $g_i^{(p)}(u)$ are positive when all $w_i > 0$ (this was a condition imposed when weights were defined). Also, the sum of all $g_i^{(p)}(u)$ is (almost) trivially one; the sum of all the numerators (top terms) from i = 0 to n is equal to the denominator (bottom term). These blending function properties enable the position of $C_r(u)$ to be interpreted as the centre of mass of masses $g_i^{(p)}(u)$ placed at control points P_i for i = 0 to n. This means that NURBS have all the local control and convex hull properties that are desirable in normal B-splines. They also have the extra local control of allocation of weights to control points. If all weights are made the same, a NURBS is identical to its equivalent B-spline. If a weight is increased, it drags the curve locally towards its associated control point, with the same form of local control that was given for movement of control points in non-rational B-splines; a change in w_i affects the curve only in the domain $[u_i, u_{i+p+1})$. This will also have a linear effect on individual points on the curve in the sense that, if a point is within the domain that moves, it will move along a straight line directly towards P_i if w_i is increased, away from P_i if w_i is decreased (fig 9.12).

It was shown above that similar local effects could be obtained by moving control points, but this would affect the convex hull of the curve, which may be critical in some applications. The difference with weight change is that local curve shape can be varied in a relatively controlled way without moving the control points, ensuring that continuity and convex hull effects are not affected adversely.

NURBS can be depicted by stepping through parameter u to draw short line segment approximations to the curve. The control polygon is an approximation to the curve; it can be reset at higher levels of subdivision to approach the curve shape in a way similar to that used for Bézier spline curves. Extra knots and control points are added by knot insertion. This will not be described here – those who wish to investigate this method should look up the more specialist literature.

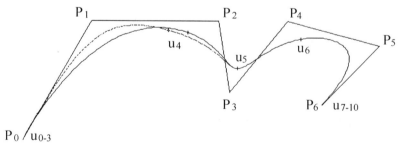

Figure 9.12 Increasing the weight of control point P_1 moves NURBS points towards P_1 within a limited knot range

Circles and Other Conic Sections

One of the main reasons for the popularity of NURBS is their ability to give exact representations of circles, ellipses, parabolae and hyperbolae. Arcs of circles (and other conics) can be created using power two NURBS with three control points. This implies that there are 7 knots. With multiplicity 3 at the end knots to maintain end point and tangent control, a balanced knot sequence is $\{0, 0, 0, 0.5, 1, 1, 1\}$. With control points P_0, P_1, P_2 arranged as in fig 9.13 so that $P_0P_1 = P_1P_2 = a$, say, and $b = P_0P_2/2$, the shorter arc of the circle between P_0 and P_2 which has P_0P_1 and P_2P_1 as tangents and lies inside the convex hull of $P_0P_1P_2$ is given by the power 2 (quadratic) NURBS with weights $w_0 = 1$ at P_0, $w_1 = b/a$ at P_1 and $w_2 = 1$ at P_2. No proof or justification of this is offered; it, and other results in this section, are merely stated as they may be useful to readers. In NURBS theory, weights are usually restricted to positive quantities in order to maintain the convex hull property. In this case, there are advantages to using a negative weight. If w_1 is set to $-b/a$, the same formulation sweeps out the longer arc between P_0 and P_2 (clearly outside the convex hull of the control points). It is convenient to hold weights w_0 and w_2 at 1, changes to w_1 enabling us to produce all forms of classical conic section curves. An arc of a hyperbola is generated for $w_1 > 1$, a parabola for $w_1 = 1$ and an ellipse if $0 < w_1 < 1$ (the circle is a special case of an ellipse). In the case of the ellipse, taking $-1 < w_1 < 0$ gives the longer of the two arcs between P_0 and P_2, lying outside the convex hull of the control points. These formulations all have axes of symmetry along that of the triangle (the vertical dotted line in fig 9.13). This symmetry is destroyed by moving the position of P_1 so that $P_0P_1 \neq P_1P_2$. It is not possible to form a circle in this case, but all other conic forms can be generated by choice of w_1 as above.

It may be useful to generate a complete circle through a single sweep from $u = 0$ to 1. In this case several circular arcs can be 'stitched together' to give the required effect. One example is given (fig 9.14) which uses multiple interior knots to effect the link. The eight control points are at the vertices and edge mid-points of a square, the start and end vertices coinciding at a mid-point. The knot sequence is

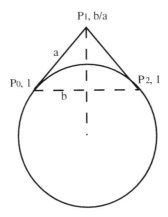

Figure 9.13 Control points and weights for an exact circular arc

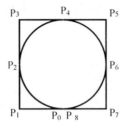

Figure 9.14 Control points for NURBS generation of a complete circle

$$\{0, 0, 0, \tfrac{1}{4}, \tfrac{1}{4}, \tfrac{1}{2}, \tfrac{1}{2}, \tfrac{3}{4}, \tfrac{3}{4}, 1, 1, 1\}$$

and the weights are

$$(1, \tfrac{1}{\sqrt{2}}, 1, \tfrac{1}{\sqrt{2}}, 1, \tfrac{1}{\sqrt{2}}, 1, \tfrac{1}{\sqrt{2}}, 1).$$

The weights are found from the value b/a in fig 9.13 as applied to fig 9.14. Similar circle representations can be devised with control points at vertices and edge midpoints of any regular polygon, the parameterization being more even as more edges are used. With n + 1 control points starting at an edge mid-point of an n-sided regular polygon, a circle is defined by knot sequence

$$\{0, 0, 0, \tfrac{1}{n}, \tfrac{1}{n}, \tfrac{2}{n}, \tfrac{2}{n}, \tfrac{3}{n}, \tfrac{3}{n}, \dots 1, 1, 1\}$$

with weights

$$(1, \tfrac{b}{a}, 1, \tfrac{b}{a}, 1, \dots 1, \tfrac{b}{a}, 1),$$

where a and b are equivalent to the lengths a and b in fig 9.13. A standard result from geometry, the exterior angle of a regular n-sided polygon is 360°/n, so the interior angle ($\angle P_0P_1P_2$ in fig 9.13) is $(180 - 360/n)°$. Now using a standard rule from trigonometry on the right-angled triangle with half this angle, we have

b/a = sin(90° − 180°/n) = cos(180°/n).

This gives an easy method for inscribing a NURBS circle inside a regular polygon of any order.

Surface Construction Using Bézier Patches

One of the most widely used methods for surface construction is the bi-cubic Bézier patch. As sections of cubic Bézier curves can be joined together to form a curve that is smoothly continuous wherever this is needed, bi-cubic Bézier patches can similarly be stitched together to form controlledly smooth surfaces. Such forms are

9. Splines: Generation of Curves and Surfaces

useful, for example, in the design of car body parts, where smoothness is largely required, but some sections may have deliberate folds or creases. A surface patch is created by sweeping a cubic Bézier curve whose four control points are themselves swept along separate cubic Bézier curves with matched curve parameters. The bi-cubic form is largely used, but higher power extensions are easily accommodated. The powers do not have to be matched; for example, the five control points of a fourth-order curve could each be swept along different cubic curves. The extension to other forms should be clear if the method for bi-cubic curves is properly understood. A cubic Bézier curve with control points P_0, P_1, P_2 and P_3 has been defined as

$$C(u) = \sum_{i=0}^{3} f_i^{(3)}(u) P_i \qquad \text{for } 0 \leq u \leq 1.$$

(The sigma notation should now be familiar.) Now, the control points P_i are found from four other Bézier curves, each with parameter v. This parameter links the four points on the secondary curves and distinguishes them from the variable u used for the swept curve. Suppose the ith secondary curve has control points P_{i0}, P_{i1}, P_{i2}, P_{i3}, for $i = 0, 1, 2$ or 3 (fig 9.15). For a given value of v, each of the control points for the sweeping curve is defined by a similar cubic Bézier curve equation,

$$P_i(v) = \sum_{j=0}^{3} f_j^{(3)}(v) P_{ij} \qquad \text{for } 0 \leq v \leq 1, i = 0, 1, 2 \text{ or } 3.$$

These values for $i = 0, 1, 2$ and 3 are shown as black squares in fig 9.15 for the arbitrary value $v = 0.25$. The formula is equivalent to the standard one given above; the summation suffix j is used as i is used to select one of the four curves. As v increases, the black squares, acting themselves as control points for the curve that sweeps out the surface, move together up their respective curves in the image. Substituting the $P_i(v)$ into the equation for the curve through them as control points gives a formula for a general point on the bi-cubic Bézier surface $S(u, v)$,

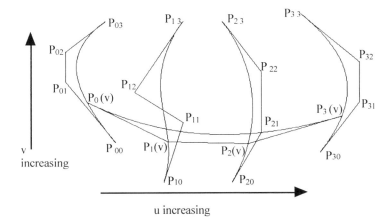

Figure 9.15 Bi-cubic Bézier patch generation: four control points sweep along four Bézier curves

$$S(u, v) = \sum_{i=0}^{3} \{f_i^{(3)}(u) \sum_{j=0}^{3} f_j^{(3)}(v) P_{ij}\}, \text{ for } 0 \le u \le 1 \text{ and } 0 \le v \le 1.$$

Here, $f_i^{(3)}(u)$ and $f_j^{(3)}(v)$ are the usual Bézier cubic *curve* blending functions. The patch is swept out as u and v both vary from 0 to 1. The double summation behaves like a pair of nested loops in a computer implementation. The order of activity in this formula is expanded algorithmically as

- For u and v from 0 to 3, clear S(u, v) to zero;
- for all i from 0 to 3
- clear term to zero;
- for all j from 0 to 3
- add $f_j^{(3)}(v) P_{ij}$ to term
- end j loop;   (term is an intermediate value)
- S(u, v) = S(u, v) + $f_i^{(3)}(u)$ *term
- end i loop;
- exit with position of S(u, v).

The inner summation is completed before the outer, which may imply asymmetry in u and v, but the summations can be merged mathematically into a symmetric formula for S(u, v):

$$S(u, v) = \sum_{i=0}^{3} \sum_{j=0}^{3} \{f_i^{(3)}(u) f_j^{(3)}(v) P_{ij}\}, \text{ for } 0 \le u \le 1 \text{ and } 0 \le v \le 1,$$

which gives exactly the same surface. This result can be reached from the original double summation formula. The 'sigma' with count index j can be considered as an algebraic bracket containing all terms of its summation, with $f_i^{(3)}(u)$ as a factor. As $f_i^{(3)}(u)$ is not affected by changes in j, it can be moved across the sigma sign to multiply all summed terms, just as a factor can be moved into a bracket. For example,

$$3 \sum_{i=0}^{3} i^2 = 3(0 + 1 + 4 + 9) = 42$$

is equivalent to

$$\sum_{i=0}^{3} 3i^2 = (0 + 3 + 12 + 27) = 42.$$

The factor of 3 can stay outside the sigma (or bracket) or can be incorporated into each term as it is not affected by the summation index i. The reverse is true – terms that are constant across summations can be allowed to move outside the sigma sign. This will be used in a moment, but first the two summations must be reordered. When the two summations are next to each other, as in the latest version of the

9. Splines: Generation of Curves and Surfaces

formula, and the limits of both are constants or are independent of each other, then the summations can be performed in opposite order without changing the result. For example, to find the sum of all elements of a matrix, there is no difference between first summing all rows and then adding the row sums, or first adding all columns and then adding the column sums. Using similar reasoning, the formula can be written (with some fairly straightforward re-ordering), as

$$S(u, v) = \sum_{j=0}^{3} \sum_{i=0}^{3} \{f_j^{(3)}(v) f_i^{(3)}(u) P_{ij}\}, \quad \text{for } 0 \leq u \leq 1 \text{ and } 0 \leq v \leq 1.$$

Now $f_j^{(3)}(v)$ is not affected by changes in i, so it can be allowed to move across the sigma sign with dummy count i. We arrive at

$$S(u, v) = \sum_{j=0}^{3} \{f_j^{(3)}(v) \sum_{i=0}^{3} f_i^{(3)}(u) P_{ij}\}, \quad \text{for } 0 \leq u \leq 1 \text{ and } 0 \leq v \leq 1.$$

The purpose of this tortuous argument is to show from the symmetry of both expressions for $S(u, v)$ that the roles of u and v are equivalent. The procedure could be restarted to give exactly the same surface by exchanging the roles of u and v. First, generate four cubic Bézier curves with control points $\{P_{00}, P_{10}, P_{20}, P_{30}\}$, $\{P_{01}, P_{11}, P_{21}, P_{31}\}$, $\{P_{02}, P_{12}, P_{22}, P_{32}\}$ and $\{P_{03}, P_{13}, P_{23}, P_{33}\}$ and curve parameter u. The four points $P_j(u)$ with parameter u on each of these curves are given by

$$P_j(u) = \sum_{i=0}^{3} f_i^{(3)}(u) P_{ij}, \quad \text{for } 0 \leq u \leq 1 \text{ and } j = 0, 1, 2 \text{ or } 3.$$

These are used as control points to define a cubic Bézier point with parameter v, giving exactly the same formula for $S(u, v)$. This time, we sweep the surface-defining curve from left to right of fig 9.15 so that its control points track along the four Bézier curves described in the paragraph above. Some texts match their grid point definitions to u and v in different ways. The equivalences established here show that this is of no consequence; it is merely a matter of personal taste whether u or v sweeps across. Consistency, however, is desirable. If an implementer starts by using one ordering, this should be stuck to throughout.

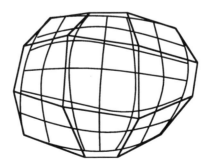

Figure 9.16 A bi-cubic Bézier patch and its control mesh

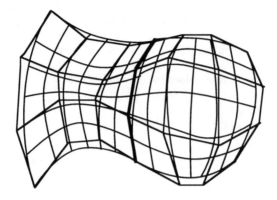

Figure 9.17 Smoothness across bi-cubic Bézier patch joins by equal length and collinearity of some mesh edges

These two processes can define a surface from a set of 16 control points. They produce exactly the same surface, so the processes are equivalent. Because of this relationship, it is usual to depict the control points as a (4 by 4) control mesh (fig 9.16). No labels are included in this figure for two reasons – it avoids clutter and allows a number of equally sensible possible labellings. Parameter u could vary either horizontally (to right or left) or vertically (upwards or downwards), with v taking an alternative direction. All these methods, however, result in exactly the same surface, for which a few of the forming cubic Bézier curves are shown.

Bi-cubic Bézier patches have many properties analogous to their equivalent curves. The surface is necessarily in contact with the control mesh only at the four patch corner points P_{00}, P_{03}, P_{30} and P_{33} (figs 9.15 and 9.16). For some meshes, the patch may cross at other points, but this is not necessarily the case. At the corner point of contact P_{00}, the lines $P_{00}P_{01}$ and $P_{00}P_{10}$ are tangential to the surface. This indicates a way of ensuring smoothness in joining one patch to another at corners and across edges. If two adjacent patches are to connect smoothly, the two sets of grid edge control points must coincide and the four grid lines converging on these from one patch must be in line with equivalent lines from the other patch. Making these line segments of equal length also gives smoothness of curvature (fig 9.17).

This construction ensures an exact fit of the two boundary curves at the junction as well as ensuring an uncreased join. For a bi-cubic Bézier patch, the four bounding curves at such joins are themselves cubic Bézier curves, defined by

- $C(u, 0)$: the curve with control points P_{00}, P_{10}, P_{20} and P_{30};
- $C(u, 3)$: the curve with control points P_{03}, P_{13}, P_{23} and P_{33};
- $C(0, v)$: the curve with control points P_{00}, P_{01}, P_{02} and P_{03};
- $C(3, v)$: the curve with control points P_{30}, P_{31}, P_{32} and P_{33}.

If adjacent patches share the same boundary control points, it is clear that their boundary curves will be coincident. Plate 9.1 shows examples, each shell having two bi-cubic Bézier patches smoothly joined at the central ridge.

Bi-cubic Bézier patches have become standard modelling tools in many 3D CAD packages. Clearly, it is more complicated for designers to have to specify 16 control points per patch compared to only four points per cubic Bézier curve and the

9. Splines: Generation of Curves and Surfaces

computational work done in creating such a patch is considerably greater. To identify a single point with parameters (u, v) on a surface involves five standard curve calculations – four to give control points and one to pinpoint the specific point. However, particularly if interactive tools are available to study the effect of varying control point locations, the bi-cubic Bézier spline and equivalent derivative spline formulations have extended the possibilities for designers who wish to create 'free form' surfaces.

If the bi-cubic form is considered inadequate, extensions to higher orders are algorithmically simple, although they are inevitably computationally more intensive. The powers of Bézier patches do not have to be equal. For example, it is possible to create a power 5 by 4 Bézier patch based on a grid of 6 by 5 points. The extension is performed by changing the upper limits on the summations and in the blending functions in the formula for S(u, v).

One of the formulae for S(u, v) from above indicates the form of bi-cubic blending functions, which have similar roles to blending functions in the 2D case.

$$S(u, v) = \sum_{i=0}^{3}\sum_{j=0}^{3} \{f_i^{(3)}(u)f_j^{(3)}(v)P_{ij}\}, \quad \text{for } 0 \le u \le 1 \text{ and } 0 \le v \le 1.$$

The multiplying factor for the control point P_{ij} is the bi-cubic Bézier blending function,

$$f_{i,j}^{(3,3)}(u, v) = f_i^{(3)}(u)f_j^{(3)}(v), \quad \text{for } i, j = 0 \text{ to } 3, 0 \le u \le 1 \text{ and } 0 \le v \le 1.$$

These are functions of the two variables u and v defined in the domain $0 \le u \le 1$, $0 \le v \le 1$. The bi-cubic Bézier curve formulation can be recast as

$$S(u, v) = \sum_{i=0}^{3}\sum_{j=0}^{3} f_{i,j}^{(3,3)}(u, v) P_{ij}, \quad \text{for } 0 \le u \le 1 \text{ and } 0 \le v \le 1.$$

It may be considered mathematically sloppy to use the same symbol 'f' for both surface and curve blending functions, but the numbers of suffixes and superfixes should be enough to distinguish which is being used.

The double suffix in the surface equation should not confuse the issue. The formula for S(u, v) can be restated as 'the sum, for all control points, of the blending function multiplied by the control point position'. A blending function for a surface is the product of two 'ordinary' blending functions for curves. With a little thought on the summations used, it should be clear that the blending functions for surfaces share many of the properties of their curve equivalents. For given values of u and v, the blending functions of a surface all add to one and they are all non-negative. (We suggest this as an exercise for the adventurous reader, but ask those who are mathematically exhausted at this stage to take it on trust, at least on first reading.) This means that the formula for S(u, v) can (as for many spline curves) be interpreted to define S(u, v) as the centre of mass of a collection of masses $f_{i,j}^{(3,3)}(u, v)$, each placed at its corresponding control point P_{ij}. This immediately gives the convex hull property for such surfaces – all points on a bi-cubic Bézier patch lie within the convex hull of the control points. A centre of mass cannot lie 'outside' its defining masses.

It is easy, from the blending function formulation, to give the exact form for a general Bézier patch of any pair of powers. A patch of power (m by n) is based on a mesh of (m + 1) by (n + 1) control points, $P_{00}, \ldots P_{m0}, P_{01}, \ldots P_{m1}, \ldots P_{0n}, \ldots P_{mn}$. The blending functions for this patch are

$$f_{i,j}^{(m,n)}(u, v) = f_i^{(m)}(u) f_j^{(n)}(v) \text{ for } 0 \leq u \leq 1 \text{ and } 0 \leq v \leq 1,$$

where $f_i^{(m)}(u)$ and $f_j^{(n)}(u)$ are blending functions for Bézier curves of power m and n respectively. The general Bézier patch formulation is

$$S(u, v) = \sum_{i=0}^{m} \sum_{j=0}^{n} f_{i,j}^{(m,n)}(u, v) P_{ij}, \quad \text{for } 0 \leq u \leq 1 \text{ and } 0 \leq v \leq 1.$$

Surface Generation Based on Other Forms of Curve

Constructions similar to those described above can create surfaces based on other forms of spline curves. The extension is reasonably straightforward. B-spline surfaces require additional information in the form of two knot sequences (one for each parameter u and v). Rational B-spline (NURBS) surfaces also require a matrix of weights whose elements are attached to control points. If interpolating surfaces are to be produced from a grid of control points, description of two surface tangents in the u- and v- directions for each control point can be conceptually difficult. These two surface tangents are orthogonal to the surface normal, so it may be convenient to define that at each control point, but the actual directions of tangents are still difficult to tease out of this information. The approach of the section above, where tangents are defined automatically from neighbouring control points, is probably the most effective and the easiest to implement. Here, the slope established from two neighbour points can be applied to each control point both 'across' and 'along' the grid. In all cases, the general blending functions for a surface can be found by multiplying the equivalent blending functions for curves. For B-spline and NURBS surfaces, with blending functions

$$f_{i,j}^{(m,n)}(u, v) = f_i^{(m)}(u) f_j^{(n)}(v) \text{ for } 0 \leq u \leq 1 \text{ and } 0 \leq v \leq 1$$

where $f_i^{(m)}(u)$ and $f_j^{(n)}(v)$ are blending functions for curves,

$$S(u, v) = \sum_{i=0}^{m} \sum_{j=0}^{n} f_{i,j}^{(m,n)}(u, v) P_{ij}, \quad \text{for } 0 \leq u \leq 1 \text{ and } 0 \leq v \leq 1.$$

These look the same as for the Bézier case, but the blending functions will themselves have different forms. The properties of such surfaces are analogous to the properties of their equivalent spline curves. For example, using the NURBS depiction of a circle or an arc of a circle across one of the parametric variables, forms such as cylinders, cones and spheres can be depicted exactly by NURBS surfaces.

Depiction of Surface Patches

Individual surface patch points are identified above. This does not solve the method of how to depict a patch. The method most commonly used is equivalent to approximating a curve by short line segments, according to a given precision. Suppose we have a patch with parameters u and v and we want to depict it to precision of k subdivisions in the parameter u direction, l (this is the character 'l' rather than the number '1') subdivisions in the parameter v. An initial array A, say, of (k + 1) points along the patch edge for v = 0 can be defined by calculating curve positions in a loop for u from 0 to 1 in steps of 1/k. Now set v = 1/l, and similarly step through u to create an array B of points for this new value of v. We have two arrays of points representing the edges of a 'strip' across the surface. An approximation to this strip can be formed by triangulating the sequence of points, taking triangles $A_i B_i B_{i+1}$ and $B_{i+1} A_i A_{i+1}$ for i = 0 to k − 1 (fig 9.18). It may be tempting to try to create quadrilateral patches of form $A_i B_i B_{i+1} A_{i+1}$, but the four points would be unlikely to lie in the same plane, giving twisted facets that would pose problems for most rendering systems.

This method of surface depiction could be applied to surfaces devised from any form of spline curve − Bézier spline, B-spline, NURBS or interpolating spline. The surface produced, however, is only an approximation to the true spline surface, so one of the purported advantages of NURBS − their exact depiction of circles − is nullified.

Ray tracing (chapter 10) is a rendering method that can give more precise depiction of curved surfaces, giving a precision equivalent to that of the display device. Rays are cast along the line of sight to interrogate which surfaces are potentially visible. Intersection of a ray with a spline surface is not easy. The convex hull property can be used to filter out obviously missing rays. If a ray is identified to pass through the surface's convex hull, probably the best way from that stage is to use numerical methods to identify an intersection point to a required degree of precision. Points on the ray can be classified as to whether they lie 'inside' or 'outside' the surface, as discussed in chapter 7 on vectors. Once a pair of inside and outside points is found on the ray (if such a pair exists), the mid-point is tested to decide if it is inside or outside. From this information, a narrower ray section containing the ray–surface intersection can be devised (from the three points, the 'inside–outside' or 'outside–inside' pair is chosen, with the 'inside–inside' or 'outside–outside' pair discarded). This rather 'messy' process is continued until the

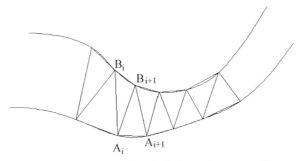

Figure 9.18 A method for depicting strips on a spline surface

required precision is achieved; convergence to an acceptably close pair of values is usually quite rapid.

An alternative method has been used by the author to produce plate 9.1. This involves densely covering the spline surface with points and rendering each point individually. This can be done by looping in u and v at small enough intervals to ensure at least one point is rendered per pixel of the image. Alternatively, and this is the method used in plate 9.1, points are scattered randomly over the surface until it is filled in a method similar to that used for IFS generation (chapter 8). At the moment, the question of how many points to create to determine if an object is filled is a matter of trial and error. This is a slow process compared to more standard scan line rendering (chapter 10), but does produce an exact surface (relative to the precision of the pixel rendering device) compared to the triangulated version normally fed to scan line renderers.

Summary

This chapter is the most mathematically detailed of the book. Long as it may seem on first reading, it has only begun to explore the available range of spline curves and surfaces, and their applications. One of the latter, although not often recognized as such, is a well-known computer graphics technique for developing solid textures, useful for creating images of solid continuous 3D effects such as wood grain or marble. Continuous textures are created through piecewise definition of spline linked 'solids'. As in moving from curves (parameter u) to surfaces (parameters u, v), so the process can be extended further to include 3D solid spaces (parameters u, v, w). This will not be explored here, but readers may find it helpful to think of such developments as no more than an extension of spline theory into a higher dimension.

The surfaces presented above are all based on a rectangular parameter structure. Triangular patches, too, can be defined and are useful in certain circumstances. For reasons of space and time, they are not developed here, but are adequately covered in specialist literature.

The literature of computer graphics is scattered with references to spline curves. This chapter is not intended to give an exhaustive coverage of all such forms. It is hoped that, having absorbed the concepts developed above, readers could interpret other manifestations of spline curves. They should also appreciate the constructive power of splines in modelling and animation. The chapter should give a better understanding to users of spline modelling systems. Some may be able to implement spline methods to suit their own needs. Even if the mathematics has been difficult to follow, the importance of the concept of spline curves and surfaces should be clear, and their basic properties understood.

10. Drawing and Rendering: How to Create Pictures

What is a 3D Drawing?

Figure 10.1, in its simple way, shows some of the difficulties involved with creating drawings of a three-dimensional object. A drawing is inherently a two-dimensional object, lying on the surface of the paper, canvas or display device. Even an illustration on a textured canvas or on the surface, for example, of a sphere is two-dimensional (we have argued this with respect to the representation of the globe in discussing dimension, chapter 3). In creating an image of a 3D object in a 2D space, some detail is lost. For example, do we really know that the rear faces of the figure in the lower right corner of fig 10.1 make it into a cube? Could there be some kind of structure cut into or protruding from the hidden rear faces? Do the figures in the top row give us a better impression? Which is the 'front' face in each of these? There is no clue in the top left figure; viewers should be able to 'flip' between two possible views, the well-known 'Necker cube' effect. The two figures on the right have one face reduced in size (a hidden face in the lower figure), so give an indication of perspective, suggesting which of the two possible faces should be at the back. This helps to give a suggestion of depth, but the top right part of fig 10.1 can still, with some thought, be flipped from one version to another. Some forms of depth cueing, with lines illustrated more faintly to indicate greater depth, or partial clipping of edges, to indicate depth priority (fig 10.2), gives visual clues, at the cost of edge integrity. The pure 'wire frame' interpretation of 3D objects is inevitably ambiguous.

Are the hidden line versions on the lower row really 'better', giving us a proper interpretation of a solid cube? We still cannot turn these in our hands. They could

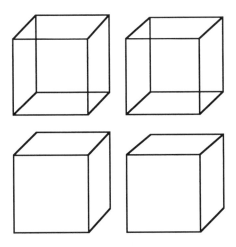

Figure 10.1 Four representations of a cube

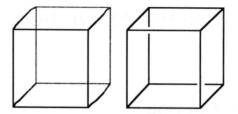

Figure 10.2 Two ways of suggesting depth in wire frame drawings

be shown in an interactive space where similar manipulation is allowed, but on the page or as a static image, this is not possible. And what are we to make of 'shapes' such as the devil's tuning fork of fig 8.6? Is this a 3D object? This shows how difficult it is to make a correct interpretation of the information lost when 3D objects are changed or 'projected' into 2D. It may be impossible to make 3D interpretations of abstract 2D structures. Good drawings of 3D objects should be relatively unambiguously interpretable as manifold objects, the types of solids that could be tangibly appreciated. Many of M.C. Escher's drawings are appreciated, but for their non-manifold features. Such interpretation is part of the study of perception, which looks at the ways humans interpret objects from visible clues.[1]

In terms of computer-generated '3D' illustrations, there is a chain of processes that is followed, which is analogous to that followed by a painter or photographer in composing and producing an image. In computer graphics, the two main components of this process are identified as modelling and rendering, composing the objects to be viewed and then creating the image from them, as discussed in chapter 1.

The scene to be depicted must be created using the black arts of geometric modelling (chapters 8, 9). Figure 10.3 shows a particularly simple set of objects: a cylinder and a pyramid. Use of a 'geometric modeller' usually involves selecting the objects to be depicted from a set of allowed 'primitives', combining objects where necessary to generate a wider range of essential shapes, manipulating these objects by changing their scales, locations and orientations to compose the scene. These

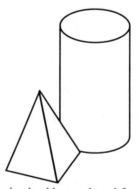

Figure 10.3 Some simple objects selected from geometric primitives

---

[1] Richard Gregory (1997) *Eye and Brain* (5th edn), Oxford University Press, is a very readable introduction to this area of study.

10. Drawing and Rendering: How to Create Pictures

objects may also be given surface characteristics, such as colour, glossiness or transparency, to be used in the rendering process discussed below. Some of the modelling techniques are extremely sophisticated; for example, to drape a cloth or to display the surface of ocean waves, 'physically based modelling' may be used. This simulates the physical laws that govern cloth draping or wave generation, giving a relatively natural effect if properly performed. Pure physical laws are not necessarily directly followed; simplified versions that give the required appearance or behaviour can be adequate. This allows the user to arrange descriptive features such as wind strength, shoreline shape and ocean depths or the cloth stiffness and the object it is draped over, allowing the rules to match the object performance to these characteristics, in the hope of obtaining a 'natural' appearance. The alternative would be for the user to locate each defining surface section individually, which would not necessarily lead to a natural effect and would involve considerable work.

Once the objects are satisfactorily arranged, the image has to be created. This involves locating a pseudo observer or camera in the space occupied by the model (fig 10.4). A 'line of sight' is defined, and the section of the scene to be depicted is framed using a 'view window', which can be considered as a picture frame hanging in space before the observer. In the photographic analogy, this is equivalent to selecting the aspect ratio of the image to be produced and using a zoom lens to frame the set of objects included in the image. Lighting is arranged with pseudo light sources as an extra set of geometric 'objects' (not shown in fig 10.4). The location of objects in the 3D scene is projected onto the view window. This is a mathematical process that converts the 3D coordinates of the 'world space' in which the model was created into a 2D coordinate system describing the location of points in the view window, as discussed in chapter 6 on matrix transformations. Perspective projection places the projected point in the position where the line from the observer to a point in the scene cuts through the view window. This is not the only method. Parallel projection is often adequate; other methods may simulate use of special lenses, for example giving a 'fisheye' lens effect by projection onto a window on a spherical view surface. Full colour rendering takes into account the effect of the lights on the perceived image – the 'rendering technique' is usually chosen from a standard repertoire of methods discussed below. The decision of what will appear in the view window involves the process of clipping, deciding which objects or parts of objects lie outside the visible region through the window. The

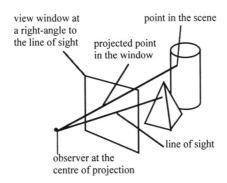

Figure 10.4 Creating a 2D image from a geometric model

hidden surface problem is dealt with naturally by the properties of light in our real physical world, but algorithms are used to decide which parts of objects obscure others in the synthesized scene. There are many hidden surface methods; some work by prioritizing objects in real valued world or object space, others work in the 'image space' of the picture as divided into pixels. One method uses distance of parts of objects, such as polygons, from the centre of projection to decide a priority order for rendering, first illustrating those that are more distant. Subsequent nearer objects are overdrawn, obscuring sections that lie behind them. This is known as the painter's algorithm, as if a painter in oils has first created a background, and then overpaints foreground figures. Another works in the opposite order, drawing foreground figures first, creating a mask or template against which to clip or block out subsequent (more distant) objects. We will discuss another method, the z-buffer, later. Finally the contents of the view window are transferred to the 'viewport', the region of the display device on which the image is created. This last stage is commonly known as the window to viewport mapping.

This is a general overview of how images are created, which for convenience repeats some of the concepts introduced in chapter 1. The discussion should explain to users why it sometimes takes quite a long time to compose and render a complex scene to a reasonable level of detail. Specific techniques may differ in the level or ordering of activities, but few would be less complex than this. In general, there are many switches of coordinate systems, with formulae to convert from one to the other. A scene is composed in a mental model in the user's brain. The computer model is built more or less directly from this, with primitive objects normally defined in 'easy' positions in a 3D 'canonical' space. These primitives will be instantiated (or instances of them will be created) in the 3D world coordinate space. Projection converts into a 2D view coordinate system within the view window, usually through an intermediate transformation into a view coordinate space (chapter 4). The image is located in its final space, the discrete 2D space of the viewport within the display device (discrete means that the image is pixellated, points are located as integer values; the coordinate systems used before this one are real valued, in continuous space). Post-processing methods, such as anti-aliasing, can be used to disguise the artefacts produced by pixellation, rather as a charcoal artist can smudge individual marks with a wet thumb.

Several geometric modelling methods have been discussed (chapters 8 and 9); rendering methods will be expanded on later. This description is intended to give an indication of the kinds of process undertaken and the kinds of specifications that need to be made by (or for by default) those generating images (of 3D scenes) by computer (on a 2D display).

Methods for Rendering

Once the objects to be depicted have been created, the process of rendering generates more or less 'realistic' images of the artificial world. Here we use realism in the sense that the artificial objects could be believed to have been photographed to produce the image, known as 'photorealism'. There is a developing tradition in 'non-photorealistic rendering', but we concentrate on the simulation of 'natural' or

10. Drawing and Rendering: How to Create Pictures

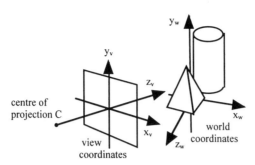

Figure 10.5 World and view coordinates

photographic effects in the following, using the established method of perspective projection with a variety of more-or-less understandable rendering methods. These include calculations of the amount of light falling onto a surface using Lambert's law[2] (a proper law of physics), as well as a number of artificial methods for masking the straight-edged boundaries of polyhedral models to give the appearance of smooth shapes.

As with modelling (chapter 8), there is more than one method for rendering. The choice of which to use depends on the balance of attributes required or the platform available to perform the rendering. As has now become familiar, speed, realism and computer memory requirements have to be traded. There is no perfect system.

The Viewing Pipeline

We look first at the process of converting object points to image points. This involves transformations across several sets of coordinate systems. In a typical system (some would arrange things differently, but here we describe a general system), the defining points of objects are described in real valued Cartesian 'world coordinates' as (x_w, y_w, z_w). These locations are converted (using translations, scalings and rotations) into 'view coordinates', (x_v, y_v, z_v) (fig 10.5). Some systems use a view coordinate origin at the position of a pseudo observer, the 'centre of projection' C, but the method illustrated here places the origin at the point where the direct line of sight or view direction from the centre of projection passes through the view plane. It is usual for the view plane to be orthogonal (at right angles) to the view direction, although some special effects can be created by varying this angle. The plane may be replaced by other geometric forms. A fish-eye lens effect can be achieved by a spherical view surface. The remainder of this discussion is constrained to the orthodox method of a view plane orthogonal to the view direction and at distance d from the centre of projection C.

The affine transformations that convert a point in the world coordinate system into its equivalent coordinates in the view coordinate system depend on the location of the origin O_v of the view coordinate system, a unit view direction vector **v** in the direction of z_v in fig 10.5 and the unit 'up' vector **u** in the direction of y_v (for simplicity, we have not included the normal 'hat' accent to indicate a unit vector). If O_v is not explicitly given, it can be found from the centre of projection C as

---
[2] German mathematician Johann Heinrich Lambert (1728–77) also proved that e and π are irrational.

$O_v = C + d\mathbf{u}$.

Here, we are being a bit mathematically sloppy in using C and O_v as vectors; we should really change to the equivalent position vectors **c** and $\mathbf{o}_v$, but the meaning should be clear through the context, as it was in chapter 9.

Note that **u** and **v** are defining vectors of the view coordinate system, themselves defined with reference to the world coordinate system. Their companion 'right' vector **w**, pointing in the x_v direction, is found from them using the vector (cross) product

$\mathbf{w} = \mathbf{v} \times \mathbf{u}$,

defining a left-handed coordinate set as in fig 10.5. This is unusual compared to the usual right-hand screw rule for coordinate direction (page 61), but early developers of computer graphics systems preferred to have their z coordinate pointing into the scene to designate depth from the centre of projection. There is nothing wrong with being unorthodox if the method is consistently used and adds to the comfort of users. **w**, **u** and **v** are the directions of x_v, y_v and z_v respectively.

The transformation to convert a point P from a location in world coordinate space to its position in view coordinate space is equivalent to conversion of the view coordinate system (x_v, y_v, z_v) into the world coordinate system (x_w, y_w, z_w). As the axes move, the point P moves with them, so its transformed position with respect to (x_w, y_w, z_w) is equivalent to its original position with respect to (x_v, y_v, z_v).

This sequence of operations can be calculated from O_v, **u** and **v** using affine transformations. This begins with a translation to superimpose O_v on O_w. Then three rotations are needed about the y_w, x_w and z_w axes to line up the original view coordinate axes with the static world coordinate axes (these rotations may take place in a different order, but this is typical). Finally, a scaling is used to reflect the left-handed view coordinates into right-handed world coordinates. This scaling may also incorporate size changes in the unusual circumstance of the world coordinates and view coordinates having different measuring scales; for example, one is in inches, the other in centimetres. Note that each of these transformations is applied in the world coordinate system; we are transforming from the view coordinates into world coordinates as a reference frame.

It is quite difficult to visualize the angles needed for the series of rotations, so we adopt a different method here, using a result from vectors. If P is the point to be referred to the view coordinate system, the component of vector O_vP in the directions of each of the view coordinate axes gives the required coordinates. Such components are found by the scalar (or dot) product (chapter 7).

Suppose the coordinates of P and O_v are given with respect to world coordinates as (x, y, z) and (x_o, y_o, z_o) respectively. Then, the vector

$O_vP = P - O_v = (x - x_o, y - y_o, z - z_o)$.

Again, purists may not like this use of notation, but the meaning should be clear. Three scalar products then give us the required view coordinates, (x_v, y_v, z_v)

$x_v = (P - O_v).\mathbf{w} = (x - x_o)w_x + (y - y_o)w_y + (z - z_o)w_z$,

10. Drawing and Rendering: How to Create Pictures

$$y_v = (P - O_v).u = (x - x_o)u_x + (y - y_o)u_y + (z - z_o)u_z,$$
$$z_v = (P - O_v).v = (x - x_o)v_x + (y - y_o)v_y + (z - z_o)v_z.$$

The relevant coordinates of **w**, **u** and **v** (w_x is the x coordinate of **w** in world coordinates, and so on) representing the directions of view coordinate axes are readily available, as is the view coordinate origin. Any point in world coordinates can be changed to its equivalent position with respect to view coordinates by these equations. However, we have stressed the usefulness of having matrix representations of affine transformations (and this is one) to enable streams of operations to be concatenated into one. We can express this in matrix form by a rearrangement of the equations,

$$x_v = w_x x + w_y y + w_z z - (w_x x_o + w_y y_o + w_z z_o),$$
$$y_v = u_x x + u_y y + u_z z - (u_x x_o + u_y y_o + u_z z_o),$$
$$z_v = v_x x + v_y y + v_z z - (v_x x_o + v_y y_o + v_z z_o).$$

The final bracketed terms in each equation,

$$k_x = -(w_x x_o + w_y y_o + w_z z_o),$$
$$k_y = -(u_x x_o + u_y y_o + u_z z_o),$$
$$k_z = -(v_x x_o + v_y y_{ow} + v_z z_o),$$

do not depend on the position of P, so can be pre-computed from definitions of the view coordinates. Thus, we can write

$$x_v = w_x x + w_y y + w_z z + k_x,$$
$$y_v = u_x x + u_y y + u_z z + k_y,$$
$$z_v = v_x x + v_y y + v_z z + k_z.$$

These may now be written as a matrix version of homogeneous coordinate equations,

$$\begin{bmatrix} x_v \\ y_v \\ z_v \\ 1 \end{bmatrix} = \begin{bmatrix} w_x & w_y & w_z & k_x \\ u_x & u_y & u_z & k_y \\ v_x & v_y & v_z & k_z \\ 0 & 0 & 0 & 1 \end{bmatrix} \begin{bmatrix} x \\ y \\ z \\ 1 \end{bmatrix},$$

so the conversion from world to view coordinates can readily be absorbed into the sequence of matrix transformations.

Increasing d in the rendering process is equivalent to the photographic effect of zooming in. As the view plane moves further from the centre of projection C (figs 10.5 and 10.6), diverging lines from C cut off longer edges on the view plane.

When objects are located relative to the view coordinate system, the stage of projection, or reduction of the 3D coordinates to the correct location in the 2D view plane, is relatively easy. Figure 10.6 shows a view directly down the y_v-axis of fig 10.5. Projected coordinates (x_p, y_p) give locations in the view plane (see also figs 6.5 and 6.6, this result was established in chapter 6, pages 172–176). Perspective projection of P is achieved by finding (x_p, y_p), the point at which a line from P

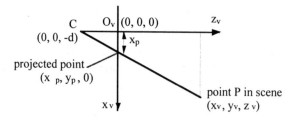

Figure 10.6 Perspective projection; view of fig 10.5 from the positive y_v-axis

expressed in view coordinates (x_v, y_v, z_v) to the centre of projection C $(0, 0, -d)$ passes through the view plane. Using coordinates as defined in figs 10.5 and 10.6, by an argument of proportions,

$$x_p = dx_v/(d + z_v)$$
and $$y_p = dy_v/(d + z_v).$$

Figure 10.6 illustrates the first equation; the latter can be accepted by analogy or by drawing an equivalent figure down the direction of the x_v-axis.

Parallel projections (giving less sense of realism) can also be used, where the projection lines are parallel to each other. This is effectively the same as moving the centre of projection a very long way away along the negative z-axis, so the projection lines become practically parallel. In this case, the projection is relatively trivial, with the values of x_p and y_p remaining equal to x_v and y_v.

The formulae for perspective projection are fairly simple, but they do not fall into the natural system of affine transformations, those used to manipulate objects and to change coordinate systems. They must be dealt with carefully; for example, trying to project a point with view coordinates $(x_v, y_v, -d)$, one that is 'level' with the centre of projection, leads to a 'zero divide' error. This problem can be eliminated by projecting only points that have $z_v > (-d + n)$, where n is a small value greater than 0. This defines a near plane a small distance n in front of the observer at the centre of projection. Only objects beyond the near plane will be projected. This is not an unnatural thing to do; we cannot clearly observe any object closer than a small distance from the eyes.

Now the question of clipping raises its ugly head. Clipping involves eliminating from consideration all objects and parts of objects that would lie outside the view window when projected. If the view window has lower left corner (x_{vl}, y_{vl}) and upper right corner (x_{vh}, y_{vh}) – suffixes l and h refer to 'low' and 'high' here – we can reject all parts outside the view window as those that satisfy

$$x_p < x_{vl}, \quad x_p > x_{vh},$$
$$y_p < y_{vl}, \quad y_p < x_{vh}.$$

Alternatively, clipping can be performed in 3D prior to performing projection. This is more complicated, clipping against the potential viewing region, the 'view frustum' (fig 10.7). This is a truncated pyramid within the near plane, far plane and the four slant planes from the centre of projection alongside the edges of the view window. A point cannot be viewed if it lies outside the view frustum, which means it lies outside any one of these six bounding planes. In circumstances where there

10. Drawing and Rendering: How to Create Pictures   303

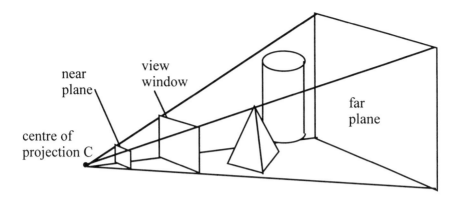

Figure 10.7 The view frustum between near and far planes

are many objects lying outside the viewing region, for example in a flight simulator where extensive terrain data is held as a geometric model, it may be better to clip before projection, thus reducing the number of points to be processed at this stage. When objects to be seen are created specially so that they can be expected to lie within the view frustum, it may be better to clip after projection. Again, we have a situation where the 'best' method depends on circumstance.[3]

The net result of the processes described above is that for every point (x_w, y_w, z_w) in the objects to be depicted that is potentially visible within the view frustum, a projected point in the view window (x_p, y_p) can be found, via the view coordinate point (x_v, y_v, z_v). One further transformation from the projected coordinate system converts into display coordinates, locating the correct point within a defined 'viewport' on the display surface (x_d, y_d). This is an integer location, typically the location of a pixel on a display screen (fig 10.8). By another argument of proportionality, the position of (x_d, y_d) within the viewport can be defined as

$$x_d = x_{dl} + (x_p - x_{vl})*(x_{dh} - x_{dl})/(x_{vh} - x_{vl})$$
$$y_d = y_{dl} + (y_p - y_{vl})*(y_{dh} - y_{dl})/(y_{vh} - y_{vl}).$$

The values of (x_d, y_d) are rounded to the nearest integer to give the pixel locations required. Many screen addressing systems have x or horizontal coordinates from left to right in the normal style, but y or vertical coordinates from top to bottom of the

---

[3] We have not described the full horrors of clipping here. For example, when considering an object edge, it is not enough to identify if the two ends of the edge are outside the visible region, as some part of the line itself may cut across that region. Clever clipping methods save overall time by using 'exclusion tests' for rapid identification of some simple cases, such as when a line is completely inside or easily identified as completely outside the view window, then performing more complex tests on the remaining inconclusive cases. Some of these will need to be split against the window in 2D or frustum boundaries in 3D if they are partly inside. The clipping of faces is done by linking clipped edges with parts of the window boundary. Standard computer graphics texts give full descriptions of various methods; we just accept for our purposes that they are available.

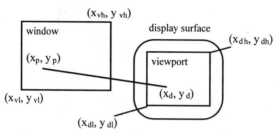

Figure 10.8 The window to viewport mapping transforms (x_p, y_p) in the view window to (x_d, y_d) in the viewport

screen. Regardless of this, the formulae above will give correct interpretations if the lower left and upper right coordinates of the viewport are properly allocated. Some systems introduce an intermediate stage between window and viewport, transforming to 'normalized device coordinates', NDC. In real valued NDC, the contents of the window are 'placed' into a unit square, with x and y coordinates ranging from 0 to 1. NDC can be seen as a standardized form of output; the process of placing this image onto the display surface of a specific device is then dependent on the device only. Manufacturers can supply their equipment with the 'device driver' software needed to convert from NDC to device coordinates.

Figure 10.8 shows a window and viewport with the same aspect ratios (width/height). This need not be maintained in a window to viewport mapping. Readers should be aware that using different aspect ratios in each window gives scaling distortions to the resulting image, as in setting a currently standard (4 by 3) television transmission to be received on a (16 by 9) wide screen. Different aspect ratios are almost invariably set as an intermediate stage in using NDC, but this is usually remedied by correct selection of the final NDC to display viewport transformation.

Hidden Surface Removal

We have seen how to transform points in object space to points in the image space of the viewport. It is necessary for rendering methods considered below to carry the depth information from view coordinates with the point location in display coordinates. Values $\{(x_d, y_d), z_v\}$ tell us where the point may be displayed on the display and its original spatial depth behind the view plane. Depth information makes it possible to decide when an object or part of an object is occluded by other objects in the process of rendering, giving vital information for the 'hidden surface problem', as mentioned in chapter 6. We will turn later to deciding the colour that an object should receive on rendering, but consider here whether an object should be rendered at all through the process of hidden surface removal.

Hidden surface removal algorithms generally fall into two types: those that operate in object space and those that operate in image space. Object space algorithms consider features of objects in view coordinates, usually ordering these features according to their depths from the centre of projection. These depths are used to prioritize the order of depiction. For B-Rep models, the types of features

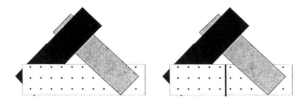

Figure 10.9 An object may need to be split to create absoulte depth priority

considered are the polygons that constitute faces of objects. The painter's algorithm renders objects in prioritized order from the furthest to the nearest, overpainting what already exists on the screen. Questions of priority are not always easy to solve, particularly when groups of features are mutually overlapping as in fig 10.9, which shows how objects may have to be split to create absolute ordering by breaking such cycles.

The z-buffer, a secure image space method, is commonly used currently due to the ready availability of memory. Although postulated some time ago, it was not feasible in times when computer memory was expensive and scarce. It does not require pre-ordering of scene objects; all objects that have potential to be rendered can be considered in any order. For each pixel of the screen, a separate array of memory known as the z-buffer holds the depth z_v of the object currently depicted in that pixel. Suppose it is decided that part of a new object to be rendered should occupy a particular pixel (using the methods of the last section). The question is whether it should take priority in the pixel's current content. Its z_v is compared with the z-buffer value for the pixel. If the new object has larger z_v than the value currently in the z-buffer, it is further away from the centre of projection than the object that currently occupies the pixel, so it is not drawn. If its z_v is less than the z-buffer value, z_v is entered into the z-buffer in this location and the current object's colour is placed into the pixel. If the z-buffer is held to adequate precision, this method is relatively error free; it even deals with interpenetrating objects.

For most hidden surface methods on B-Rep models, back face culling is used as a first trawl through the set of objects to be depicted, eliminating all faces that are oriented away from the centre of projection. This is easily done in the view coordinate system by eliminating faces whose surface normal has positive z coordinate (this means it points broadly away from the centre of projection, where an observer would be situated). Alternatively, the face direction through the clockwise or anticlockwise ordering of vertices can be considered (as discussed in

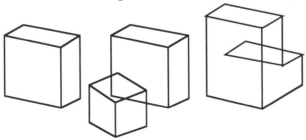

Figure 10.10 Back face culling is adequate for single convex objects, but may not be for multiple objects or for objects with concave features

chapter 8 on B-Rep modelling). Back face culling is a complete hidden surface method only for single concave objects (fig 10.10), but it is only a partial solution when multiple objects may occlude each other or when convex objects occlude parts of themselves. However, it is a useful first stage for such scenes as a great deal of the data set is eliminated from further consideration by a fairly simple initial test.

Flat or Lambert Shading

Having decided what pixel should be occupied by a part of a model, we have to consider the colour that should be placed in that pixel. Colour is not just a property of the object being depicted, it depends on the quality of the lighting and the relative locations of object and light source. The brown appearance of red vehicles under yellow sodium street lighting is familiar. We must take into account the surface colour of an object, colours of light sources and the angles between surfaces and the light direction. A page traps more light and thus appears brighter if it is held at right angles to the rays from a bright lamp compared to one at a more oblique angle. Figure 10.11 illustrates this principle, although the proximity of the light source to the page means the angle of incidence of light is variable in this case. The brightest possible illumination occurs when the rays are parallel to the surface normal, illustrated as an arrow in the left image of fig 10.11. With the orientation of the right-hand image, none of the rays fall at right angles to the surface, and this will appear less bright. Another way of thinking about this is to observe that the object will 'cut off' more light and thus be brighter if it casts a larger shadow.

In physics, the distance from light source to object is also important; the amount of light received depends upon the inverse square of distance. The effect of this is relatively small in the context of most computer graphics scenes, so this characteristic is not often modelled.

To calculate the apparent colour of a surface, we need to know the angle of incidence of a light source on a surface: the angle between the light direction and the surface normal. We mentioned in chapter 8 that a unit surface normal, a direction at right angles to a surface, could be stored as part of the data structure for a face in a B-Rep model. Formulae also can be used to define surface normal directions in CSG modellers, but the following description is particularly related to rendering B-Rep forms. A unit surface normal **n** can be stored in vector form as (x_n, y_n, z_n) (fig 10.12), where the values of x_n, y_n and z_n represent the displacements in the x, y and z directions respectively undertaken by tracing a line of unit length orthogonal to the face concerned. If the unit vector from a surface point towards the light is similarly recorded as $l = (x_l, y_l, z_l)$ ('l' is italicized to make sure it is distinguished from 'I' below), then the cosine of the angle θ between the light direction and the surface

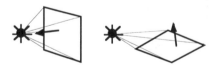

Figure 10.11 Objects that trap more light appear brighter

10. Drawing and Rendering: How to Create Pictures

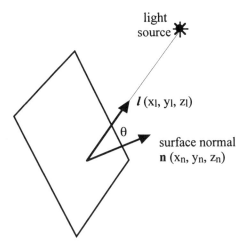

Figure 10.12 Angle of incidence of light on a surface

normal (fig 10.12) is given as the scalar product of the two vectors

$$\cos(\theta) = \mathbf{l}.\mathbf{n} = x_n x_l + y_n y_l + z_n z_l.$$

The meaning of scalar product and the method for finding unit vectors in particular directions were discussed in chapter 7 on vectors. Lambert's law states that the intensity of colour at the surface due to this light source is proportional to $\cos(\theta)$, so we take

$$I = C_l C_s \cos(\theta)$$

where C_l and C_s are values indicating the intensities of colours of the light source and the surface respectively. Representations of colour were discussed in chapter 3; the additive (red, green, blue) is that underlying the meaning of colour in this section. The angle θ varies for all points on the surface, but if the light source is very distant (for example, the sun), for all practical purposes light rays are parallel within a reasonable sized surface, and this angle can be considered to be constant. Using a point near the centre of the surface as representative gives reasonable results, particularly when the surface is relatively small compared with the light distance. This assumption of constant illumination across a surface may be adopted in flat shading for simplicity; methods to be discussed later give alternatives.

Those who find the above argument difficult to understand mathematically should not despair; the point in developing it is to show that a formula to give the effective colour of a surface can be developed from known characteristics of the surface and the light. Also, the meaning of a colour intensity has not been discussed since chapter 3; some reminders appear below.

If there are several light sources, all such terms for each source must be summed. It is usual to add an 'ambient' light effect, simulating the way in which indirect light bounces off surrounding objects. Without an ambient light term, all objects not directly lit would appear intensely black. Thus the colour intensity at a surface is calculated as

$$I = \Sigma C_l C_s \cos(\theta) + AC_s ,$$

where A is a factor taken to simulate the ambient light effect and the meaning of Σ should now be known. For colour images, the terms C_l and C_s have three components, one for each of the red, green and blue components. For a colour computer display, the colour intensity I also has three components I_{red}, I_{green} and I_{blue} resulting from one application of the formula for each colour component. Calibration of the intensity of the light source C_s has to be considered carefully to avoid 'washing out' parts of the image with very bright light. This can occur if the value of I becomes greater than the full 'bright white' intensity of the display. Also, the overall intensity should be enough to enable darker sections of the image to be seen; those who have adjusted the 'brightness' and 'contrast' controls of a television should be familiar with this problem. It may be necessary to re-run an image generation, with light source intensity changed to get the required effect, as it may be necessary to change lighting when a photograph is taken, or to change the exposure of a photographic film to accord with existing lighting. Some computer image synthesis systems can simulate the automatic adjustment of modern cameras. An absolute representation of the full range of screen colours can be found without truncating over bright colours that go outside the permitted scale, then all pixel contents may be scaled down to match the colour performance of the system in use.

This method is commonly known as Lambert shading or flat shading; all parts of a face have the same surface normal. Faces appear separate. This may not be wanted when B-Rep faces approximate a smooth surface, when differences between faces can be 'smudged' using Gouraud or Phong shading. These are discussed below. Plate 10.1 gives examples of 'cylinders', approximated by octagonal prisms in three cases, rendered using different shading methods.

Scan Line Methods

Many popular rendering methods are scan line based, where a horizontal row of pixels in a viewport is considered as a scan line. Efficiency is gained by passing on the attributes of one scan line to the next, and by considering clusters of pixels with similar attributes rather than just individual cases. Figure 10.13 shows a typical scan line in a viewport where three overlapping triangles are to be drawn – most computer graphics scenes contain considerably more polygons than this. For each scan line, an ordered set of 'spans' is considered – these are regions cut off from left to right of the scan line between polygon edges. Spans are updated using an 'active edge list', another example of a linked list, as the scan line sweeps down the viewport. This stores those edges intersected by the scan line. For the scan line shown in fig 10.13, the active edge list is {AB, DE, AC, GH, DF, IH}. GI was previously active, but was replaced by IH when the scan line passed beyond I. As the scan line moves down the viewport, EF will replace DE and BC will replace AB. When the scan line passes through point H, GH and IH are discarded from the

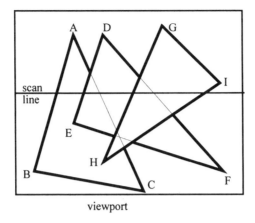

Figure 10.13 A scan line intersects objects within a viewport

active edge list; after passing through F and C, the active edge list will be empty. The spans for the scan line shown in fig 10.13, in order, lie between the left edge of the viewport, AB, DE, AC, GH, DF, IH and the right edge of the viewport. Fast updating of active edge lists can be performed using integer-based algorithms equivalent to line drawing algorithms to update the pixel positions of span ends, and to swap span end locations at crossing points.

The basic algorithm for scan line shading is

```
for each scan line
   for each span
      identify the visible polygon
      calculate span colour
      fill span with colour
   end for (span)
end for (scan line)
```

For each span, the decision as to which polygon is visible is made using a hidden surface algorithm based on the z_v depth of each polygon. If polygons penetrate each other, extra span divisions must be introduced using new edges that split polygons at these intersection lines. These edges are identified using vector methods (chapter 7). For flat shading, each span is given the same colour intensity (we use 'colour' to be synonymous with 'colour intensity' below). With Gouraud and Phong shading, there is variation of colour across each span, giving smooth shading (plate 10.1). These two methods will be discussed in some detail in the following sections.

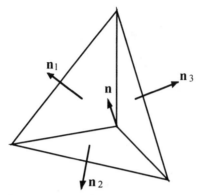

Figure 10.14 The normal **n** at a vertex is taken as the average of neighbouring face normals n_1, n_2 and n_3 for 'smooth shading'

Gouraud Shading

Smooth shading is used when it is desirable to smudge the boundaries between faces of an object, for example when a collection of plane faces is used to approximate a curved surface. Suppose three or more faces come together at a vertex and that the surface normal for each face is stored in the face data structure (fig 10.14). An average normal for the vertex can be calculated from face normals, and can be stored in the vertex data structure. This involves averaging the x, y and z coordinates of the face normals, then normalizing the resulting vector to unit length, as explained in chapter 7 on vectors.

Suppose this process has been done for each vertex of the polygon under consideration, taken as a triangle ABC in fig 10.15, with n_A, n_B and n_C being adjusted normals at A, B and C respectively. A scan line cuts the triangle's edges AB and AC at points M and N as illustrated (AB and AC are in the active edge list). The colour to be used at each vertex A, B and C is calculated from the Lambert method with an ambient component,

$$I = \Sigma C_l C_s \cos(\theta) + AC_s ,$$

giving three colour intensities I_A, I_B, I_C, where n_A, n_B and n_C are used to calculate the relevant value of $\cos(\theta)$ at each vertex. If adjacent faces at these vertices have different surface colours C_s, a strategy for deciding an averaged colour at each vertex must be adopted. Smooth change of colour from I_A to I_B as we move from A to B is achieved by linear interpolation of colour along the line AB. Let s be AM/AB, the fraction of the 'journey' from A to B where M is located. This is easily found as

$$s = (y_M - y_A)/(y_B - y_A),$$

10. Drawing and Rendering: How to Create Pictures

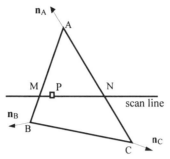

Figure 10.15 Intensity at P is interpolated from vertex intensities

where y_A is the vertical pixel address of A, and so on. There is no worry about zero divide here as AB is assumed to have a non-zero vertical component, otherwise it would not be in the active edge list as intersecting a scan line. If s is explicitly needed, the scan line by scan line increment of s is pre-calculated as $1/(y_B - y_A)$, and this is added or accumulated into s for each successive scan line, for speed of calculation. The colour at M is calculated as

$$I_M = (1 - s)I_A + sI_B.$$

This form of linear interpolation was discussed in chapter 9 on splines. Colour change can also be accumulated by pre-calculating the scan line to scan line change as $(I_B - I_A)/(y_B - y_A)$ and adding this colour into successive values of I_M. This is a calculation on colour, so each of the red, green and blue components of I_M must be evaluated in this way. As the parameter s varies from 0 (at A) to 1 (at B), I_M varies from I_A to I_B, giving smooth variation of colour. We deal similarly with the colour at N. Let t = AN/AC. Then,

$$I_N = (1 - t)I_A + tI_C.$$

Now finally, suppose that u = MP/MN. The colour at pixel P is given as

$$I_P = (1 - u)I_M + uI_N.$$

Horizontal lines are not entered into an active edge list, but are dealt with by this equation as their pixel representations coincide with scan lines through rounding of their end coordinates to integer pixel locations in the window to viewport mapping. All three colour interpolations are efficiently implemented by accumulation of step by step changes as described above for I_M.

Gouraud shading gives smooth colour variation across the surface of the triangle. This matches colours across neighbouring faces due to their use of the same vertex colours. The method is described here for triangle rendering, but it can be applied to 'well-behaved' polygons (they do not self-intersect or 'wrap over' themselves) with arbitrary numbers of edges. Each span to be rendered depends on two active edges; the intensity at each of the bounding vertices of these edges is known. For example, in fig 10.16, the intensity at P is found by a series of interpolations, between B and C for I_M, between E and D for I_N and between M and N for I_P. By this method, an

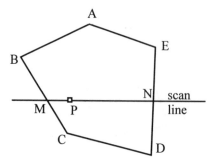

Figure 10.16 Interpolation along active edges of a general polygon

image of a polyhedral model is given smooth transition of colours across internal boundaries, although its geometric model is held as a collection of plane faces. In drawing a polyhedral approximation of a cylinder, for example, edges between the curved surface and the end circles are sharp, so edges are labelled as susceptible or not to smooth shading, with averaging of vertex normals accordingly adjusted.

Unnatural artefacts may be seen due to the underlying shapes of polygons. For example, if polygon ABCDE (fig 10.16) is viewed at a different orientation without changing its surface normal or the location of point P (fig 10.17), we would expect the colour I_P of point P to remain the same. This is not necessarily the case. In fig 10.17, interpolated colour I_M is calculated from I_A and I_B, I_N from I_D and I_C, so the colour at P depends on vertex colours at A, B, C and D. In fig 10.16, the colour at P depends on vertex colours at B, C, D and E. Now the colours at A and E are independent of each other; they depend on the orientation of polygons impinging on these vertices that can be changed without affecting the structure of polygon ABCDE. Thus, the use of A rather than E will almost certainly affect the colour given to P, although in a physical sense it should remain the same. This effect may be noticed in animation sequences, although it is less of a problem in still images. The underlying polygonal structure may still give undesirable artefacts in still images, where silhouettes appear angular in contrast to the internally smooth rendering, so although Gouraud shading (named after Henri Gouraud) is a relatively fast way of achieving reasonable smooth shading, it can be improved.

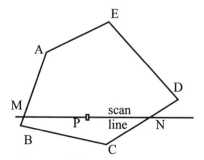

Figure 10.17 Reorientation of ABCDE makes colour at P dependent on different vertices

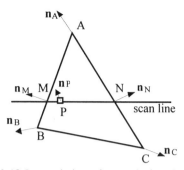

Figure 10.18 Interpolation of normals for Phong shading

Phong Shading

Phong shading is conceptually similar to Gouraud shading, but instead of being based on the linear interpolations of three component colour intensities, interpolation is performed on vertex normals, such as $\mathbf{n}_A$, $\mathbf{n}_B$ and $\mathbf{n}_C$ in fig 10.18. The normal at M is found from

$$\mathbf{n}_M = \text{norm}\{(1-s)\mathbf{n}_A + s\mathbf{n}_B\},$$

where 'norm' is a function that normalizes a vector by reducing it to unit length (see chapter 7). This must be done at each stage of interpolation to ensure balanced results. There is no way to calculate an incremental change in the vector, as colours were interpolated in Gouraud shading. For example, dividing the overall change from $\mathbf{n}_A$ to $\mathbf{n}_B$ into pixel by pixel incremental steps as vector quantity $(\mathbf{n}_B - \mathbf{n}_A)/(y_B - y_A)$ gives a linearly interpolated non-unit vector at intermediate point M. This must be set at unit length to 'balance' with vectors in subsequent interpolations, so a non-linear interpolation method must be used. The time savings possible in Gouraud shading by incremental calculation of colours are not possible for vectors. Even worse, each normalization involves calculation of a square root, making vector interpolation even more time consuming. The best that can be done is to calculate the interpolated value of parameter s using increments $1/(y_B - y_A)$. As $s = AM/AB$ varies from 0 (at A) to 1 (at B), $\mathbf{n}_M$ varies from $\mathbf{n}_A$ to $\mathbf{n}_B$, rather like a rod with one end hooked to line AB, while it changes its direction smoothly from that of $\mathbf{n}_A$ to $\mathbf{n}_B$. We deal similarly with the normal at N. Let $t = AN/AC$.

$$\mathbf{n}_N = \text{norm}\{(1-t)\mathbf{n}_A + t\mathbf{n}_B\}.$$

Now finally, suppose that $u = MP/MN$. The normal at pixel P is defined as

$$\mathbf{n}_P = \text{norm}\{(1-u)\mathbf{n}_m + u\mathbf{n}_N\}.$$

$\mathbf{n}_P$ is used to calculate the value of $\cos(\theta)$

$$\cos(\theta) = \mathbf{n}_P.\mathbf{n}_l,$$

where $\mathbf{n}_l$ is a vector from P towards the light source and θ is the angle between the calculated surface normal and the direction from P towards the light source. This is used in the now familiar shading equation

$$I = \Sigma C_l C_s \cos(\theta) + AC_s,$$

giving a value for the colour to occupy the pixel at P. Extension of the method to general polygons with more than three edges is done equivalently to Gouraud shading. This is not described explicitly here, but reference to fig 10.16 with normals rather than intensities being interpolated should enable readers to understand the concept. This method is known as Phong shading, after its inventor Phong Bui-Tong, whose sad death occurred in a car accident shortly after the method became widely used.

Phong shading is considerably more time consuming than the Gouraud method, where the three colour components of a colour intensity are interpolated at each stage. In Phong shading, the necessary normalization to unit length of each interpolated vector involves time-consuming square rooting and there are three extra divisions per interpolation to find the required value of each vector component. Then, the intensity calculation is performed for each pixel lying within the represented triangle, compared to just one intensity calculation per polygon vertex in Gouraud shading. Vertex intensities are also shared between faces, so this is a further time improvement in Gouraud shading. The number of light sources obviously has a more serious effect on Phong shading. Phong shading does, however, diminish the effects of some of the artefacts produced by Gouraud shading and, in an extension discussed below, allows shiny objects to be simulated. Gouraud shading is suitable only for matt surfaces.

Both these smooth rendering scan line methods (Gouraud and Phong) simulate the effects of smooth curved surfaces from approximate polygon models by varying the intensity of colour displayed within visible regions of objects. However, they do not smooth the visible boundaries of objects. Cylinders rendered by these techniques have smooth polygon boundaries across the interior 'curved' boundaries (the two central objects of plate 10.1), but a perceptibly jagged boundary is seen on the polygonal ends, which should appear circular. Alternatively, surfaces can be represented as greater numbers of polygons giving the usual trade-off between the quality of representation, the amount of memory required and the speed of representation. Increasing the polygon count slows rendering time and increases memory requirements.

Exact Object Rendering

More advanced techniques can be used to trace smooth boundaries (the right-hand object of plate 10.1). The author uses a method of distributing points across the surfaces in object space, in a way similar to the IFS method described in chapter 8.

10. Drawing and Rendering: How to Create Pictures

All surfaces are two-dimensional, so they can be described parametrically using two defining parameters. By taking small enough steps in each of these parameters, or by scattering points randomly across the domain of these parameters, enough points can be generated to fill the image of the surface. The curved surface of the cylinder, described in chapter 5 for example (fig 5.25), has angle θ varying between $0°$ and $360°$ and z value varying between 0 and c. From values of θ and z generated in these ranges, individual surface points and normals can be calculated from the formulae of chapter 5. To create the circular ends, a point within a circle of radius r_0 can be represented in polar form by ranging the radius from 0 to r_0 and the angle θ from $0°$ to $360°$, and then generating values of x and y from these as

$$x = r \cos(\theta), \qquad y = r \sin(\theta).$$

For the end $z = c$, the normal is in the z-axis direction, $(0, 0, 1)$; for $z = 0$, it takes the opposite direction, $(0, 0, -1)$.

Points and normals created in this way can be transformed to the required positions in world coordinates and entered into the shading formula to generate an even smoother effect. The decision of when enough points have been created is done in an ad hoc way by this technique. Although it gives 'exact' curved surfaces to the precision of the pixel display, it is considerably slower than the more standard scan line methods for generating Gouraud and Phong rendering. Programming the method is considerably easier than standard methods as the organization of devices needed to maintain active edge lists, decide on span priorities, arrange the scan sequences, to perform clipping and so on, is greatly eased as only one point at a time is rendered by the technique. It is easy to decide if a single point is visible or not by an inclusion test against the view window and a z-buffer to decide on depth priorities. The technique was used to generate plate 10.4, where all objects are displayed by point scattering. The spherical ball, conical and cylindrical surfaces of the plant pot are defined using two variables after the methods of chapter 5, and passed into the rendering equation. The plants are 3D IFSs. Reflected or specular highlights are seen in this image, calculated due to the availability of surface normals at each point depicted. These will be considered later, after we have discussed another feature of this image, shadows.

Shadows

Both Gouraud and Phong shading give matt surface effects. There are no reflective 'specular highlights' and no shadows. Shadows may be created by adding 'shadow volumes' to the geometric model. First, pyramids are created with vertices at the light source and generating edges through the vertices of objects that cast shadows (fig 10.19). Then the sections of these pyramids that extend behind back faces when viewed from the light source create the shadow volume. The intersections of these shadow volumes with other objects in the scene are calculated to create new edges that give the boundaries of shadows. These edges are combined carefully with the edges of existing faces to define new shadow faces as parts of these existing faces. The process of defining shadow areas is equivalent to hidden surface removal with

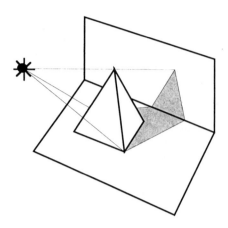

Figure 10.19 New edges added to create shadows

the light source as the centre of projection. This adds to execution time, particularly if there is more than one light source. Shadow faces do not receive light from the source under consideration, so this component of the light equation is not included when these faces are rendered. If no light sources penetrate a shadow region, only the ambient term is used.

This method gives crisp-edged shadows from single light sources, quite unlike the soft shadows observed in real situations, such as distributed light sources (strip lights or light panels, for example) or the kind of diffuse light that enters through a window on a dull day. This kind of shadow can be simulated in these shading techniques only by adding several light sources, for example using a row of point lights to simulate a strip light or a grid of point lights to simulate a light panel. This brings us back to trade-offs. This still gives only a set of discrete light sources rather than a single distributed source, so several overlapping shadows are seen rather than a smooth penumbra. We have already seen that calculating a single shadow is computationally expensive. Increasing the number of light sources may increase realism, but adds considerably to computation time. An alternative way of generating soft shadows, 'radiosity', is described below. This, too, is computationally intensive, but gives, arguably, the most correct approximation to the laws of physics.

A more memory-intensive, but conceptually easier and reasonably robust, method of dealing with shadows is to use a shadow buffer. We commented above that the process of generating shadows is related to hidden surface removal with the centre of projection set at the light source. The z-buffer is a popular hidden surface method; a memory location is allocated to hold the depth of the object currently displayed in each pixel in the image. If the object currently under consideration for display is closer to the centre of projection than the z-buffer depth, the new object replaces that currently shown in the pixel, with the z-buffer being updated accordingly. Otherwise, the object is hidden behind the currently displayed object, and is not drawn.

Now imagine that we set up a shadow buffer for each light source. This is independent of the pixel display, so may be of different spatial precision. Suppose we are about to draw an object into the image. Regardless of whether it may be seen from the centre of projection, it may cast a shadow (consider the back left face of the

10. Drawing and Rendering: How to Create Pictures 317

pyramid of figure 10.19), so all objects, regardless of whether they have 'passed' the z-buffer test, must also be subjected to a shadow buffer test. If an object is closer to the light source than the existing one in the shadow buffer, it is fully lit and the shadow buffer is updated. Thus, if the object is to be drawn, it is given the full rendering component for this light source. If the object is further from the light than the shadow buffer content, it is not lit by this source, so is rendered (if visible according to the z-buffer) without the added component from this light source. The scheme works if objects closer to the light source are rendered first, so those further away will be given the correct shadow information. However, with more than one light source, there may be conflicting priorities that cannot be satisfied. A way around this is to process all objects in the scene twice: once to fill the shadow buffer, the second time to render the objects using the z-buffer and information already held in the shadow buffer. This method works well, but it requires more memory and effectively undertakes the full view transformation process once for each light source. It is also advisable to have greater spatial precision in the shadow buffer than the z-buffer to reduce the incidence of Moiré patterns that may be observed on surfaces at oblique angles to the observer. Points at each edge of a pixel's spatial span have different depths within the shadow buffer, so another computer graphics 'fudge' that helps with this problem is to render a narrow range of depths behind the shadow buffer as fully lit, as though there is a thin translucent film on top of each object. Shadows add greatly to ability to interpret an object in a 3D sense, so the extra work needed to create them is valuable in the search for photorealism.

Specular Highlights

Shadows may be added to all the rendering methods featured above, but specular highlights are usually added to Phong and exact object methods only as the surface normal for each rendered point is readily available, as needed for the calculation of specular highlights. Faces or surfaces can be designated as susceptible to reflection or not, and the degree of shininess can be controlled by a gloss factor. All visible objects reflect light. Those that are matt do so 'diffusely', with light rays scattered in many directions. Very shiny surfaces, such as mirrors or highly polished metals, reflect incoming rays directly back along a single ray, perhaps with some light being absorbed to give a change of colour.

For a point P on a reflective face, the reflected ray direction PR is calculated as shown in chapter 7 on vectors. Suppose the angle between PR and the direction PC from P to the centre of projection C is α (fig 10.20). If α is small, a reflected highlight should be seen. This can be achieved by adding a specular highlight term to the colour intensity formula for the pixel occupied by P,

$$I = \Sigma C_l C_s \cos(\theta) + AC_s + \Sigma k_r C_l \cos^n(\alpha).$$

k_r is a constant of proportionality that may be a function of the surface indicating how much light of each colour component is absorbed. C_l is the colour of the light source as used for the other rendering components, the power n to which $\cos(\alpha)$ is

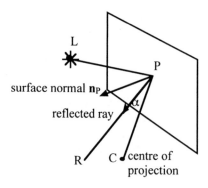

Figure 10.20 A reflected ray PR directed fairly close to the centre of projection

raised is a 'gloss factor' and Σ indicates that the respective terms are summed for all light sources. The value of $\cos(\alpha)$ can be found as the scalar product of unit length vectors in the directions of the reflected ray (x_r, y_r, z_r) and the centre of projection C (x_c, y_c, z_c) as

$$\cos(\alpha) = x_r x_c + y_r y_c + z_r z_c.$$

Now $\cos(\alpha) = 1$ when $\alpha = 0°$, and $\cos(\alpha) = 0$ when $\alpha = 90°$, having the desired property of taking its largest value when the reflected ray and PC are in line, and taking the value 0 when they are as far out of line as possible, the fall in value between these terms being gradual. With $n = 1$ in the specular term, this gives a bright spot where the reflected ray and PC are coincident, the brightness dying off as α increases, when an observer moves away from the line of the reflection. For α between 0° and 90°, $\cos(\alpha)$ must lie between 1 and 0, so $\cos^n(\alpha) < \cos(\alpha)$ when n is greater than 1 (think of the effect of raising a fraction such as 1/2 to a power; half squared is a quarter, ... raising the power gives smaller values). For the pixel which has $\alpha = 0°$ (or close to 0), the value of n has no (or minimal) effect as $\cos(0) = 1$ and $\cos^n(0) = 1$ for any value of n. As the value of α increases away from 0°, the specular reflection will diminish more rapidly with increasing n, giving a tighter reflected spot, as if the surface is highly glossy. An 'eggshell' finish highlight can be found using small values of n (typically $n = 1$ to 5). With $n = 5$ or more, the effect is of a more highly glossy surface with a near single point reflection.

In plate 10.2, all curved objects are depicted using exact object rendering. The gloss factor for the rearmost sphere is 10, and that for the foreground cone and other sphere are 5; the effects on the levels of glossiness are apparent. Plate 10.2 also shows how the z-buffer copes well with interpenetrating objects, and how the shadow buffer casts shadows effectively across complex combinations of curved and separate surfaces, with one minor defect caused by a computer graphics 'fudge'. The 'foreground layer' method is used to avoid Moiré patterns as described when discussing the shadow buffer; a narrow layer of a surface is considered to be fully lit. The effect of this is seen on the foreground cone: the shadow of its vertex, where it touches the ground, is seen to occur just behind its proper position.

Textures

In the real world, faces of objects do not always have constant colour. In this section, a few of the ways of introducing texture to computer graphics are described briefly.

Random textured effects can be shown by accumulating random deviations from the calculated surface normal in Phong shading. Although the location of the surface is not changed, this artificial change of orientation scatters light in different directions, as though from a series of surface bumps. This method is known as 'bump mapping', and is one of the many computer graphics techniques developed by 'pioneer' Jim Blinn.

Texture mapping uses photographically or procedurally generated texture, such as tree bark. Procedurally generated patterns may be subjected to random variation or 'noise' to make them appear more natural, as in the simulated tree bark and leaf patterns in plate 10.3 (this image was generated by the author's former student, Jon Thum). If a photograph of a large section of tree bark is scanned in or digitized into yet another space, 2D texture space, sections of this space can be mapped onto surfaces to be depicted, maintaining as much continuity as possible. Coordinates in 2D are mapped into polygons or into parametric representations of 2D surfaces, in a method similar to that used in exact object depiction. If a digitized photograph of bark is texture mapped onto a simulated tree trunk, like wallpapering a cylinder with bark patterned paper, surprisingly realistic results can be obtained. A major problem occurs when the back of such a tree trunk is observed – the join in the wrapped texture is almost invariably seen, as would be the join in poorly designed wallpaper. Also, as textures are held to limited precision in pixel form, pixel structures may be seen if the texture is enlarged before depiction. Pixels from texture space are mapped onto sections of faces or surfaces in world space, which are eventually mapped into pixels in the image space. There is no direct correspondence between pixels of texture space and pixels in image space; they will almost invariably overlap each other. A good renderer will calculate an appropriate average of the texture space pixels that occupy each pixel in image space, but this is another time-consuming activity and is not always done. In spite of these drawbacks, this method is much used in animation and virtual reality systems, as it can produce reasonably realistic scenes with fairly low rendering times.

Plate 10.4 uses procedurally generated textures imposed on the exact object depiction of the beach ball, allocating colours within set ranges of the angle θ (equivalent to the angle of longitude) used to generate a sphere (fig 5.24, chapter 5). This allows the surface colour C_s to change locally, giving different contributions to the rendering equation across the surface. Procedural texturing is also applied to the check floor and patterned wallpaper. For the latter, colour is allocated by interpolation between light and mid blue colours using a cosine function of the horizontal component of the wall direction.

'Solid texture' creates a texture as a set of colour patterns in a 3D texture space, often using random effects in the generation phase. A standard method effectively creates a spline solid, controlling the variation of texture by randomly affected values and direction vectors at 3D grid vertices, the locations of points within the grids being interpolated smoothly using cubic spline expansions. The extension to 3D spline solids is equivalent to the switch from spline curves to surfaces. If the texture space is mapped into the 3D space 'occupied' by a 3D object, the value of the texture at the object's observed surface points may be used as the surface colour C_s or to create variation in a given surface colour, which thus changes continuously across points in an object face. This method can be used successfully in modelling solid characteristics, such as the grains of marble or wood, when the spline can exhibit regular periodic effects subject to random variations. It avoids the problem of 'wallpaper joins', and gives convincing results if a model is split or carved, as the texture is continuous in 3D space. The method also allows continuous bump mapping, the object's surface normal being perturbed continuously by the interpolated tangent of the texture spline. This technique was developed by Ken Perlin, amongst others.

Ray tracing

Lambert, Gouraud and Phong shading are the major scan line rendering methods. Ray tracing operates pixel by pixel in image space, rather than the scan line by scan line of the previous methods.

In ray tracing, the track of the light that reaches an observer at the centre of projection is reversed. Rather than following the light from its source to the centre of projection, a set of reverse light rays, at least one per pixel, is sent out to search the set of objects that comprise the scene to be viewed, to identify which features generated that light. In fig 10.21, for example, a ray from centre of projection C passes through a pixel equivalent position in the view window (not shown in fig 10.21 to simplify the diagram) and intersects objects in the scene at points P, Q, R and S. P is the nearest – if the ray is defined as a parametric vector line, the smallest positive value of parameter t (see chapter 7 on vectors) can be used to identify this. Thus, the pixel should be given the colour identified at point P. This depends on the surface colour C_s of P itself and the light sources that reach P. To interrogate these, secondary rays are cast with P as source. Each of these secondary rays is treated as if P has replaced the centre of projection C. There are different forms of secondary rays. Rays towards light sources may intersect other objects before the light source is reached, identifying that P is in shadow with respect to this source, otherwise P is fully rendered with respect to this source using the standard rendering

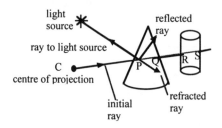

Figure 10.21 The first stage of ray splitting in ray tracing

equation. Shiny objects have reflected rays that are tested for intersections with other surfaces. Transparent surfaces have refracted rays, as defined in chapter 7 on vectors, which are similarly tracked to find the next ray/surface intersection. At all ray/surface intersections, the light component from the surface is calculated similarly by further subdivision. Each splitting of a ray contributes a diminishing contribution to the colour of the pixel under consideration. This process can continue down several levels in a tree data structure, the colour of the pixel being found as the sum of a set of these terms. The process can be terminated when none of the lower-level terms under consideration achieves a particular threshold, or until a certain level of subdivision is reached, in practice often as low as two or three levels or subdivision 'down' the tree. It is clear that this is a time-consuming effort, each split sub-ray requiring as much computing power as the original ray.

Some systems allow environment mapping, where the region to be depicted is surrounded by a box whose inside is painted with an image equivalent to photographs taken in all directions from the object location. If a reflected ray from a shiny object intersects this surrounding box without any other objects intervening, its shiny source will receive a contribution as though reflected from a surrounding scene. The surrounding box is not seen directly, but will affect reflections of objects from behind, to the side, and so on, of the centre of projection

The method of ray tracing produces very glossy images (too often of shiny balls on chequer-boards). Shadows, which are created relatively simply by the method, are almost invariably unnaturally hard edged, as single light sources are much more easy to process than distributed light sources. These may be dealt with as an array of single lights, and thus would add enormously to the time taken to render. Ray tracing is particularly suited to depicting bright and shiny scenes with hard-edged shadows.

Radiosity

Radiosity, more correctly 'radiation exchange', is the closest approximation to the true laws of physics in computer graphics rendering, but it is generally more time consuming than any of the methods mentioned above and gives only matt surface effects. Only a very sketchy description of the method is given here.

Surfaces are divided into sections, each of which receives light from its surroundings. A section traps some of this light; it is this property that gives surfaces colour, and is then assumed to re-transmit the remainder diffusely, with the light spread evenly in all possible directions from the surface, this diffusion giving the matt effect. From real surfaces, relatively less light is radiated in directions close to the surface, more light close to the surface normal, but due to difficulty this feature is not usually modelled. A blue surface will release predominantly blue light, trapping a large proportion of light with other wavelengths, slightly heating the surface with this absorbed energy. Suppose we have a point P on a face A that radiates light uniformly. This will travel through all parts of a hemisphere sitting on the surface with P as centre (fig 10.22). The fraction of light radiated from P that arrives at another surface B can be found by evaluating the area projected from B towards centre of projection P onto this hemisphere, and dividing it by the full area

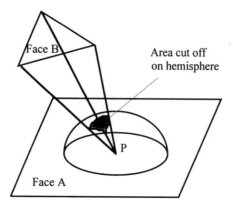

Figure 10.22 Solid angle that contributes to the 'form factor' in radiosity

of the hemisphere, $2\pi r^2$, where r is the radius of the hemisphere. Full or partial occlusion by surfaces that lie between A and B must be taken into account in the area calculation, making this an even more difficult task. If the surfaces involved are small enough to consider that they radiate approximately the same amount of light from each point, the overall fraction of light that leaves A to illuminate B is found by averaging such fractions for all points in A, effectively moving P and its hemisphere over all of A and repeating the area calculation each time. The mathematically precise way of doing this is to sum a double integral (as explained in chapter 5) over face B for each point in A, so this requires a double integral over A. This leads to a quadruple integral, which is usually mathematically intractable. There are various ways of approximating this calculation; for example, a representative sample of points in A can be tested against a similar sample in B, or a large sample of randomly directed rays from A is tested against all other surfaces in the environment. Several techniques substitute less difficult surfaces, such as surrounding boxes with plane faces, for the hemisphere. We will not explore the many ways of achieving this result further, but comment that this is an extremely difficult and time-consuming process, whichever way is chosen.

The result of this complicated evaluation is to find the form factor F_{A-B} for A with respect to B: the fraction of light leaving A that arrives at surface B. This must be found for all surfaces with respect to all other surfaces. Assuming we know the amount of light arriving onto A, we can extract the light absorbed by its surface characteristics to evaluate the amount of light leaving A. This is done by multiplying the incoming light by a reflectivity factor ρ_A indicating the proportions of red, green and blue components released by the surface, giving the surface its intrinsic colour. Then the form factor indicates how much of this diffusely reflected light arrives at surface section B. B then reflects some of this, so affecting the amount of light arriving at other surfaces, which ... . This may seem to be getting hopelessly circular, but the effects can be formulated as a set of equations of form

$$L_A = E_A + \rho_A \Sigma f_{B-A} L_B.$$

L_A is the intensity of light leaving A; the summation is taken over all surface sections B and represents the total amount of light to arrive at surface A from all

10. Drawing and Rendering: How to Create Pictures

sections in the scene. The self-referenced form factor f_{A-A} is zero in most cases except when A is concave, when it may receive some of the light transmitted by itself. It is included for completeness in the development below, although it may be left out if it is known that there are no concave elements in the scene. In such cases the sum over all B surface elements will include A. The amount of light leaving A is the amount of light received by A multiplied by the reflectivity ρ_A plus a value E_A representing light directly emitted by A. If A is not a light source, this is set to zero. If A is a light source, such as a ceiling illumination panel or a window, E_A represents the amount of light it independently adds to the environment, again having three red, green and blue components. If all surface elements have zero values of this component, the scene will not be illuminated. There must be some light sources to create vision.

The radiosity equation can be set out as

$$L_A - \rho_A \Sigma f_{B-A} L_B = E_A,$$

which is equivalent to

$$L_A - \rho_A f_{A-A} L_A - \rho_B f_{B-A} L_B - \rho_C f_{C-A} L_C - \ldots - \rho_C f_{Z-A} L_Z = E_A,$$

or $\quad (1 - \rho_A f_{A-A}) L_A - \rho_B f_{B-A} L_B - \rho_C f_{C-A} L_C - \ldots - \rho_C f_{Z-A} L_Z = E_A,$

where A, B, C, are separate surface sections in the scene. This contains all surface section emissions, L_A, L_B, L_C, ... L_Z (Z represents the last surface element) in its left-hand side. This equation for surface section A is one representative equation for all surfaces in the scene; there are as many equations as there are surface sections. We want to know terms L_A, L_B, ... L_Z as these are the light intensities leaving each surface section – what we need to know to picture the scene. We can set the E_A terms on the right-hand side; indeed we can vary these if we wish as if turning on or off lights. The full set of equations can be cast as a matrix equation,

$$\begin{bmatrix} 1 - \rho_A f_{A-A} & -\rho_B f_{B-A} & \ldots & -\rho_Z f_{Z-A} \\ -\rho_A f_{A-B} & 1 - \rho_B f_{B-B} & \ldots & -\rho_Z f_{Z-B} \\ \ldots & \ldots & \ldots & \ldots \\ -\rho_A f_{A-Z} & -\rho_B f_{B-Z} & \ldots & 1 - \rho_Z f_{Z-Z} \end{bmatrix} \begin{bmatrix} L_A \\ L_B \\ \ldots \\ L_Z \end{bmatrix} = \begin{bmatrix} E_A \\ E_B \\ \ldots \\ E_Z \end{bmatrix}.$$

or $\quad ML = E$

say. In most cases, diagonal terms will be 1; the version here allows for concave faces that cast light upon themselves. This can be solved for the L terms by inverting the square matrix M to give

$$L = M^{-1} E,$$

or by a suitable method such as Gaussian elimination (see chapter 6 on matrices) or an 'off the shelf' equation-solving routine. With high-precision definition of the scene to be viewed, this can lead to a matrix M of many thousands of rows. It may be necessary to operate with reduced scene complexity.

This process has taken two stages: evaluation of form factors and calculation of light emissions. The scene still has to be rendered using the usual projection from a centre of projection to a window and so on. However, the location of an observer does not affect the light emitted, unless the observer is considered a light transmitter/receiver. Once the initial calculations have been performed, movement around the scene can be evaluated rapidly, provided objects within the scene are not rearranged, which would necessitate recalculating form factors (the most time-consuming part of the exercise) and hence rebalancing the set of equations. If lights are turned on and off in a scene, the form factors and hence matrix M are not changed, but initial values in the E matrix will change, so the equation solution process will need to be recalculated.

Radiosity gives good matt shading effects, with convincing soft shadows. For best effect, the sizes of surface regions should be relatively small, but this adds to the complexity. It few surface patches are used to save on rendering time, a post-processing phase can blur light distributions across large patches and their boundaries, in a way similar to anti-aliasing, discussed below.

There are no specular highlights or glossy surfaces in most radiosity images, as surfaces are all assumed to be diffuse radiators, scattering light equally in all directions. A hybrid method which combines the best features of radiosity and ray tracing can give very good effects, but exacerbates the worst features of each in requiring considerable amounts of memory and very high rendering times.

Anti-Aliasing

The effect of the 'jaggies' inherent in a pixel-based scheme can be reduced by anti-aliasing. The upper part of fig 10.23 shows how the pixel structure of an image can produce a staircase effect at the boundary of two coloured regions. Pixels are coloured black if more than half of their area lies within the original black zone. The lower part of the figure has been anti-aliased by colouring each pixel with the average colour. Thus, if three eighths of the pixel area is black, the remainder white, its colour is set to a 'three eighths grey'. A large pixel size is used here to illustrate the effect. With more realistic small pixel sizes, the effect is to present an apparently smoother edge to the viewer, the effect of the 'jaggies' being visibly diminished, as in fig 10.24 which shows an anti-aliased receding chequer-board pattern.

Pixel average values may be calculated directly from the geometry of the data set when first evaluating pixel intensities. With scan line methods, this entails calculating fractional edge/scan line intersections exactly as real values, adding to the time taken for an operation normally performed in integer arithmetic using Bresenham's algorithm. For ray traced images, a typical anti-aliasing method sends several rays (typically five or nine) through the same pixel, then sets the pixel intensity to the average for those rays. It should be obvious that this adds enormously to execution times. This can be done by distributing the rays regularly inside the pixel shape or by randomly selecting ray directions to fall within the pixel shape. A less precise form of anti-aliasing may also be done as a 'post-processing' stage in image space as a filtering process, setting each pixel's intensity to an average intensity of itself and its neighbours.

10. Drawing and Rendering: How to Create Pictures

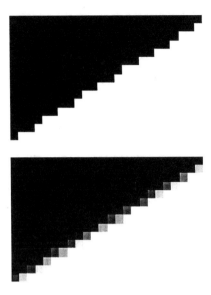

Figure 10.23 Anti-aliasing using grey scales to represent average content of a pixel

In showing still images from motion sequences, a similar technique may be used for temporal anti-aliasing. The content of each pixel may be considered to show a representative short time span, with samples of the pixel content within this time being averaged. Typically, the time could be one twenty-fifth of a second, equivalent to one frame of film, giving the effect of 'motion blur'.

Figure 10.24 and plate 10.4 show some examples of stochastic anti-aliasing using the exact object depiction method. Textures on the floor, beach ball and wall are subject to this. More that one point per pixel are randomly scattered onto the surfaces concerned, and the average intensity calculated from each point rendered is entered as the pixel colour. This reduces the pixellization of interior colour boundaries and makes shadows more precise, as seen in the shadows of the plants.

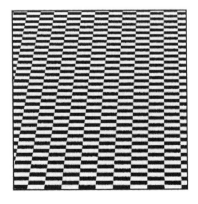

Figure 10.24 An anti-aliased chequer-board

Summary

This review of 3D rendering algorithms ends with a summary of the main operational properties. Scan line methods are relatively fast, processing rows of pixels at a time through identifying spans between the depicted edges of scene polygons. They operate in the 'computer graphics pipeline', in which a sequence of space transformations leads from conceptual, through world and view spaces to the display space, objects being clipped to the boundaries of the visible region during this process. Lambert shading gives a quick, convincingly solid, but obviously polyhedral image. Gouraud shading gives a smoother effect; Phong shading is even more smooth and can easily be adapted to include specular highlights. All these methods can be enhanced by adding texture mapping and shadows (use of an ambient light term ensures that regions not directly lit are not totally black). Similar effects can be added to exact surface rendering, avoiding artefacts seen through polygonal approximation of curved surfaces.

Ray tracing works pixel by pixel, identifying the nearest visible object and tracing higher-level rays (towards light sources, reflected and refracted) to sources of light to evaluate colour intensity. Texture and environment mapping add realism and true mirror effects can be achieved. Clipping is trivial, only pixels within the viewport are interrogated, but ray tracing is slower than scan line methods. It can operate on B-Rep models and is the most suitable method for direct rendering of CSG models – although algorithms for scan line rendering of CSG models have recently been developed, models are usually converted to B-Rep before being scan line rendered. Ray tracing produces unnaturally glossy images with very sharp shadows.

Soft shadows and subtle lighting effects are created by radiosity, which simulates the way light is exchanged between surfaces. This is more time consuming than previous methods and gives matt surface images with soft shadows. Hybrid methods have been developed from the mid 1990s to blend some of the best features of ray traced and radiosity rendering, again at the cost of rendering time.

The development in this chapter would probably not give enough detail for systems developers to create their own renderers; there are plenty of specialist texts that give such information. It should enable researchers to understand technical texts discussing the nature of standard rendering systems, and to appreciate the reasons for the apparent slowness of some rendering engines.

The first seven chapters, with a few peeps ahead, review a range of mathematical techniques used in computer graphic generation of images. Cover is sometimes brief and the selection is not exhaustive (although it may seem exhausting). Among the omissions are, for example, the generation of random numbers, the proof of the binomial expansion and logarithms. Research papers on innovative ways of using mathematics to generate images are published daily. However, the selection should be enough to give readers an understanding of the main processes of computer graphics as described very briefly in the last three chapters. To fill in some gaps and to find alternative descriptions that may clarify some obscurities, suggestions for further reading follow.

The author hopes that readers have reached a good level of understanding and will gain as much diversion as he has from the generation of computer graphic images, such as that of plate 10.4, which illustrates many features described in the book.

Suggestions for Further Reading

The following is a selection from my own bookshelf of works that have influenced me throughout my developments in mathematics and computer graphics. They range from journalistic or biographic descriptions of general areas to books with technical detail sufficient to enable the implementation of methods. Many are classics with relatively recent editions indicated where available, which is intended to be personal and opinionated rather than exhaustive. A few are out of print, but readers may be able to track them down through libraries or web searches. There is an imbalance towards areas that I have found particularly interesting. I certainly have no intention to condemn any book by omission. The list concentrates on authored books rather than conference collections and research papers.

I have made my own arrangement of works into areas that might appeal to different categories of reader. Books are listed below my brief descriptions of them. Non-mathematicians might find it useful, for example, to try to 'get into the mind' of a mathematician. The first category is of this kind. Other categories are for those who would like to develop mathematical skills and understanding, to cover the general issues of computer graphics, of geometric modelling and to look at the ways in which fractals can be used to model natural phenomena.

Mathematical and Scientific Minds

For non-mathematicians, Doxiadis' novella gives a superb insight into the compulsions and fears that drive great and near great mathematicians, as well as the day-to-day activities of the more mundane kind. The book is a historical novel framed within the history of mathematics in the mid 20th century, with many of its prominent practitioners such as Turing and Gödel having cameo roles.

Boyer and Merzbach have produced a hefty and readable account of the history of mathematics up to the mid 20th century. This book has enormous interest in a historical sense, placing mathematical development against the frame of world history, but also gives a sound insight into the processes of mathematics itself.

Gleick's *Genius* is the biography of mathematical physicist Richard Feynman. This is a fine, readable work in a journalistic style that describes the eccentric life and genuine originality of a great thinker. Feynman committed relatively little of his work to print, but several transcriptions of his public lectures exist and are worth reading. Of these, *The Meaning of it All* makes sense of interpretation of scientific results in a few short essays. Fewer devastatingly erroneous interpretations of scientific information would have been made if more politicians had read and understood this.

Gardner's description of right and left handedness is typical of his works, presenting a puzzle with great clarity (any of his books of mathematical puzzles are recommended), and that's what mathematics really is, the solving of puzzles – or problems if you want to be more formal.

Gregory's wonderful work on perception is an intriguing read. This applies to any of Gregory's other works on the topic. Anyone involved in the production and/or analysis of images should read this.

Boyer E, Merzbach UC (1991) A History of Mathematics, 2nd edn. John Wiley and Sons
Doxiadis A (2000) Uncle Petros and Goldbach's Conjecture. Faber and Faber
Feynman RP (1999) The Meaning of it All. Penguin Books
Gardner M (1990) The New Ambidextrous Universe, revised edn. W.H. Freeman
Gleick J (1996) Genius, new edn. Abacus
Gregory RL (1997) Eye and Brain, 5th edn. Oxford University Press

Understanding of Mathematics

As a child, I was given Hogben's *Man Must Measure*, an illustrated book on the basis of mathematical thought. I still have my original copy, although it is out of print, and this has recently influenced a colleague undertaking a PhD on the description of spatial concepts.

Hogben's renowned classic in popularization of mathematics for an adult audience is *Mathematics for the Million*. This sat on my bookshelf alongside the work intended for children.

Before entering university to study mathematics, I was greatly enthused by an early edition of the wonderful book by Courant and Robbins (now updated by Stewart, who is a successor of Gardner in writing on 'mathematical recreations' for *Scientific American*; these columns are also much recommended). Its first edition was published in 1941, so it predates digital computers, let alone computer graphics. The book, however, includes much that is used in computer image synthesis. For example, it describes projective geometries, homogeneous coordinates and shows fractal patterns from multiple reflections, although its publication was before Mandelbrot coined the word 'fractal'. Stewart's contribution makes good this omission. The book gives a genuine flavour of the nature of mathematics. If I have achieved a small fraction of the crispness of its presentation, I will count myself successful.

Another book I enjoyed in my youth and am pleased to see still in print is the work by Cundy and Rollitt on making mathematical models. It gives practical instructions on how to make a variety of real models as well as explaining background information on solid geometry and topology.

Abbott's *Flatland* is still read over 100 years after its first appearance in 1882. Although in some senses dated – its treatment of women's roles in society is certainly not endorsed by me – it still gives an amusing insight into the nature of dimension through describing a civilization of beings occupying a two-dimensional world. By analogy, Abbott makes clear extensions to the third and higher dimensions.

Baylis and Haggerty have produced a good introduction to number theory and subsequent developments. The book, subtitled 'a student's guide to the enjoyment of higher mathematics', is suitable for those who may have missed out on early learning of mathematical foundations.

Hart's *Guide to Analysis* is another technical book written in an accessible style. It covers much of the same introductory work but is then directed towards study of the calculus.

Abbott EA (A Square) (1999) Flatland, new edn. Penguin Books
Baylis J and Haggerty R (1988) Alice in Numberland. Macmillan
Courant R, Robbins H and Stewart I (1996) What is Mathematics?, new edn. Oxford University Press
Cundy HM and Rollitt AP (1997) Mathematical Models, 3rd edn. Tarquin Publications
Hart M (1988) Guide to Analysis. Palgrave
Hogben L (1955) Man Must Measure. Rathbone Books
Hogben L (1989) Mathematics for the Million, 4th edn. Merlin Press

Computer Graphics in General

The most comprehensive work is that of Foley, van Dam, Feiner and Hughes. It has a sound mathematical base and has an excellent mathematical appendix. It packs a wide coverage of the topics of 'classical' computer graphics into its 1100+ pages.

Hearn and Baker produced a smaller, but just as clear, work that is much easier to carry around. Some readers find this more approachable than that of Foley et al., and it's just as sound in a mathematical sense.

Watt and Watt concentrate more on the rendering and animation side of computer graphics; this gives a good and well-illustrated review of available methods. Watt and Policarpo, too, extend this to a wider consideration of computer image generation and analysis.

Mortenson's earlier book is good on the basic mathematics of computer graphics, but does not extend to colour rendering.

Foley JD, van Dam A, Feiner SK and Hughes JF (1995) Computer Graphics: Principles and Practice, 2nd edn. Addison-Wesley Longman
Hearn D, Baker MP (1996) Computer Graphics: C Version, 2nd edn. Prentice Hall
Mortenson ME (1989) Computer Graphics: An Introduction to the Mathematics and Geometry. Heinemann Newnes
Watt A and Policarpo F (1998) The Computer Image. Addison-Wesley
Watt A and Watt M (1992) Advanced Animation and Rendering Techniques. Addison-Wesley

Geometric Modelling

The works of Escher do not comprise a teaching text on geometric modelling, but for the sake of including them they are put here. They give examples of non-manifold objects, of the ways in which 2D images can show objects that may appear, in some sense, to be three-dimensional, but do not make sense when interpreted in the 3D world.

Mäntylä's book on solid modelling concentrates on the achievement of valid B-Rep models using his 'geometric workbench' method to constrain constructions to avoid the creation of non-manifold models.

Farin's work is a developing description of techniques used in the generation of spline curves and surfaces. It is a sound basis for anyone wishing to work in or to understand this area.

Salomon gives a good overview of computer graphics techniques, with over half the book's main text taken up with descriptions of curves and surfaces, so its emphasis is on this aspect of geometric modelling

Escher MC (1992) M.C. Escher: The Graphic Work. Benedikt Taschen
Farin G (1996) Curves and Surfaces for Computer Aided Geometric Design, 4th edn. Academic Press
Mäntylä M (1988) An Introduction to Solid Modeling. Computer Science Press
Salomon D (1999) Computer Graphics and Geometric Modeling, Springer

Fractals and Related Issues

A good introduction to this area is given in Gleick's journalistic overview, now updated since its first edition.

Levy's very readable book is similarly descriptive of major developments in artificial life, which is enjoyable but perhaps peripheral to the main concerns of this book. It does contain a good account of synthetic plant development using fractal methods. This, and Gleick's *Chaos*, could sensibly have been included in the first section as descriptive of the scientific mind.

Mandelbrot's book is a definitive work on fractals. It gives a mathematically complex development compared to the two more descriptive overviews mentioned above.

The two works by Peitgen with (separately) Richter and Saupe explain and give superb colourful examples of the Mandelbrot set, as well as other fractal methods through essays by contributors such as Mandelbrot, Voss and Barnsley.

Barnsley's own book is a mathematically advanced description of IFS as used for fractal object generation; this is also beautifully illustrated. The books by Fisher and Lu describe the use of IFS for image compression.

The definitive work on simulation of growing plants is that of Prusinkiewicz and Lindenmayer; the underlying theory of L-systems is thoroughly developed in this. Several subsequent research papers by Prusinkiewicz and his collaborators, particularly those published in the ACM SIGGRAPH series, have extended the use of L-systems further. It would be good to see these developments absorbed into any future editions. A softback edition of the original was released in 1996.

Kaandorp develops a related method, working more directly in geometry, for the modelling of 'marine sessile organisms' such as corals and sponges. He relates his theoretical models directly to observations in the field, so the effects of different currents and nutrient distributions can be used to validate model veracity.

Barnsley MF (2000) Fractals Everywhere, 3rd edn. Morgan Kaufmann
Fisher Y (1994) Fractal Encoding – Theory and Applications to Digital Images, Springer
Gleick J (1996) Chaos, new edn. Minerva
Kaandorp JA (1993) Fractal Modelling: Growth and Form in Biology. Springer

Levy S (1993) Artificial Life, new edn. Penguin Books
Lu N (1997) Fractal Imaging. Academic Press
Mandelbrot BB (1982) The Fractal Geometry of Nature, updated edn. W.H. Freeman
Peitgen H-O and Richter PH (1986) The Beauty of Fractals. Springer
Peitgen H-O and Saupe D (1988) The Science of Fractal Images. Springer
Prusinkiewicz P and Lindenmayer A (1990) The Algorithmic Beauty of Plants. Springer

Journals

Technical articles in journals extend the capabilities of computer graphics week by week. Readers who have reached this far should have adequate mathematical underpinning to attack some of these papers with confidence. The following are just three examples of journals in computer graphics and related issues.

ACM Computer Graphics (ACM SIGGRAPH, an ACM publication)
Computer Graphics Forum (Eurographics, published by Blackwell)
Virtual Reality (the Virtual Reality Society, published by Springer)

Index

Names

Abbott, E.	58, 60, 328–329
Argand, J.R.	132–134
Baker, M.P.	329
Barnsley, M.	238, 253–254, 330
Baumgart, B.G.	228
Baylis, J.	7, 328–329
Bernstein, S.N.	265, 277
Bézier, P.	259, 262–272, 274, 275–277, 282, 284, 286–293
Blinn, J.	319
Boole, G. (see also 'Boolean' in subject index)	28
Born, M.	149
Boyer, C.B.	7, 239, 242, 327–328
Bresenham, J.	105, 106, 324
Briggs, P.	253
Bruneleschi, F.	59
Cantor, G.	236–237, 241
de Casteljau, P. de F.	259, 262, 264, 269–270
Catmull, E.	273
Cayley, A.	149
Courant, R.	7, 56, 59, 328
Cundy, H.M.	328
van Dam, A.	329
Descartes, R. (see also 'Cartesian' in subject index)	38
Doxiadis, A.	327–328
Dürer, A.	59
Escher, M.C.	218–219, 329
Euclid	50
Euler, L.	60, 230, 259
Farin, G.	330
Fatou, P.	249
Feiner, S.K.	329
Ferguson, J.	259, 272
Feynman, R.	132, 327–328
Fisher, Y.	255, 330
Foley, J.D.	329
Galileo, G.	44
Gardner, M.	61, 327–328
Gauss, C.F.	156, 259, 323
Gleick, J.	132, 327–328, 330
Gödel, K.	327
Gouraud, H.	5, 189, 308, 310–315, 320
Gregory, R.	296, 328
Haggerty, R.	7, 328–329
Hart, M.	329
Hausdorff, F.	234, 242
Hearn, D.	329
Heisenberg, W.	149
Hermite, C.	272
Hilbert, D.	239
Hogben, L.	44, 328
Hughes, J.F.	329
James, M.	252
Jones, H.	253
Julia, G.	243–246, 248–249
Kaandorp, J.A.	253, 330
von Koch, H.	235–237, 241
Koestler, A.	44
Kronecker, L.	7
Kutta, M.W.	135
Lambert, J.H.	5, 299, 306–308, 320
Leibniz, G.W.	116
Levy, S.	330–331
Lindenmayer, A.	250, 252, 330–331
Lobaschevsky, N.I.	259
Lu, N.	255, 330–331
Maclaurin, C.	130–131
McPheeters, C.	234
Mandelbrot, B.B.	36, 234, 240, 242, 247–249, 328, 330–331
Mäntylä, M.	231, 330
Mech, R.	252
Minkowski, H.	187
Mercator, G.	40, 60
Merzbach, U.C.	7, 239, 242
Monge, G.	259
de Morgan, A.	34
Mortensen, M.E.	329
Newton, I.	116, 134–135
Peano, G.	239, 241
Peitgen, H.-O.	243, 330–331
Perlin, K.	320
Phong, B.-T.	5, 189, 308, 313–315, 320
Pickover, C.A.	249
Poincaré, H.	56, 230
Policarpo, F.	329
Prusinkiewicz, P.	250, 252, 330–331
Pythagoras	17, 40, 49, 54, 55, 165, 190, 192
Reeve, D.E.	249
Richardson, L.F.	242
Richter, P.H.	243, 330–331
Robbins, H.	7, 56, 59, 328
Rollitt, A.P.	328

Rom, R.	273
Runge, C.	135
Salomon, D.	330
Saupe, D.	243, 300–331
Sierpinski, W.	238–239, 241–242
Smith, A.R.	250
Stewart, I.	7, 56, 59, 328
Sylvester, J.J.	149
Taylor, C.	130, 135
Tunbridge, A.	253
Turing, A.	327
da Vinci, L.	59
Watt, A.	135, 329
Watt, M.	135, 329
Wyvill, B.	2, 234
Wyvill, G.	2, 234

Subject Index

A

absolute error 29–30
accuracy 29
acceleration 135
active edge list 308–309, 311–312, 314
adder 22
addition
  of bits 28
  of matrices 149–150
  of numbers 9, 10, 12, 15, 16, 20, 22
  Boolean 33–34
  of vectors 186–188, 192
additive colour 67–70
adjacent side 48
affine invariance 267, 272, 282
affine transformation 79–93, 100, 103, 147, 160–171, 178, 255, 267, 282, 299–302
algebra 11, 22
ambient light 308
analogue computer 26
analysis 122
and 33
angle 44–46
animation 3, 4, 103, 134–135, 260, 268–269, 274–275, 312
  physically based 134–135
anti-aliasing 298, 324
arccos 65, 98–99
arcsin 98–99
arctan 53–54, 65, 98–99
Argand plane 132–134, 243–249
argument of a complex number 132–134
arithmetic mean 15
arithmetic operation 10
array 177–188, 215–216, 219–220, 305
aspect ratio 297, 304
Aspergillus nidulans 252
associative law 10
asymptote 136, 139
attractor (of an iterated function system) 253–255
average (or mean)
  of numbers 16
  of unit vectors 208
  weighted 196, 261, 265

axes
  of coordinate systems 38–39
  right-handed set 61
axiom (of L-system) 250–251

B

B-Rep (see boundary representation)
B-spline 275–284, 292–293
Babylonian number 25
back face culling 305–306
base-exponent form 31
basis function (see blending function)
Bernstein polynomials 265, 277
Bézier
  curve 259, 262–272, 274, 275–277, 282, 284, 286–292
  surface or patch 286–293
binary counter 26, 28
binary digit (see bit)
binary number 25–28, 31–32
binary operator 9, 15, 20, 149
binary tree 217–218, 225–226
binomial expansion 118
biomorph 249
bit 26, 28–30
blending function 261–262, 264–265, 271, 27–277, 279–284, 287–288, 291–292
Boolean
  algebra 2, 28, 32–35, 222, 225–227
  variable 33–34, 219
bound (upper) 8, 12
boundary representation (B-Rep) 2, 215, 223, 227–231, 256, 305, 306
Bresenham's algorithm 105, 106, 324
bump mapping 319
byte 29–30

C

c, m, y (cyan, magenta, yellow) 70
CAD (computer aided design) 1, 227, 259, 282, 290
CAD/CAM (computer aided design/manufacture) 227
calculus
  differentiation 114–125, 129–130
  integration 124–129
camera (see centre of projection)
cancelling 15, 17, 202

canonical space 79, 224–225, 298
Cantor set 236–237, 241
carry bit 28
Cartesian coordinates
  2D 38–44, 46, 54, 59–60, 82, 105–107, 132–134, 189–191, 297–298, 301
  3D 2, 61–66, 87–89, 142, 191–192, 196–198, 215, 229, 260, 282, 297–301
de Casteljau's algorithm 264
de Casteljau trellis 269
Catmull-Rom spline 273
cellular decomposition 2
centre of mass 261, 265, 277, 284, 291
centre of projection 172, 174, 297, 299, 301–303, 305–306, 316–318, 320
chain rule for addition of vectors 188
chaos game 253
child node 217–218
circle 137–139, 282, 285–286, 292–293
clipping 4, 175, 178, 213, 297, 302–303, 315
closed interval 37
closed set 44
closure 9, 11, 14, 16, 20, 22
codomain 73–76
collage theorem 255
colour 3, 35–36, 66–71, 297, 304–305, 306–314, 319, 321, 325
  additive 67–68
  c, m, y 70
  r, g, b 67–70, 307–308, 306–313
  h, s, v 69–70
  subtractive 69–70
column (of a matrix) 148, 150–152
commutative law 9, 10, 11, 13, 14, 152, 187, 222
comparator 8
components of a vector 190–192, 194
computer languages 74
computer memory (see memory)
complement 33–34
completeness 20, 24, 37
complex conjugate 22
complex number 20–24, 36, 110, 132–134, 244–249
concatenation of transformations 93, 160, 165–166, 179–180
cone (mathematical) 2, 137, 143–144, 225 259, 282, 292, 315
cones (in the eye) 66–67

conformability (of matrices)
  for addition 149
  for multiplication 150
conic section 137–141, 282, 284
connected set 245, 249
constraint 61–62
constructive solid geometry (CSG) 2, 215, 222–227, 306
  CSG tree 225–226
context sensitive (L-system) 252
contiguous memory 216
continuity 116, 123, 239, 260, 274, 284
  of curvature 268
  piecewise 259, 267–269
continuous numbers 12, 16, 35
continuous variable 26
contractive transformation 253
control mesh (of a spline surface) 290
control point (of a spline) 259, 262, 264, 269–270, 273, 276–277, 279, 281–285, 287–292
control polygon 262, 272, 275
convergence 244, 248
convex hull 265, 277, 284–285, 291
coordinate systems (see Cartesian, polar)
cosine (cos) 46–56, 97–99, 119, 121–122, 130–134, 191–194, 207, 286, 307, 310, 313–314, 317–319
cosine rule 55
countability 16
cross product (definition) 194–195
CSG (see constructive solid geometry)
cube 223–224, 228–229, 295
  Necker 295
cube of a number 11, 17
cube root 17
cylinder 2, 142–143, 211, 224, 259, 282, 292, 296, 314–315
cylindrical polar coordinates 63–64, 66, 142–143, 315

D

data compression 253
data structures 215–218, 225, 310
decimal number 25
decimal place (correct to) 29
definite integral 126
degrees (angle) 44–46, 131
degrees of freedom 60–63, 158
denary number 25–27, 29, 31–32
dendrite 245–246, 249
denominator 14, 201

dependent variable	73–74
depth cueing	295–296
derivative	116–124, 130
second	119–120, 124, 130
partial	122–123
determinant (of a square matrix)	153, 165–166, 204, 212
diagonal matrix	148–149
difference (Boolean)	34, 222, 225–226
differential	116–124
differentiation	114–124, 129–130, 265
partial	122–123
digit	25
digital computer	26
dimension	4, 56–61, 167, 191, 239, 260, 294, 295–296, 315
fractal	234–235, 239–243
Hausdorff–Besicovitch	234
discrete numbers	12, 16, 35, 298
discrete variable	26
display coordinates	303–304
distributive law	10
divergence	244, 248
division	
by zero not allowed	13, 201–202, 208, 212–213, 276, 280–281, 302, 310
of numbers	11, 16, 20, 22
domain	73–74, 94–95, 98–99, 189, 269, 272, 276–277
dot product (definition)	193–194
double integral	128–129
double precision	30, 32
drawing	295–298
dust	238, 245

E

e (see exponential)	
eccentricity of an ellipse	139
eggshell finish	318
element	
of a matrix	147
of a set	8
ellipse	41–42, 137–139, 282, 285
enumeration	8
environment mapping	321
environmentally sensitive (L-system)	252
equality	
of matrices	149
of vectors	185
equations	41–44, 62–63, 79, 154–159 198

error	29–30, 32, 35–36, 328
Euler formula (topology)	230
Euler–Poincaré formula	230
exact object rendering	314–315, 317
exponent	31
exponential (e)	18, 119, 130–132, 136, 232–233
extrapolation	260
eye	66–67

F

factorial	130
factorization	24
falsity	32–33
Ferguson curve	272
field	
scalar	231–234
vector	231
finite element method	2
fish-eye lens	297, 299
flat shading (see Lambert shading)	
Flatland	59
floating point number	30–32, 35
floating point overflow	32
floating point underflow	32
fluxions	116
focus of an ellipse	139
form factor	322–324
fractal	2, 36, 60, 134, 234–258
dimension	234–235, 239, 240–243
fraction	14–16
frustum of a cone	144
functions	73–103, 215
fundamental theorem of algebra	23, 133
fungus	252
fuzzy logic	33

G

Gaussian elimination with partial pivoting	156, 323
geometric modelling	1, 4, 215–234, 296, 298, 315
geometric work bench	231
geometric series	236–238
gloss	5, 297, 324
gloss factor	317–318
Gouraud shading	5, 189, 308–315, 320
grad	123
gradient	105–106, 114–124

graph (of a function)   50–52, 76–78,
    95, 100–103, 105–108, 111–115,
    121–122

H

h, s, v (hue, saturation, value)   69–70
half-space   43–44, 63, 224–225
Hausdorff–Besicovitch dimension 234
Hermite polynomials   272
hexadecimal   28
hidden line methods   4, 294
hidden surface methods   4, 174,
    297–298, 304–306, 315
homogeneous coordinates   147,
    162–176, 282
hyperbola   137, 139–141, 282, 285
    rectangular   140–141
hyperbolic paraboloid   256
hypercube   58–59
hypotenuse   47–50, 54, 192

I

i   21
identity
    number   12
    matrix   152
IEEE standards   29, 31–32
IFS (see iterated function system)
image set   74–75, 94, 98–99
image space   305
imaginary number (i)   21, 132
imaginary part of a complex number
    22, 132
incident ray (of light)   205–207
independent variable   73–74
index   11, 15
inequality   8, 41–44, 63, 204,
    223–225, 232
infinity   8, 13, 32, 98
inflection (see point of inflection)
inner product   151
input (of a function)   74, 95
inside   197–198, 200, 223, 225, 229,
    293
integer   12–14, 16, 24, 35
    short   30
    long   30
integration   124–129, 135
integral   126
    double   128–129, 322
intensity of colour (see colour)

interpolating spline
    curve   271–275
    surface   292–293
interpolation 196, 260–262, 313–314
    linear   261–263, 311–313
intersection
    Boolean   33–35, 222
    of a line with a plane 200–205, 321
    of two planes   211–213
    of polygons   309
    of two straight lines   208–210
    of sets   33
interval   19, 37, 43, 276
inverse
    of a square matrix   153, 159–160,
        170–171, 323
    of affine transformations   94,
        170–171, 225
    of functions   53, 95–99
irrational   18–19
isosurface 2, 122–123, 215, 231–234
iterated function system   2, 238,
    253–255, 314–315

J

Julia set   243–246, 248–249

K

key framing   134
key point (of a field)   232–233
knot   275, 278, 282
    insertion   284
    multiplicity   278
    sequence 275–278, 281–282, 292
    span   276–277, 279
Koch curve   235–237, 241

L

L-systems   2, 250–253
Lambert's law   299, 307
Lambert shading   5, 306–310, 320
leading diagonal (of a matrix)   148
leaf node   217–218, 220, 225–226
left-handed coordinates   300
light/lighting   4, 66, 297, 306, 314,
    316–317, 320–322
    distributed source   316
limit   116, 127
Lindenmayer systems (see L-systems)
line (see straight line)
linear equations   154–159
linear independence   158

Index 339

linear interpolation 261–263, 311–313
linked list 216–217, 228, 308
list data stricture 216–217
locus 41, 105, 136
logarithm 241–243
logical circuit 28
logical value 29
LOGO 46, 250
long integer 30
lower triangular matrix 148–149

M

Maclaurin series 130–132, 135
magnitude of a vector 186, 188–192
Mandelbrot set 36, 243, 247–249
Manhattan grid distance 41
manifold objects 219, 224, 230, 296
mapping 73–76
mantissa 31
marching cubes algorithm 234
matrix 89, 147–183, 225, 255, 289, 292, 301–302, 323–324
  computer implementation of 176–182
matt surface 314–315, 317, 321, 324
maximum 119–120
Mayan number 25
mean (see average)
medical modelling 219–221
membership of a set 8
memory (computer) 3, 28–29, 219–221, 229–230, 298, 305, 316
  contiguous 216
Mercator projection 40, 60
mid-point displacement 256
minimum 119–120
Minkowski's inequality 187
modelling (see geometric modelling)
modulus of a complex number 132–134
modulus of a number 41, 189
Moiré pattern 317
monotonic 95–99, 274, 275
motion blur 325
multiplication
  of a matrix by a scalar 150
  of a vector by a scalar 188–192
  of bits 28
  of matrices 150–152
  of numbers 10, 15, 16, 20, 22
  of vectors, the scalar (dot) product 193–194
  of vectors, the vector (cross) product 194–196

N

nand 34
nanometre 66
nappe 137, 143, 225
natural numbers 7–12, 16, 24
NDC (normalized device coordinates) 304
near plane 173–175, 302–303
necessary condition 230
Necker cube 295
negation 33–34
neighbourhood 242–243
Newton's laws 134–135
non-affine transformation 176
non-commutativity
  of matrix multiplication 152
  of the vector product 195
non-singular matrix 153
non-uniform rational B-spline (see NURBS)
nor 34
normal (to a surface) 123, 142, 144, 189, 195, 197–198, 205–207, 229, 292, 305–306, 310–315
normalization of a vector 208, 310, 313
normalized device coordinates 304
normalized representation (of floating point number) 31
not 33–34
nought 12
null pointer 216
number 7, 35
  computer representation of 25–32
  representation of 25–28, 35
number base 25–28
numerator 14
numerical methods 111–112, 126, 293
NURBS
  curve 259–260, 282–287
  surface 292–293

O

object space 305
observer (see centre of projection)
octal 28
octree 218, 220
one-to-one 20, 37, 94–95
open interval 37

open set 44
opposite side 48
or 33
order (of a matrix) 148, 150–152
origin (of coordinate systems) 38
orthogonal 58, 194–195, 208, 292, 299, 306
output (of a function) 74, 95
outside 197–200, 223, 225, 229, 293

P

painter's algorithm 298, 305
parabola 137, 139–140, 282, 285
parallel lines, distance between 210–211
parallel projection 175–160, 297, 302
parallel vectors (test for) 195
parallelogram law for addition of vectors 186–188
parametric
 curve 121–122, 138, 142, 264
 line 196–197
parent node 217–218, 225–226
partial derivative 122–123
partial differentiation 122–123
Peano curve 239–241
perception 4, 66, 296
period 50
perspective projection 58, 61, 79, 147, 172–176, 180–182, 282, 295, 297–299, 301–303
Phong shading 5, 189, 308–309, 313–315, 317, 320
photon 66
photorealism 298, 317
physically based animation 134–135
physically based modelling 297
pi (π) 18, 44–46
pixel 3, 68, 244, 298, 305, 308–314, 317, 320, 324–325, 328
plane
 equation of 198–200, 211–212
 vector definition of 197–203
point of inflection 116, 120
point set topology 235
pointer 216–218, 228–229
polar coordinates
 2D 46–47, 54, 60, 63,82, 84, 132–134
 3D cylindrical 63–64, 66
 3D spherical 63–66, 141–142
polygon 142, 286, 305, 308, 310, 312

polyhedron 2, 142, 231, 299
polyline 262, 277
polynomial 23, 107–114, 120, 233, 265, 272, 284
pop (off a stack) 216, 226, 234, 250
position vector 191–192, 261, 300
post-multiplication (of a matrix) 152
post-order traversal 218
power 11
precision 54
pre-multiplication (of a matrix) 152
pre-order traversal 217–218
primitive 222, 225, 296
product
 of bits 28
 of numbers 10
production (of L-system) 250–252
 context sensitive 252
 environmentally sensitive 252
 stochastic 251
program 28
projection 4, 58, 61, 79, 296–298, 301, 303
(see also perspective and parallel)
projective invariance
proof 11, 50
push (to a stack) 216, 226, 234, 250
pyramid 296, 303, 315–316
Pythagoras' theorem 17, 40, 49, 54, 55, 165, 190, 192

Q

quadrant (of a plane) 48
quadratic
 equation 109–111
 function 108–111, 139
quotient 14

R

r, g, b (red, green, blue) 67–70, 307–308, 322–323
radians 44–46, 131
radiosity 5, 316, 321–324
random (see stochastic)
range 74–75, 99
raster device 3, 327
rate of change 105–106, 135
rational number 14–17, 19–20, 24
ray tracing 5, 293, 320–321, 324
real number 17–20, 22, 24, 35, 298
real number line 20
real part of a complex number 22, 132
rectangular hyperbola 140–141

Index

recurring decimal 19
reductio ad absurdum 17
reflection
   of light 205–207, 317–318, 320–321
   reflectivity 322–323
   transformation 81, 96
   (see also specular highlight)
refraction of light 205–207
relative error 29–30, 32
rendering 185, 197, 295–326
resultant of two vectors 186
retina 66–67
right-hand screw 61, 87, 191, 195
right-handed coordinates 61, 300
right vector 300
rods (eye) 66
Roman numerals 25
root
   of an equation 23–24, 43. 109–111, 120
   of a complex number 132–133
   of a number 17, 19
   of a tree data structure 217–218, 225–226
rotation 147, 253
   2D 79, 82–84, 100, 102–103, 160–164
   3D 87–89, 144, 167–171, 177–180, 300
rounding error 36
row (of a matrix) 148, 150–152
Runge–Kutta method 135

S

saturation (of a colour) 69
scalar 150, 185, 188–189, 202, 283
   field 231–234
scalar product (definition) 193–194
scaling 79–81, 86, 100–103, 138, 144, 147, 160–171, 177–179, 253, 300, 304
scan line 308–309, 315
seahorse valley 247–249
second derivative 119–120, 124
self-similarity (of a fractal) 235
   statistical 241–243
semi-open interval 38
series expansions 129–132
set 8, 24
set theory 33
shadow 315–317, 320, 325
   shadow buffer 255, 316–318

shadow volume 315
soft shadow 316, 324
shear 85–86, 160–161, 178–179
shiny surface 314, 317–318
short integer 30
SI number representation 30
Sierpinski tetrahedron 239, 241
Sierpinski triangle (or gasket) 238–239, 241–242
sigma (Σ) 127–128, 151, 276, 287–289, 317–318
significand 31–32
significant figure (correct to) 29
sine (sin) 46–56, 97–99 119, 121–122, 130–132, 195, 207, 286
sine curve 52
sine rule 55
single valued 73, 95
singular matrix 153
skew lines (distance between) 208
slope 105–106, 114, 140
smooth shading (see Gouraud shading, Phong shading)
soft object modelling (see isosurface)
solution (of an equation) 43
space-filling curve 239–240
span 308–309, 315
specular highlights 4, 315, 317–318, 324
speed 135
sphere 2, 141–142, 224, 259, 282, 292, 295, 315
spherical polar coordinates 63–66, 141–142
spline 2, 231, 259–296, 319–320
square matrix 148
square of a number 11, 17
square root
   of a complex number 133–134
   of a number 17, 19, 313–314
   of a negative number 21
stack data structure 216, 226, 234, 250
straight line 41–42, 105–106, 108, 140, 204
   vector representation of 196–197, 205–208, 210–211, 320
stochastic 249, 251, 254–255, 319–320, 324
   production (L-system) 251
subdivisibility (of a fractal) 235, 241
subset 13
subtraction 9, 11, 16, 20, 22
   of vectors 189
subtractive colour 69–70

sufficient condition	230
sum	
of bits	28
of numbers	9
sum bit	28
surd	18
swept models	2
symbolic logic	33
symmetric matrix	148–149

T

tangent (to a curve)	266, 268, 271–275, 292, 320
continuity of	267–269
tangent (tan in trigonometry)	46–55, 97–99
Taylor series	130, 135
terrain modelling	256–257, 303
tesseract	58–59
texture	
solid	5, 294, 319–320
2D	5, 319
three dimensions (3D)	61–66, 141–145, 296
topology	2, 219, 228–230, 259
point set	234–235, 242
transformation	36, 147, 225, 253, 299
affine	79–93, 85–86, 100, 103, 160–171, 178, 225, 255, 282, 299–301
contractive	253
non-affine	176
of a graph	100–103
perspective	172–176, 180–182
2D	79–84, 85–86, 160–166
3D	86–89, 166–183, 176–182, 300
using matrices	160–183
translation	58, 79–80, 86, 100–103, 144, 147, 160–171, 177–178, 253, 300
transpose (of a matrix)	148
traversal (of a tree data structure)	217–218
tree data structure	217–218, 225–226, 321
triangle law for addition of vectors	186–188, 190
triangulation	
in cartography	55
of geometric surfaces	293
trigonometry	46–56, 84–85, 97–99, 132
truth	32–33
turtle graphics	46, 56, 250

twisted facet	292
two dimensions/2D	38–41, 46–41, 56–63

U

unary minus	13
unary operator	33
union	33–35, 213, 222, 225–226
unit matrix	152–153, 323
unit vector	190–192, 205–208, 210, 299–300, 306–307, 310, 313–314, 318
unsigned	30
up vector	299
upper triangular matrix	148–149

V

value (of a colour)	69
vanishing point	176
variable	
dependent	73–74
independent	73–74
vector	62, 123, 135, 185–213, 260, 293, 299–300, 306–307, 309, 308–314, 318, 320
field	231
product (definition)	194–195
velocity	135, 266, 268–269, 274–275
Venn diagram	34–35
view coordinates	4, 172, 298–302, 304–305
view direction	299
view frustum	303
view plane/window (see window)	
view vector	299
viewing pipeline	299–304
viewport	4, 79, 298, 308–309
visualization	61
voxel modelling	2, 215, 218–221, 227

W

waves	52, 103
light	66–67
weight (in NURBS)	282–286, 292
weighted average/mean	196, 261, 265
window	4, 79, 297–299, 301, 303
window to viewport mapping	39, 79, 303, 311
wire frame illustration	5, 294
word (of computer memory)	29
world coordinates	39, 79, 298–301

X

xor 34

Z

z-buffer 255, 297, 305, 316, 318
zero 12–13, 32
zero divide (see division by zero not allowed)
zoom lens 297, 301